Advances in Intelligent Vehicle Control

Advances in Intelligent Vehicle Control

Editor

Juan A. Cabrera

MDPI • Basel • Beijing • Wuhan • Barcelona • Belgrade • Manchester • Tokyo • Cluj • Tianjin

Editor
Juan A. Cabrera
University of Málaga
Spain

Editorial Office
MDPI
St. Alban-Anlage 66
4052 Basel, Switzerland

This is a reprint of articles from the Special Issue published online in the open access journal *Sensors* (ISSN 1424-8220) (available at: https://www.mdpi.com/journal/sensors/special_issues/intelligent_vehicle_sensors).

For citation purposes, cite each article independently as indicated on the article page online and as indicated below:

LastName, A.A.; LastName, B.B.; LastName, C.C. Article Title. *Journal Name* **Year**, *Volume Number*, Page Range.

ISBN 978-3-0365-6009-0 (Hbk)
ISBN 978-3-0365-6010-6 (PDF)

Cover image courtesy of Juan A. Cabrera

Contents

About the Editor

Juan A. Cabrera

Juan A. Cabrera is currently a Full Professor in the Department of Mechanical Engineering at the University of Malaga. His field of research has always been linked, on the one hand, to the study of mechanisms and design of machines and, on the other, to vehicles, specifically to the study of tire models, vehicle dynamics and control of active safety systems. Dr. Cabrera wrote his doctoral thesis on intelligent braking systems. In 2006, he undertook a postdoctoral stay at the Cornell Creative Machines Lab at Cornell University under the supervision of Professor Hod Lipson. He has published more than 80 works that have featured in prestigious journals, book chapters and at national and international congresses. He has supervised 7 doctoral theses and has directed 10 research projects and 10 technology transfer contracts. He has also published eight patents and six of these patents are in the field of vehicles. He is an active member of the international scientific community, acting as a reviewer for various journals in the field of mechanisms, vehicles and optimization.

sensors

MDPI

Editorial

Advances in Intelligent Vehicle Control

Juan A. Cabrera

Department of Mechanical Engineering, University of Malaga, 29071 Malaga, Spain; jcabrera@uma.es

Citation: Cabrera, J.A. Advances in Intelligent Vehicle Control. *Sensors* **2022**, *22*, 8622. https://doi.org/10.3390/s22228622

Received: 3 November 2022
Accepted: 4 November 2022
Published: 9 November 2022

Publisher's Note: MDPI stays neutral with regard to jurisdictional claims in published maps and institutional affiliations.

Advanced intelligent vehicle control systems have evolved in the last few decades thanks to the use of artificial-intelligence-based techniques, the appearance of new sensors, and the development of technology necessary for their implementation. Therefore, a substantial improvement in vehicle safety, comfort, and performance has been achieved. The appearance of new vehicles with new technologies incorporated in them requires new control strategies that will continue to increase handling, stability, and energy efficiency.

In recent years, intelligent vehicle control has been widely investigated from different points of view. Many researchers have studied active safety systems, advanced driver assistance systems, autonomous driver systems, etc., through strategies incorporating aspects of artificial intelligence, making them adapt and learn from situations never explored before. To achieve this, it has been necessary to develop increasingly precise dynamic vehicle models and incorporate new intelligent sensors and sensor fusion techniques to learn the vehicle's state accurately. However, it is important to observe not only the state of the vehicle where these systems are incorporated but also those of vehicles around it that can influence the vehicle's behavior. This requires communication between vehicles and developing architectures that enable smart transportation. On the other hand, the incorporation of electric vehicles (EVs) in recent years has enabled a new way of focusing on vehicle control systems, fundamentally due to the incorporation of new systems that must be studied differently.

Today, there are still many challenges in this field of research that have to be solved or improved, and that is why the Special Issue, "Advances in Intelligent Vehicle Control" in the journal *Sensors*, has compiled 11 works that have tried to provide an answer to the initial issues raised in it, such as:

- Development of intelligent control algorithms for active safety systems [1,2].
- Smart sensors: development of advanced strategies using future smart sensor technology and intelligent sensor fusion for the measurement and estimation of vehicle states, tire and road conditions, situation awareness assessment, environment mapping, fault diagnosis, and driving conditions [3–6].
- Intelligent and efficient driving: advanced vehicle control systems for assisted and autonomous driving and vehicle navigation by incorporating new sensors and measurement systems to develop new strategies to avoid critical driving situations and save energy [7–11].

The article titled "Investigation on the Model-Based Control Performance in Vehicle Safety Critical Scenarios with Varying Tyre Limits" [1] was intended to investigate the possibility of physical model-based control to consider the variations in terms of the dynamic behavior of the systems and of boundary conditions. Different scenarios with specific tire thermal and wear conditions were tested on diverse road surfaces, validating the designed model predictive control algorithm in a hardware-in-the-loop real-time environment and demonstrating the augmented reliability of an advanced virtual driver aware of available information concerning the dynamic limits of the tire.

In the article "Nonlinear Ride Height Control of Active Air Suspension System with Output Constraints and Time-Varying Disturbances" [2], addressed the problem of nonlinear height tracking control of an automobile active air suspension with output state constraints and time-varying disturbances. The proposed control strategy guaranteed that the ride height stayed within a predefined range and converged close to an arbitrarily small neighborhood of the desired height, ensuring uniform ultimate boundedness. The authors designed a nonlinear observer to compensate for time-varying disturbances caused by external random road excitations and perturbations, achieving robust performance. Co-simulation showed the efficiency of the proposed control methodology.

The article, titled "Roll Angle Estimation of a Motorcycle through Inertial Measurements" [3], deals with a method to estimate the roll angle in a motorcycle. They developed a multibody motorcycle model and used an observer based on a Kalman filter to estimate the roll angle. The multibody model is a seven-element assembly without closed kinematic loops, where six elements belong to motorcycle parts, and one of them represents the torso of the driver. This model used 12 Degrees of Freedom (DOF). Six DOF from the chassis rigid body condition; five revolute joints from the two wheels, swingarm, steer, and torso roll movement; and one prismatic joint between the fork bars and fork bottles. One of the most important parts of this model is the tire behavior and properties. The authors used a toroidal tire defined with an outer radius, which represents the undeformed outer radius of the tire, and the torus tube radius, which should be selected to represent the tire curvature near the contact patch in the most accurate way. To test their roll angle estimation algorithm, they performed maneuvers in six different scenarios. From these maneuvers, some measurements were obtained, mimicking the properties of actual sensors by adding some white Gaussian noise. These measurements were used to verify the performance of the state observer.

In the article "Semantics Aware Dynamic SLAM Based on 3D MODT" [4], the authors proposed a framework to solve the dynamics of Simultaneous Localization and Mapping (SLAM) problems. They used a Visual-LIDAR based on Multiple Object Detection and Tracking (MODT) to handle the dynamic regions of the scene. The framework was tested on a dataset developed for LIDAR-based autonomous driving and evaluated and contrasted with state-of-the-art SLAM algorithms. The results suggest that the proposed dynamic SLAM framework can perform in real time with budgeted computational resources. In addition, the fused MODT provides rich semantic information that can easily be integrated into SLAM.

The article, titled "A Redundant Configuration of Four Low-Cost GNSS-RTK Receivers for Reliable Estimation of Vehicular Position and Posture" [5], proposed a low-cost sensor system composed of four GNSS-RTK receivers to obtain accurate position and posture estimations for a vehicle in real time. The four receiver antennas are positioned so that each combination of three antennas is optimal for obtaining the most accurate 3D coordinates with respect to a global reference system. The redundancy provided by the fourth receiver allows further improvement of the estimates and maintains accuracy when one of the receivers fails. They carried out successful experiments with a ground rover on irregular terrain. Angular estimates similar to those of a high-performance IMU were achieved in dynamic tests.

In [6] "Deep Transfer Learning Based Intrusion Detection System for Electric Vehicular Networks", the authors proposed a deep-transfer-learning-based Intrusion Detection System (IDS) model for an In-Vehicle Network (IVN) along with improved performance compared to several other existing models. The unique contributions included effective attribute selection, which is best suited to identify malicious CAN messages and accurately detect normal and abnormal activities, designing a deep-transfer-learning-based model and evaluating capacity considering real-world data. To this end, an extensive experimental performance evaluation was conducted. The architecture, along with empirical analyses, showed that the proposed IDS greatly improves detection accuracy over mainstream machine learning, deep learning, and benchmark deep transfer learning models and demonstrated better performance for real-time IVN security.

The articles, titled "Service-Centric Heterogeneous Vehicular Network Modeling for Connected Traffic Environments" [7], deals with connected vehicles using mobile networks.

In this article, a heterogeneous network model for heterogeneous vehicular communications is presented. After developing the network model, the following conclusions were reached:

- Network cooperation supports cloud computing on big traffic data to realize intelligent traffic services.
- A heterogeneous network coordinator and gateway are the key to proper connection management.
- Service-oriented traffic applications become smarter with increased traffic data mastery and processing power.
- Practical simulation verified higher message diversion and flow utilization and a lower rate of message loss and delays for traffic services when implementing heterogeneous vehicular communications.
- The mathematical modeling of a service-oriented network prioritization and implementation of content-centric services in a heterogeneous vehicular environment was also presented to theoretically support the implementation of the heterogeneous vehicular network.

In the paper titled "Overcoming Challenges of Applying Reinforcement Learning for Intelligent Vehicle Control" [8], the authors apply Reinforcement Learning (RL) in intelligent vehicle control. They analyze the implications of RL in path-planning tasks. Concretely, first of all, they discuss the role of Curriculum Learning (CL) in structuring the learning process of intelligent vehicle control, not showing the learning examples randomly but organized instead in a meaningful order that gradually illustrates more concepts and gradually more complex ones. Second, they study a method to transfer RL policies from simulation to reality to make the agent experience situations in simulation so that it knows how to react to them in reality. To achieve this, they used 2D discrete grid environments with four possible states for each cell in the environment. Additionally, in the second set of experiments, they used the traffic junction environment as the learning environment. Finally, for the physical platform, they assembled multiple robots controlled by the Arduino Yún microcontroller. From the result of their experiments, they concluded the following: as the complexity of the environment influences the learning time and the performance of the agents when using intelligent vehicle control tasks such as path planning through reinforcement learning, several problems can arise, such as the existence of many possible states that the agent could experience, the existence of multiple agents, difficulty in representing states, or how we should formulate these safety-critical tasks to be solved by trial-and-error.

In [9] "On–Off Scheduling for Electric Vehicle Charging in Two-Links Charging Stations Using Binary Optimization Approaches", the authors dealt with the problem of scheduling charging periods of electrical vehicles (EVs) to satisfy users' demands for energy consumption as well as to optimally utilize available power. They proposed a new model for on–off scheduling of EV charging, assuming that every three-phase charger is equipped with two ports that can be served alternately. The scheduler considers individual charging rates and maximal currents that supply the entire farm separately for each phase. For this model, the authors analyzed several algorithms, and they concluded that binary quadratic programming solved with successive linear approximation algorithms satisfied the most important criteria and constraints in all statistical tests performed. This algorithm ensured smooth charging profiles but had a relatively long delay and was not the fastest. For this reason, the authors suggested that the issue of the algorithmic approach is still open, and further research in this area will be performed in the future.

The authors of the article "Advanced Driver Assistance Systems (ADAS) Based on Machine Learning Techniques for the Detection and Transcription of Variable Message Signs on Roads" [10] aimed at avoiding distractions when drivers pay attention to traffic signs. The authors developed a prototype of a Variable Message Sign (VMS) reading system using machine learning techniques. The assistant consists of two parts: one that recognizes a signal on the street and another that extracts its text and transforms it into speech. They used a neural network to recognize the VMS in an image and indicate its location with a confidence percentage.

Finally, in [11] "A Bidirectional Versatile Buck–Boost Converter Driver for Electric Vehicle Applications", the authors presented a novel dc-dc bidirectional buck–boost converter between a battery pack and an inverter to regulate the dc bus in an electric vehicle powertrain. The converter was based on the versatile buck–boost converter, which has shown excellent performance in different fuel cell systems operating in low-voltage and hard-switching applications. The theoretical analyses were validated using simulations and experimental tests performed on a 400-V 1.6-kW prototype. The authors concluded that:

- The current controller regulated the traction of the dc bus during motoring and regenerative brake conditions.
- The system presented zero steady errors and a fast-transient response in the start-up for dc bus voltage reference changes and under realistic conditions using an EV powertrain system emulation.

Acknowledgments: I would like to thank all authors who have submitted their manuscripts for consideration in this Special Issue and the reviewers for their hard work during the review process. Furthermore, I wish to express my deepest gratitude to the assistant editor of *Sensors*, for her kind help and support.

Conflicts of Interest: The author declares no conflict of interest.

References

1. Sakhnevych, A.; Arricale, V.M.; Bruschetta, M.; Censi, A.; Mion, E.; Picotti, E.; Frazzoli, E. Investigation on the Model-Based Control Performance in Vehicle Safety Critical Scenarios with Varying Tyre Limits. *Sensors* **2021**, *21*, 5372. [CrossRef] [PubMed]
2. Zhao, R.; Xie, W.; Zhao, J.; Wong, P.; Silvestre, C. Nonlinear Ride Height Control of Active Air Suspension System with Output Constraints and Time-Varying Disturbances. *Sensors* **2021**, *21*, 1539. [CrossRef] [PubMed]
3. Maceira, D.; Luaces, A.; Lugrís, U.; Naya, M.; Sanjurjo, E. Roll Angle Estimation of a Motorcycle through Inertial Measurements. *Sensors* **2021**, *21*, 6626. [CrossRef] [PubMed]
4. Sualeh, M.; Kim, G.-W. Semantics Aware Dynamic SLAM Based on 3D MODT. *Sensors* **2021**, *21*, 6355. [CrossRef] [PubMed]
5. Morales, J.; Martínez, J.L.; García-Cerezo, A.J. A Redundant Configuration of Four Low-Cost GNSS-RTK Receivers for Reliable Estimation of Vehicular Position and Posture. *Sensors* **2021**, *21*, 5853. [CrossRef] [PubMed]
6. Mehedi, S.; Anwar, A.; Rahman, Z.; Ahmed, K. Deep Transfer Learning Based Intrusion Detection System for Electric Vehicular Networks. *Sensors* **2021**, *21*, 4736. [CrossRef] [PubMed]
7. Khasawneh, A.M.; Abu Helou, M.; Khatri, A.; Aggarwal, G.; Kaiwartya, O.; Altalhi, M.; Abu-Ulbeh, W.; AlShboul, R. Service-Centric Heterogeneous Vehicular Network Modeling for Connected Traffic Environments. *Sensors* **2022**, *22*, 1247. [CrossRef] [PubMed]
8. Pina, R.; Tibebu, H.; Hook, J.; De Silva, V.; Kondoz, A. Overcoming Challenges of Applying Reinforcement Learning for Intelligent Vehicle Control. *Sensors* **2021**, *21*, 7829. [CrossRef] [PubMed]
9. Zdunek, R.; Grobelny, A.; Witkowski, J.; Gnot, R.I. On–Off Scheduling for Electric Vehicle Charging in Two-Links Charging Stations Using Binary Optimization Approaches. *Sensors* **2021**, *21*, 7149. [CrossRef] [PubMed]
10. De Las Heras, G.; Sánchez-Soriano, J.; Puertas, E. Advanced Driver Assistance Systems (ADAS) Based on Machine Learning Techniques for the Detection and Transcription of Variable Message Signs on Roads. *Sensors* **2021**, *21*, 5866. [CrossRef] [PubMed]
11. González-Castaño, C.; Restrepo, C.; Kouro, S.; Vidal-Idiarte, E.; Calvente, J. A Bidirectional Versatile Buck–Boost Converter Driver for Electric Vehicle Applications. *Sensors* **2021**, *21*, 5712. [CrossRef] [PubMed]

sensors

MDPI

Article

Investigation on the Model-Based Control Performance in Vehicle Safety Critical Scenarios with Varying Tyre Limits

Aleksandr Sakhnevych [1,*], Vincenzo Maria Arricale [1], Mattia Bruschetta [2], Andrea Censi [3], Enrico Mion [3], Enrico Picotti [2] and Emilio Frazzoli [3]

1 Department of Industrial Engineering, University of Napoli Federico II, 80125 Naples, Italy; vincenzomaria.arricale@unina.it
2 Department of Information Engineering, University of Padova, Via Gradenigo, 6/B, 35131 Padova, Italy; mattia.bruschetta@dei.unipd.it (M.B.); picottie@dei.unipd.it (E.P.)
3 Institute for Dynamical Systems and Control, ETH Zurich, 8092 Zurich, Switzerland; acensi@idsc.mavt.ethz.ch (A.C.); enmion@ethz.ch (E.M.); efrazzoli@ethz.ch (E.F.)
* Correspondence: ale.sak@unina.it

Abstract: In recent years the increasing needs of reducing the costs of car development expressed by the automotive market have determined a rapid development of virtual driver prototyping tools that aims at reproducing vehicle behaviors. Nevertheless, these advanced tools are still not designed to exploit the entire vehicle dynamics potential, preferring to assure the minimum requirements in the worst possible operating conditions instead. Furthermore, their calibration is typically performed in a pre-defined strict range of operating conditions, established by specific regulations or OEM routines. For this reason, their performance can considerably decrease in particularly crucial safety-critical situations, where the environmental conditions (rain, snow, ice), the road singularities (oil stains, puddles, holes), and the tyre thermal and ageing phenomena can deeply affect the adherence potential. The objective of the work is to investigate the possibility of the physical model-based control to take into account the variations in terms of the dynamic behavior of the systems and of the boundary conditions. Different scenarios with specific tyre thermal and wear conditions have been tested on diverse road surfaces validating the designed model predictive control algorithm in a hardware-in-the-loop real-time environment and demonstrating the augmented reliability of an advanced virtual driver aware of available information concerning the tyre dynamic limits. The multidisciplinary proposal will provide a paradigm shift in the development of strategies and a solid breakthrough towards enhanced development of the driving automatization systems, unleashing the potential of physical modeling to the next level of vehicle control, able to exploit and to take into account the multi-physical tyre variations.

Keywords: model-based control; vehicle dynamic potential; tyre thermodynamics; tyre wear; weather influence; vehicle safety; double lane change; safety optimization

Citation: Sakhnevych, A.; Arricale, V.M.; Bruschetta M., Censi, A.; Mion, E.; Picotti, E.; Frazzoli E. Investigation on the Model-Based Control Performance in Vehicle Safety Critical Scenarios with Varying Tyre Limits. *Sensors* **2021**, *21*, 5372. https://doi.org/10.3390/s21165372

Academic Editor: Juan A. Cabrera

Received: 21 June 2021
Accepted: 2 August 2021
Published: 9 August 2021

1. Introduction

The information concerning the vehicle's non-linear physical limits depending on the thermal and wear states of tyres, the pavement characteristics, and the boundary conditions (wet or icy ground, under-inflated or worn tyre, etc.) represents a fundamental additional value for the optimal behavior of safety- and performance-oriented control logics [1–3].

Virtual driver prototyping is becoming an increasingly exploited tool, allowing the car manufacturer to perform the majority of the testing campaign already in the design phase of the vehicle. Specific prototyping choices can be reproduced and evaluated in any condition within the virtual environment, also at the limit of performance, minimizing the time-to-market and connected costs [4,5].

In this field, closed-loop control strategies have been widely studied in past years to address the problem of path following for autonomous driving cars. Examples can be found in [6], where a nested PID steering control has been designed for the lane-keeping task, and more recently in [7], where a pure pursuit controller has been specifically developed for path tracking. The most recent VD implementations rely on a vehicle controller based on a non-linear model predictive control (NMPC) technique, which is a model-based control strategy able to compute the optimal sequence of control inputs over a prediction horizon, by minimizing a tailored cost function [8,9]. The control technique is applied in a receding horizon mode and is capable of handling constraints and the intrinsic non-linearities of the vehicle model [10].

The main advantages of the NMPC approach are the capability of the controller of handling all significant features of the process dynamics directly: in this way, the constraints on variables involved in the task (track limits, actuator constraints) can be easily integrated into the optimal control problem, hence guaranteeing the maximal exploitation of vehicle capabilities. Moreover, it is a predictive technique that allows optimizing the vehicle behavior over a future horizon in time, and therein system states and controls. In this way, the controller is allowed to retrieve information about future vehicle behavior and about possible dangerous situations, aiming at anticipating actions and providing suitable controls for challenging vehicle handling.

The objective of the work consists in the integration of the information concerning the tyre dynamic limits within the definition of a virtual driver (VD), implemented as a vehicle controller aiming at testing the vehicle behavior at limit of handling condition, and demonstrating the advantages in terms of both enhanced active safety and optimized performance. An interesting VD definition that addresses the problem of real-time obstacle avoidance on low-friction road surfaces has been proposed in [11], where the code generation tool ACADO [12] has been used to define and solve the NMPC problem. Another similar implementation of such a controller for an autonomous ground vehicle has been proposed in [13], where the controller has been also validated in co-simulation with a hard real-time dSPACE DS1005 Autobox system. The vehicle model employed in the both implementations has consisted of a four-wheel vehicle, where tyres have been described by means of a linear tyre model and Fiala tyre model for longitudinal and lateral dynamics, respectively [14,15]. The inputs are the steering angle and the front/rear braking ratios, while the bounds are defined through specifically defined spatial constraints. A different virtual driver definition, especially designed for high performance vehicles, has been developed in [16]: here, the vehicle model integrates longitudinal load transfer and Pacejka's lateral tyres forces model. The controller implementation has been tested in a real-time co-simulation with a commercial software VI-CarRealTime (VI-CRT) within a double lane change (DLC) maneuver, where the abilities of the controller have been demonstrated with high speed operating conditions and a challenging track geometry. The NMPC strategy has also been applied in racing environment as the autonomous vehicle controller for handling 1:43 scale, RC electric vehicles [17] and autonomous racecar [18], with the specific purpose of achieving aggressive maneuvering and lap time minimization.

In this work, the authors aim to investigate the possibility to employ the model-based strategies to control the non-linear time-dependent system, i.e., the full vehicle model with temperature and wear sensitive tyres operating in completely different environmental conditions. To perform the study, the standardized DLC maneuver, currently employed for the validation of virtual driver and advanced driving assistance systems (ADAS) [16,19,20], is implemented in Matlab/Simulink virtual environment. The vehicle and tyre models have been characterized and validated for a reference GT vehicle, identifying the requisite complex tyre–road coupled phenomena concerning the temperature, wear, and road pavement dependencies [21,22]. Four different roads, i.e., dry, wet, snowy, and icy, two diverse tyre mileages, i.e., new and worn, and three thermal tyre conditions are combined and analyzed in the study to understand which could be the advantages of the employment of the model-based controllers, aware of the tyre instantaneous characteristics, boundary operating and weather conditions, and overall vehicle dynamic potential [23]. The model-based control logics, able to make use of the additional information concerning the dynamic limits of the system, is tested in co-simulation with a 14 degrees-of-freedom vehicle plant model, where the tyres are described by means of a Pacejka's magic formula (MF) model. The vehicle controller is based on a robust and computationally effective non-linear model-predictive-control (NMPC), implemented in the open-source NMPC software MATMPC [24], able to take into account the additional instantaneous information concerning the varying adherence potential and the vehicle non-linearities. The information concerning the vehicle non-linear physical limits depending on the thermal and wear states of tyres, the pavement characteristics and the boundary conditions (wet or icy ground, under-inflated or worn tyre, etc.) represents a fundamental additional value for the optimal behavior of safety- and performance-oriented control logics [25–27], as it allows to maximize the potential to avoid obstacles and to reduce the severity of collisions [28].

The authors aim to lay the foundation of the future advanced driving systems, sensitive to environmental conditions and adaptive to continuously varying characteristics of the underlying non-linear system. Being currently mainly based on mere empirical calibration, the physical model-based estimation can represent a crucial factor towards the improvement of the pedestrians' and passengers' active safety, enabling the management of the activation threshold ranges on the basis of the instantaneous operating and the environmental boundary conditions [29,30]. This can be already employed in the current ADAS to communicate to the driver the necessity to co-act in specific situations, but it also constitutes a fundamental root for the future driving automatization [31,32].

The paper is organized as follows: Section 2 introduces the problem description concerning complex phenomena linked with the tyre–road interaction and their influence on the overall vehicle dynamics; Section 3 describes the advanced methodologies developed to characterize, model, and reproduce the dynamic behavior of the real system in the virtual environment, and introduces the adopted model-based control, evidencing the peculiarities of the designed cost function; in Section 4 the outputs of the conducted simulations employing different road surfaces, in adverse boundary conditions and with diverse states of the tyres are discussed, addressing particular attention towards the control strategies. Finally, in Section 5, a discussion on the next developments and the conclusions are presented.

2. Problem Description

A proper understanding of the tyre dynamic behavior and of its multiple intrinsic dependencies is a crucial topic for tyre manufacturers, to improve tyre performance and durability, for users, to set the optimal working conditions, and for researchers, to develop computationally efficient mathematical models able to represent the experimental behavior with a high degree of accuracy. Friction phenomenon, arising at the tyre–road interface, originates from three physical contributions: the adhesive term relative to molecular Van der Waals links arising between the two counter surfaces in mutual contact, the hysteretic term linked to the deformation losses within the elastomeric material, and the wear term [33,34]. All of them are deeply interconnected and dependent on the specific tyre working conditions, in terms of sliding velocity, temperature, and pressure distributions, arising at the tyre contact patch as a result of different excitation spatial frequency spectra, representative of diverse types of road pavement [35]. Furthermore, tyres may deeply modify their dynamic behavior over time due to ageing effects, influencing the dynamic potential of the overall vehicle [22].

The enrichment of the vehicle state with the information concerning the tyre instantaneous and potential friction will allow, taking into account the tyre multi-physical variations (Figures 1 and 2), represents a key point in the development of control logics, able to adapt to sudden variation in boundary conditions in order to guarantee the vehicles higher stability in critical scenarios . Indeed, in the Figure 2 it is possible to observe how the adherence ellipse changes in three tyre thermal ranges.

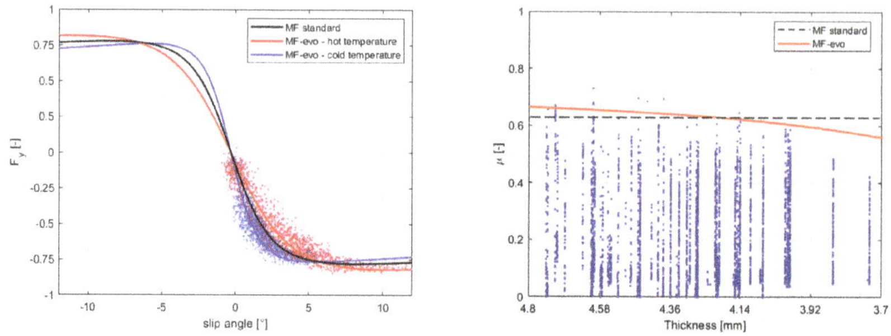

(**a**) (**b**)

Figure 1. Tyre behavior variations. (**a**) Compound temperature influence on the characteristic interaction shape. (**b**) Wear effect on available grip.

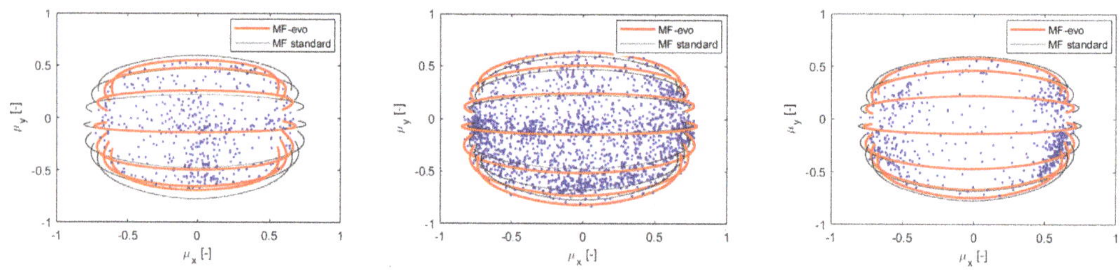

(**a**) (**b**) (**c**)

Figure 2. MF-based standard and evo tyre models compared with the experimental points in three thermal ranges: under-heating condition (**a**), optimal temperature (**b**), over-heating condition (**c**), (camber angle = −2 deg | vertical load = 3000 *N*).

The aim of the proposed adaptive control will be to avoid collisions and to minimize risks in any environmental condition, validating all the scenarios on interest in a highly accurate simulation environment. The information concerning the vehicle non-linear physical limits depending on the thermal and wear states of tyres, the pavement characteristics and the boundary conditions (wet or icy ground, under-inflated or worn tyre, etc.) represents a fundamental additional value for the optimal behavior of safety- and performance-oriented control logics. The non-linear model predictive control approach is employed to integrate the tyre varying dynamic parameters within the definition of physical constraints of the vehicle, guaranteeing the stability of the system and allowing to achieve the optimal solution for the defined vehicle instantaneous dynamic limits.

3. Physical Model, Physical Model-Based Control, and Virtual Scenario

To parametrize the vehicle and the tyres' model, the authors have collected data with a chosen GT vehicle in a specific test session on track. Due to a non-disclosure agreement with the industrial research partner, the vehicle and the track will not be specified.

The track session has consisted of handling tests in the widest possible range of tyre operating conditions in terms of temperature, pressure, and wear level. Following the vehicle model parametrization and the tyre parameters' estimation procedures described in [36,37], the vehicle non-linear system has been completely characterized in all the conditions of interest, being able to faithfully reproduce the experimental data in the virtual environment.

3.1. Vehicle Parametrization

The 14 degrees of freedom (DoF) vehicle model, based on the mathematical representation described in [38], has been modeled in a `MATLAB/Simulink` environment as follows:

- 6 DoF to reproduce longitudinal, lateral, vertical, pitch, roll, and yaw motion of the vehicle body;
- 4 DoF concerning the wheel rotation and 4 DoF for the wheel normal displacement, with the hypothesis that the degrees of freedom to the relative motion between the wheel and the vehicle body can be neglected along the longitudinal and lateral directions, allowing only the independent rotational and vertical displacements.

Furthermore, the parameterized vehicle is rear-wheel drive with front steering and internal combustion engine. The tyre model is described by Pacejka's magic formula model, whose parameters have been characterized for different conditions of temperature, pressure, and wear. Per each road surface under study (dry, wet, snowy, and icy), the tyre-road friction coefficient has been supposed constant and is applied as an additional scaling factor of the λ_{μ_x} and λ_{μ_y} parameters [39], linearly combining the tyre characteristics identified on a reference road with the ones potentially achievable on diverse pavement surfaces.

The vehicle dynamic behavior in the reference tyre conditions has been validated in a slow-ramp-steer maneuver, whose parameters are summarised in the Table 1 and outputs are illustrated in the Figure 3, feeding the model with the steering input presented in the Figure 4a):

Table 1. Slow-ramp-steer inputs.

Description	Value	Unit
start time	13.26	s
end time	20.3	s
initial velocity	27.9	m/s
initial gear	3	-
ramp duration	7.04	s
initial steer	0	deg
slope steer	−22.29	deg/s

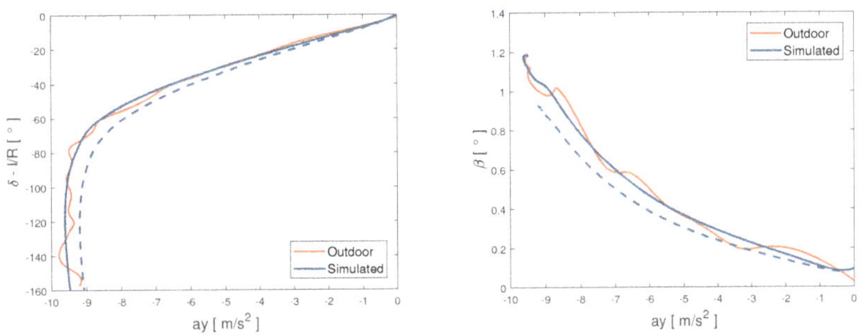

(a) (b)

Figure 3. Comparison between outdoor acquisitions and simulation output. (**a**) Steering angle vs. lateral acceleration diagram. (**b**) Sideslip angle vs lateral acceleration diagram.

For the validation purpose, lateral acceleration a_y, steering angle δ, side slip angle β have been compared for the same inputs. Figure 3 shows the comparisons between experimental data and model outputs shown on the classic $a_y - \delta$ and $a_y - \beta$ diagrams. An aspect that is worth pointing out is the difference between the black dashed and continuous lines: the first one is obtained using the starting parameters provided by the research partner, the second one is obtained employing the calibration procedure described in [37]. In particular, the starting under-steering characteristics (dashed lines) have been revised better identifying the parameters linked to the anti-roll bars stiffness and the steering maps, leading to a less under-steering behavior within the handling diagram, in agreement with the experimental data.

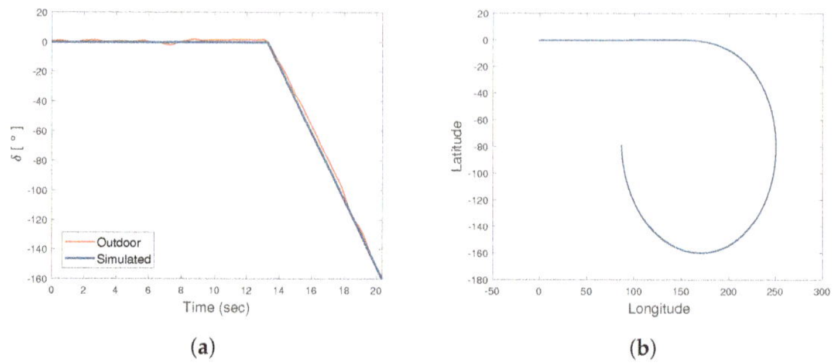

Figure 4. Example of lateral maneuver's input reproduction. (a) Experimental and simulation steering angle comparison. (b) Slow-ramp-steer trajectory in virtual environment.

The enhanced parametrization has led to a higher slope in the linear section (Figure 3a), but also higher lateral grip and side-slip angle values, related to the rear axle behavior (Figure 3b). Once the vehicle and the tyres' subsystems have been properly characterized in the specific range of temperature, pressure, and wear, the validity range of the MF tyre model has been extended adopting the MF-Evo one, described in [40]. In particular, the tyre model calibration process can be summarized in three fundamental steps: the first one is related to the pre-processing of the experimental data (which allows to discern useful information contained in the acquired data and to eliminate the non-physical outliers); the second one concerns the identification of the standard MF micro-coefficients in a specific range of temperature, pressure, and wear; the third step aims at the calibration of the additional multi-physical analytical formulations, taking into account of the entire dataset and, thus, extending the tyre model towards thermal and degradation phenomena.

The calibration results are visible in terms of adherence ellipse in the Figure 1, where the experimental data have been compared towards the MF and MF-evo outputs within different temperature working ranges of the tyre. Finally, the parameters of the MF-evo model have been further modified to extend the applicability of the tyre model on different road surfaces, modifying the identified friction factors towards the pavement characteristics, as reported in the Table 2. The resulting interaction characteristics for different tyres, in diverse thermodynamic conditions and in contact with different road surfaces have been summarized in Figures 5 and 6.

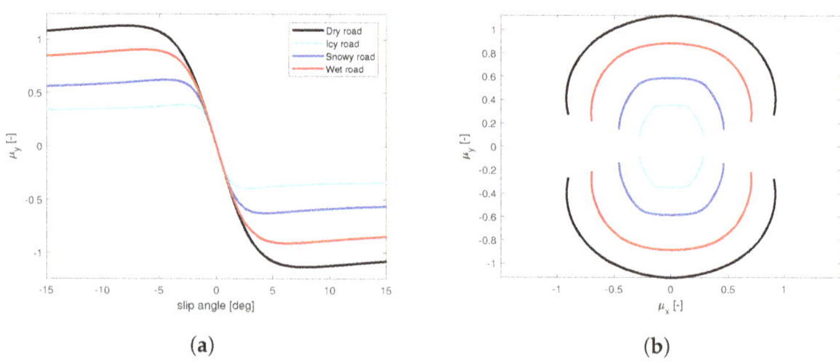

Figure 5. New tyre in optimal thermal condition in contact with different road surfaces. (a) Lateral interaction characteristics. (b) Adherence ellipse.

In steady-state conditions, the global force exerted by the tyres is in a dynamic equilibrium with the centrifugal force, as a function of the longitudinal velocity of the vehicle

v and the instantaneous cornering radius R, relating the lateral acceleration a_y and the longitudinal velocity v of the vehicle's center of mass (CM) by the equation:

$$a_y = \frac{v^2}{R}; \tag{1}$$

To demonstrate the potential influence of the road surface characteristics on the overall vehicle behavior, a set of simulations has been conducted with different tyre parameters described in Figure 5 in a steady-state lateral slow-ramp-steer (SRS) maneuver. The maximum achievable value of the forward velocity v for a given curvature and $a_y - \delta$ characteristics are reported for dry, wet, snowy, and icy pavement conditions in the Figure 7a,b, respectively.

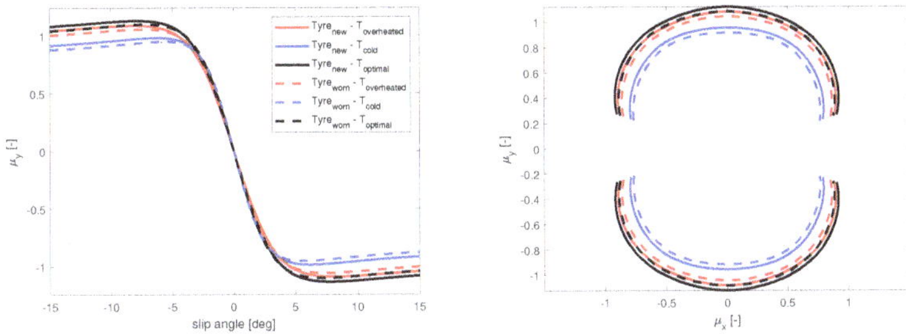

Figure 6. New and worn tyres in diverse thermal conditions in contact with the dry road. (**a**) Lateral interaction characteristics. (**b**) Adherence ellipse.

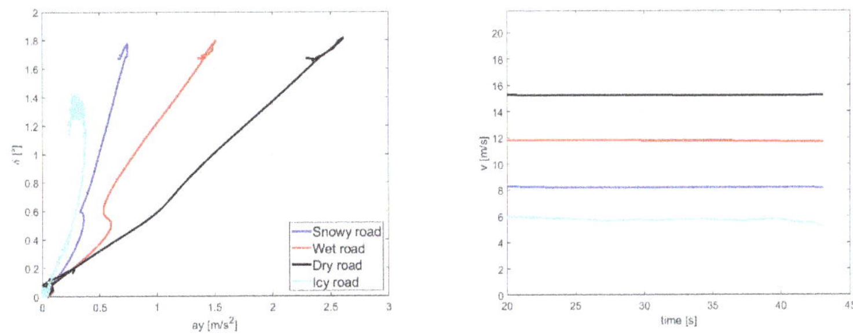

Figure 7. SRS maneuver on different road surfaces. (**a**) Vehicle understeer characteristics. (**b**) Maximum velocity achieved.

Table 2. Summary of the velocity maximum values assumed for each road scenario.

Friction Coefficient μ [−]	Lateral Acceleration a_y [m/s^2]	Longitudinal Velocity v [m/s]
0.35	0.35	5.92
0.55	0.70	8.37
0.80	1.50	12.2
1.00	2.52	15.9

3.2. Internal Vehicle Model

A four-wheel vehicle model based on the description in [16] has been used as the internal model for the NMPC controller. Specific characterization of load transfers, gear shift predictions, longitudinal force saturation, and an ellipsoidal tyre friction constraint have been also introduced in the model definition to improve the overall prediction capabilities of the controller. Finally, the model dynamics have been reformulated in spatial coordinates with respect to the curvilinear abscissa s along the track. In this way, track constraints can be defined with respect to space and the time can be considered as a minimization variable, as already highlighted in previous works [16,41,42].

The continuous-time dynamics model is described as

$$\dot{\xi} = \phi(\xi(t), u(t); p(t)), \tag{2}$$

where the state is represented by $\xi(t) \in \mathbb{R}^{n_x}$, $u(t) \in \mathbb{R}^{n_u}$ is the input, whereas the time-varying parameter vector is $p(t) \in \mathbb{R}^{n_p}$.

$$\begin{aligned} &\ddot{x} = \dot{y}\dot{\psi} + \frac{1}{m}\left(\sum_{i,j} F_{x_{i,j}} - F_x^d\right), \quad \ddot{y} = -\dot{x}\dot{\psi} + \frac{1}{m}\left(\sum_{i,j} F_{y_{i,j}}\right), \\ &\ddot{\psi} = \frac{1}{I_z}\left[a\left(\sum_j F_{y_{f,j}}\right) - b\left(\sum_j F_{y_{r,j}}\right) + c\left(\sum_i F_{x_{i,r}} - \sum_i F_{x_{i,l}}\right)\right], \end{aligned} \tag{3}$$

where longitudinal and lateral positions are x, y, while ψ is the yaw angle. m and I_z are the mass and the inertia around the vertical axis of the vehicle, respectively. a, b, c are the vehicle dimensional parameters, front wheels to CM longitudinal distance, rear wheels to CM longitudinal distance, and wheels to CM lateral distance, respectively. $F_{\{x,y\}_{\{i,j\}}}$ are the lateral and longitudinal forces on the wheels and F_x^d is the longitudinal drag force in the vehicle's reference frame. Subscripts $i \in \{f, r\}$ refer to front or rear wheels, $j \in \{l, r\}$ left or right wheels. Figure 8 illustrates the physical quantities involved and the reference systems chosen. δ_f is the steering angle of the front wheels, assumed the same for the both front tyres, and $\beta_{f,j}$ is the side slip angle of the f, j-th tyre. The projection of cornering and longitudinal forces in the vehicle frame, the position and the dynamics of the vehicle's CM in the inertial frame X, Y, and the vehicle side slip angle β are described in [16], whereas the longitudinal drag force and the down-force are modeled as [43] pp. 97–98.

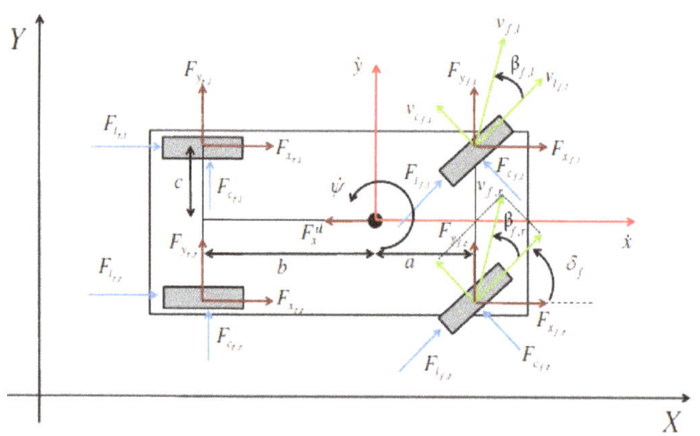

Figure 8. Internal vehicle model for control.

Differently from [16], the longitudinal tire forces in each wheel reference frame are computed as

$$F_{l_{i,j}} = f_{\mathrm{eng}_{i,j}} - f_{\mathrm{brk}_{i,j}}, \tag{4}$$

where the engine and braking forces are

$$f_{\text{eng}_{i,j}} = \text{sat}\left(\frac{\tau_{\text{eng}_i}}{r_w}, \mu F_{z_{i,j}}\right), \quad f_{\text{brk}_{i,j}} = \text{sat}\left(\frac{\tau_{\text{brk}_i}}{r_w}, \mu F_{z_{i,j}}\right), \tag{5}$$

where μ is the tyre friction coefficient, r_w is the wheel radius and the saturation function is defined in (8). Then, the engine and braking torques at the wheels are:

$$\tau_{\text{eng}_i} = \gamma_t \left(\tau^{\max}_{\text{eng},i} - \tau^{\min}_{\text{eng},i}\right) + \tau^{\min}_{\text{eng},i}, \quad \tau_{\text{brk}_i} = \gamma_b \, \tau^{\max}_{\text{brk},i}, \tag{6}$$

where $\gamma_{t,b}$ are the normalized throttle and braking efforts, $\tau^{\max}_{\text{brk},i}$ is the maximum torque given by the braking system to front/rear wheels, $\tau^{\max}_{\text{eng},i}$ and $\tau^{\min}_{\text{eng},i}$ are the maximum and minimum torque values expressed by the engine at front/rear wheels at a given gear and are changed as a time-varying parameter to the actual model gearshift. To compute the torques in the prediction horizon, an iterative strategy predicting the engine rpm, and, hence, gearshift, based on the predicted velocity is used [16]. Specifically, the engine rpm quantity is computed as

$$\text{rpm}^{\text{pred}} = \frac{v_x^{\text{pred}}}{r_w} \frac{\text{diff}_{\text{ratio}}}{\text{gear}_{\text{ratio}}} \frac{60}{2\pi}, \tag{7}$$

where $\text{diff}_{\text{ratio}}$ and $\text{gear}_{\text{ratio}}$ are the input/output torque ratios at the differential and at the gearbox (in a specific gear), respectively. The dependence of $\tau^{\max,\min}_{\text{eng},i}$ w.r.t. the engine rotational velocity has been neglected. Finally, the saturation function is defined as:

$$\text{sat}(f_a, f_b) = \frac{f_b}{1 + \exp(-5(\frac{f_a}{f_b} - \frac{1}{2}))}. \tag{8}$$

The normal forces $F_{z_{i,j}}$ are modeled considering the load transfer in steady-state condition as described in [44]. The algebraic loop in the model has been avoided by considering F_x^{sat0} (total longitudinal force expressed in the vehicle frame saturated at nominal F_z) and F_y^{static} (the sum of the lateral forces computed at nominal F_z on each wheel) as the forces used for the load transfer dynamics.

Finally, the lateral forces model is based on the simplified MF model described in [39] pp. 187–188, expressed by means of the macro-parameters B, C, D, E.

3.3. NMPC Algorithm

The goal of the NMPC controller for the virtual driver is to compute a reliable sequence of steering, throttle, brake commands in a prediction horizon, given a tailored cost function. The NMPC algorithm is based on `MATMPC` [10,45], an open source software built in `MATLAB` for real-time NMPC solution.

In `MATMPC`, a non-linear programming problem (NLP) is formulated at sampling instant i by applying direct multiple shooting [46] to an optimal control problem (OCP) over the prediction horizon $S = [s_0, s_f]$, which is divided into N *shooting intervals* $[s_0, s_1, \ldots, s_N]$, as follows

$$\min_{\xi_{\cdot|i}, u_{\cdot|i}} \sum_{k=0}^{N-1} \frac{1}{2} \| h_k(\xi_{k|i}, u_{k|i}) \|_W^2 + \frac{1}{2} \| h_N(\xi_{N|i}) \|_{W_N}^2 \tag{9}$$

$$s.t.\, 0 = \xi_{0|i} - \hat{\xi}_0, \tag{10}$$

$$0 = \xi_{k+1|i} - \phi_k(\xi_{k|i}, u_{k|i}; p_{k|i}), \; k \in [0, N-1], \tag{11}$$

$$\underline{r}_{k|i} \leq r_k(\xi_{k|i}, u_{k|i}) \leq \overline{r}_{k|i}, \; k \in [0, N-1], \tag{12}$$

$$\underline{r}_{N|i} \leq r_N(\xi_{N|i}) \leq \overline{r}_{N|i} \tag{13}$$

where $\xi_{\cdot|i} = (\xi_{0|i}^\top, \xi_{1|i}^\top, \ldots, \xi_{N|i}^\top)^\top$, and $u_{\cdot|i} = (u_{0|i}^\top, u_{1|i}^\top, \ldots, u_{N-1|i}^\top)^\top$, while $\hat{\xi}_0$ represents the measurement of the current state. System states $\xi_{k|i} \in \mathbb{R}^{n_\xi}$ are defined at the discrete arc-length point s_k for $k = 0, \ldots, N$ and the control inputs $u_{k|i} \in \mathbb{R}^{n_u}$ for $k = 0, \ldots, N-1$ are piece-wise constant. Their definitions are given in (14) and (15). Here, (12) is defined by $r(\xi_{k|i}, u_{k|i}) : \mathbb{R}^{n_\xi} \times \mathbb{R}^{n_u} \to \mathbb{R}^{n_r}$ and $r(\xi_{N|i}) : \mathbb{R}^{n_\xi} \to \mathbb{R}^{n_l}$ with lower and upper bound $\underline{r}_{k|i}, \overline{r}_{k|i}$. Equation (11) refers to the *continuity constraint* where $\phi_k(\xi_{k|i}, u_{k|i}; p_{k|i})$ is a numerical integration operator that solves (16) with initial condition $\xi(0) = \xi_{0|i}$ and returns the solution at s_{k+1}. The time has been included as a state variable with the following ODE $\dot{t} = \frac{1}{\dot{s}}$ to fulfil the minimization of the travel time over the prediction horizon. The full state vector is then given by:

$$\xi = [\dot{x}, \dot{y}, \dot{\psi}, e_\psi, e_y, \delta_f, \gamma_t, \gamma_b, t]^T, \tag{14}$$

where e_ψ, e_y are orientation and lateral error of the vehicle with respect to the center-line of the path, respectively. The input computed by the algorithm is then:

$$u = [\dot{\delta}_f, \dot{\gamma}_t, \dot{\gamma}_b, \epsilon_{\text{slip}}, \epsilon_{\text{err}}, \epsilon_{\text{gg}}]^T, \tag{15}$$

where $\dot{\delta}_f$, $\dot{\gamma}_t$, $\dot{\gamma}_b$ are the derivatives of the actual input to the vehicle and ϵ are slack variables. This formulation allows a smooth action of the controller and avoids too aggressive, unrealistic behaviors.

The dynamics equation of the model used in the NMPC algorithm can be compactly written as

$$\xi' = \phi(\xi(s), u(s); p(s)), \tag{16}$$

where $p(s) = \left[\zeta(s), \tau_{\text{eng},i}^{\text{MAX,min}}(s)\right]^T$.

The real-time iteration scheme (RTI) [47] is employed to reduce the time required to solve the (9) problem. Moreover, a non-uniform grid strategy [48] has been used for lowering the computational burden and let the controller predict a sufficiently long horizon (chosen 400 m in advance for the specific vehicle).

The cost function for the NMPC is defined as:

$$\begin{aligned} h_k(\xi_k, u_k) &= [\beta, \gamma_t \cdot \gamma_b, \zeta \cdot \gamma_t, t, \dot{\delta}_f, \dot{\gamma}_t, \dot{\gamma}_b, \epsilon_{\text{slip}}, \epsilon_{\text{err}}, \epsilon_{\text{gg}}]^\top, \\ h_N(\xi_N) &= [\beta, \gamma_t \cdot \gamma_b, \zeta \cdot \gamma_t, t, e_y - e_y^{\text{ref}}, \dot{e}_y, e_\psi - e_\psi^{\text{ref}} + \\ &\quad + \beta, \dot{e}_\psi]^\top. \end{aligned} \tag{17}$$

The penalty on the vehicle side slip β is used to limit the sliding behavior of the vehicle; simultaneous throttling and braking are penalized by the cost $\gamma_t \cdot \gamma_b$. The $\zeta \cdot \gamma_t$ cost is included to make the controller accelerate smoothly during the final phase of the track corner exit. The objective variable time t is added to minimize the time on the prediction horizon. Smooth control actions are ensured by the objective terms on the inputs. The three slack variables are also adopted to define the *soft constraints* [49], which increase the robustness of the overall procedure. Finally, the terms related to errors e_y and e_ψ, used only as terminal objective variables, are introduced to integrate information about the trajectory over the prediction horizon.

The constraints are defined as

$$\begin{aligned} r_k &= [\delta_f, \gamma_t, \gamma_b, \dot{\delta}_f, \dot{\gamma}_t, \dot{\gamma}_b, \epsilon_{\text{slip}}, \epsilon_{\text{err}}, \epsilon_{\text{gg}}, \beta + \epsilon_{\text{slip}}, \\ &\quad e_y + \epsilon_{\text{err}}, (\mu_x \frac{\ddot{x}_{\text{ext}}}{g})^2 + (\mu_y \frac{\ddot{y}_{\text{ext}}}{g})^2 + \epsilon_{\text{gg}}]^\top, \\ r_N &= [\delta_f, \gamma_t, \gamma_b]^\top, \end{aligned} \tag{18}$$

where the constraints on δ_f, γ_t, and γ_b are intrinsic bounds of the actual vehicle controls, while those on $\dot{\delta}_f$, $\dot{\gamma}_t$, and $\dot{\gamma}_b$ are added in order to improve the smoothness of the computed inputs and can be used to easily tune the aggressivity of the NMPC driving commands. Additionally, the slack variables have been constrained in order to help the optimization procedure restricting the *search space* of the inputs. The slack variables are used for defining the soft constraints: the first one is introduced on the side-slip of the vehicle and helps the controller to regain control of the vehicle in case of high skidding; the second one is used to correct the trajectory when the vehicle is out of track; the third one instead is designed to make the controller respect the required *gg diagram*, which represent the maximum combined longitudinal-lateral acceleration that can be induced by the combined longitudinal-lateral behavior of the specific tyre spec [50]. μ_x and μ_y are the longitudinal and lateral friction coefficient of the tyres, respectively, whereas the considered accelerations on the vehicle are

$$\begin{aligned} \ddot{x}_{\text{ext}} &= \frac{\sum_{i,j} F_{x_{i,j}} - F_d^x}{m}, \\ \ddot{y}_{\text{ext}} &= \frac{\sum_{i,j} F_{y_{i,j}}}{m}. \end{aligned} \tag{19}$$

At the i-th sampling instant, considering that the QP solution is Δu^{i^*}, the control input is updated by

$$u^{i^*} = u^{i-1^*} + \Delta u^{i^*}, \tag{20}$$

The first sample of u^{i^*} is applied to the vehicle, the prediction horizon is shifted forward and the optimization procedure is repeated with updated state measurement.

3.4. Co-Simulation Environment

The co-simulation platform, represented in the Figure 9, is composed of the following subsystems:

- Plant model: a 14 DoF vehicle model reproducing the overall vehicle dynamics behavior;
- Road pavement: a boundary condition module concerning the asphalt condition and computing the tire-road friction coefficient to reproduce dry, wet, snowy, and icy contact;
- Tyre: MF-Evo tyre model reproducing the tyre dynamic behavior in different thermal and wear conditions;
- Path reference: the track geometrical representation defined by the specific maneuver and employed to compute the cost function.

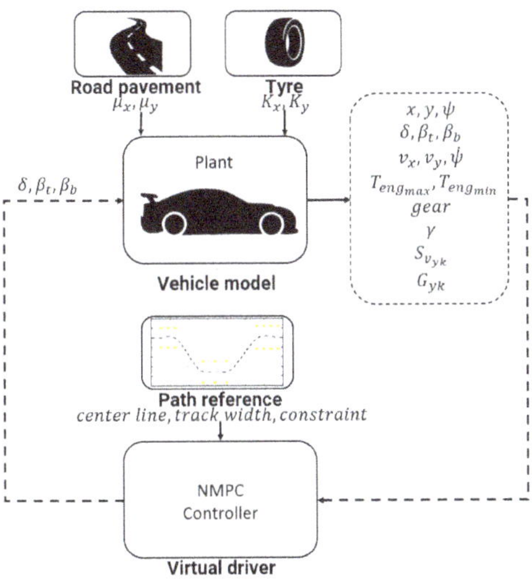

Figure 9. Co-simulaton platform.

The maneuver chosen for the current study is the emergency double-lane-change maneuver, generally performed on the highway to overtake another vehicle [51]. The test is commonly adopted because it correlates the ability of controlling the vehicle at the limits of handling with an enhanced safety for the vehicle occupants in scenarios concerning the presence of obstacles on the path [52]. Given the criteria for ideal lane-change path, prescribing a minimal length path with a smooth and continuous curvature at a given vehicle forward velocity, the trajectory of the DLC maneuver is computed without violating the track boundaries and assuring that all the tyres remain always in contact with the road surface (possible lift motions are avoided with constraints modelled within the maximum load transfers, as described in [9]).

The co-simulation is conducted in `MATLAB/Simulink` environment, coupling the plant model with NMPC controller and performing the dynamic simulation of the plant model at f_{sim} = 1000 Hz, while the control action is updated by NMPC at f_{ctrl} = 100 Hz. The simulations have been computed on a `Windows 10` machine with Intel(R) Core(TM) i7-7700HQ @ 2.80GHz CPU.

4. Analysis and Results

The knowledge of the instantaneous and potential grip directly on board and in real-time potentially allows the vehicle control logic to maximize the probability of avoiding obstacles and reducing the severity of collisions. To investigate the possible outcomes of a model-based control within a vehicle safety-linked scenario, the authors have performed within the DLC maneuver a complete design of experiment comprehending:

- *Case A:* the adoption of two different sets of NMPC weights (best and global) in the definition of the cost function.
  The best NMPC set of weights addresses the maximum achievable performance of the underlying vehicle plant model, specifically calibrated for a new tyre working in the optimal thermal range in contact with the dry road, whereas the global NMPC set of weights represents the trade-off solution to guarantee ability of the vehicle to complete the DLC maneuver in the worst proposed dynamic scenario, i.e., a worn cold tyre in contact the icy road surface. In this case, the parameters of the plant and the controller models are the same for each simulation;
- *Case B:* the analysis of the vehicle dynamic response in case of different tyre thermal and ageing conditions on the same road and in case of the tyre with a specific thermal and wear state on different pavements. In this case, the parameters of the plant and the controller models are the same for each simulation;
- *Case C:* the possibility to employ the non-linear model predictive controller calibrated with the average set of weights in conditions where the parameters of the controller model can be updated in real-time on the basis of the actual state of the plant model or can be constant and with an estimation on the friction value affected by a percentage error respect the real value.
  This particular scenario has been conducted to highlight the importance of the correct estimation of the parameters of the controller model, potentially aware of the actual knowledge of tyre-road friction. The simulation outputs with average tyre parameters within the controller model have been compared towards the ones obtained with the instantaneous parameters of the co-simulated vehicle plant to put in evidence the importance of the correct information concerning the tyre friction and stiffness for the vehicle dynamics control.

The simulation outputs have been compared in terms of the vehicle trajectory, the forward velocity, the vehicle side slip angle, and yaw angle.

4.1. Case A

In this section, the impact of two possible sets of weights, defined within the NMPC cost function, is investigated. Both the plant and controller models share the same model parameters of a new tyre in the optimal thermal window in contact with the dry road.

The *best* set of NMPC weights represents the most suitable solution to perform the DLC maneuver with both the plant model and the controller model in the maximum performance conditions of the tyre, corresponding the the maximum dynamic limits of the vehicle. The *global* set of NMPC weights stands for the conservative trade-off solution, calibrated to guarantee the accomplishment of the maneuver in all the possible tyre-linked and boundary conditions, in which the plant and controller models share the same physical parameters (i.e., the performance of the vehicle controller is limited by the worst possible dynamic scenario of a cold and worn tyre in contact with the icy road).

In the Figure 10a the trajectories of the vehicle with the *best* (red) and *global* (black) sets of NMPC weights are compared. It is easy to observe that the optimized set of weights allows the vehicle performing at a larger trajectory and achieving significantly higher velocities both in the first part of the curves and at the end of the DLC maneuver (Figure 10b). It is worth highlighting that the *best* set also demonstrates higher side slip and yaw angles (Figure 11a,b), because it is specifically optimized to perform in the scenario of a new optimal tyre in contact with the dry road, therefore allowing the vehicle to reach the actual friction limits. Furthermore, the *best* set allows the vehicle to approach to the DLC manuever and to end the scenario 6.62 and 8.34 s before, respectively (Figure 11c).

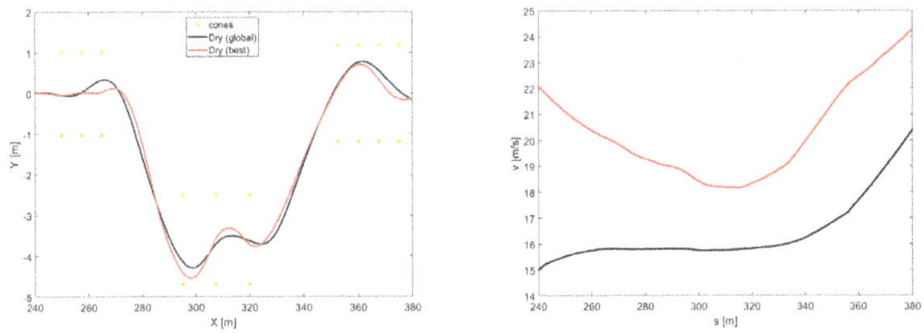

(a) (b)

Figure 10. (**a**) Vehicle trajectory performed in the DLC maneuvers in a different road surface (dry in black, wet in red, snow in blue, and icy in light blue), but with the same tyre (new tyre in optimal range temperature) for a NMPC tuned to better perform the maneuver in all road surface, tyre, and temperature condition. (**b**) Vehicle velocity.

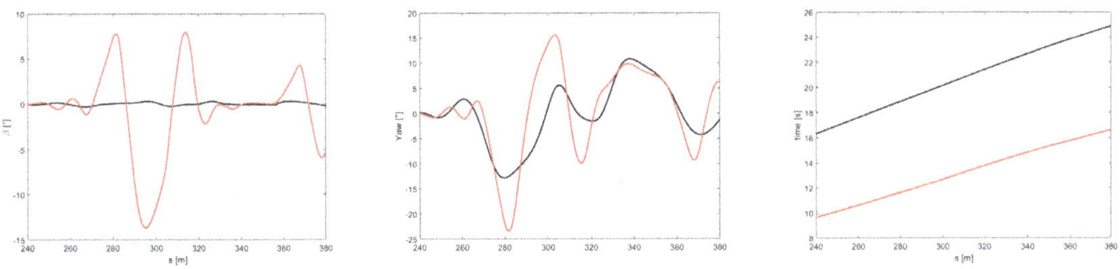

(a) (b) (c)

Figure 11. (**a**) β angle. (**b**) Yaw angle. (**c**) Time.

4.2. Case B

In this section, only the *global* set of NMPC weights has been employed to compare the dynamic response of the vehicle in two scenarios: (1) different road characteristics (dry, wet, snowy, and icy) with the new tyre within the optimal thermal range, and (2) different tyre thermal and ageing conditions in contact with the dry road. The plant and the controller models share the same physical parameters for each iteration.

- *Scenario B1*

  In the Figure 12a it is possible to observe how the vehicle maneuver characterized by the highest friction coefficient (dry pavement) performs the DLC with a largest trajectory and the highest velocity Figure 12b in minimum amount of time Figure 13c and Table 3.

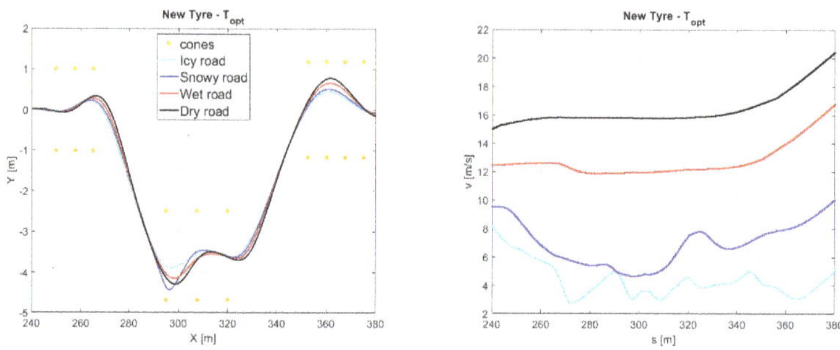

(**a**) (**b**)

Figure 12. (**a**) Vehicle trajectory. (**b**) Vehicle velocity.

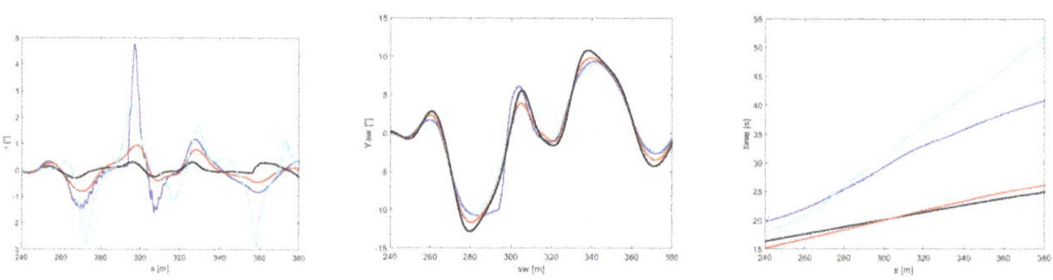

(**a**) (**b**) (**c**)

Figure 13. (**a**) Side slip angle. (**b**) Yaw angle. (**c**) Time.

Table 3. Summary of time's maneuver for each scenario.

Road Surface	Time [s]
Dry	24.8
Wet	26.2
Snowy	40.8
Icy	51.7

Since the *global* NMPC set is limited by the most critical dynamic condition (worn cold tyre in contact with the icy road), the Figure 13a shows higher values in terms of side slip angle for snowy and icy road surfaces, foreseeing the possibility to perform the maneuver in more aggressive way for dry and wet road conditions.
Such a conservative behavior can be motivated by the fact that the *global* set of weights is a result of a trade-off between completely different dynamic scenarios in the respect of vehicle maneuverability and safety.

- *Scenario B2*
  The comparison between a same road condition (dry) performing with different tyre condition (new or worn, in the optimal temperature range, cold or overheated) are shown in the following figure. Regarding the analysis of trajectories, shown in the Figure 14a it is possible to observe how they are too similar each other due to the same road pavement, however in the new tyre condition a little largest trajectory has been carried out to achieve an highest velocity Figure 14.

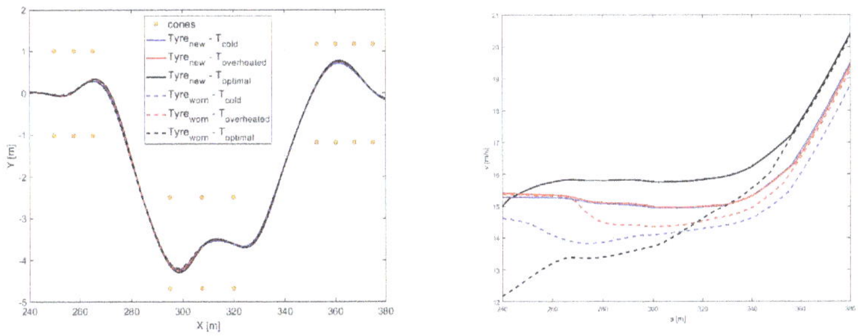

(**a**) (**b**)

Figure 14. (**a**) Vehicle trajectory performed in the DLC maneuvers in a dry road, with different tyre condition (New tyre (continuous lines) and worn tyre (dashed lines) in optimal (black), cold (blue), and overheated (red) temperature range. (**b**) Vehicle velocity.

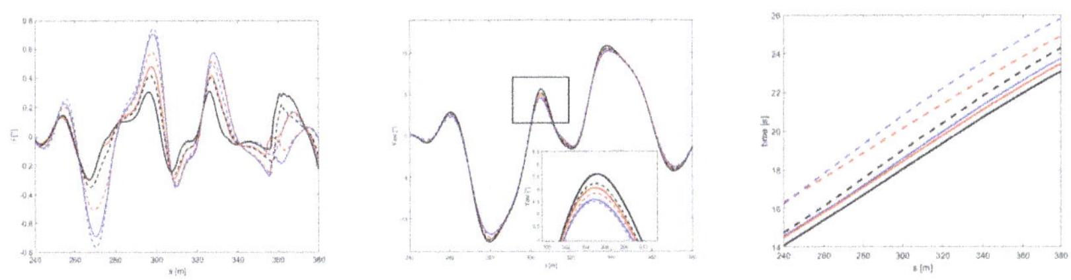

(**a**) (**b**) (**c**)

Figure 15. (**a**) Side slip angle. (**b**) Yaw angle. (**c**) Time.

Table 4. Summary of time's maneuver for each scenario.

Tyre Condition	Time s
$New - T_{opt}$	23.0
$New - T_{cold}$	23.7
$New - T_{overheated}$	23.44
$Worn - T_{opt}$	24.24
$Worn - T_{cold}$	25.8
$Worn - T_{overheated}$	24.8

The analysis side slip angle show a dependence of β angle with the tyre stiffness, indeed the highest value of β has been performed to highest cornering stiffness Figure 15a. Finally, in the Table 4 are shown the performing time for each condition.

4.3. Case C

- *Scenario C1*

  In this paragraph the aim of the authors is to argue the following query:

  If the plant and the controller do not share the same model parameters, i.e., the parameters of the controller model are not updated by a specific co-simulated estimator of the vehicle parameters and state, and of the tyres' and the road conditions are not known a priori, how a controller model with an average "parameters'" configuration could perform with different plant model employment scenarios within the DLC maneuver?

  With this purpose, the controller model has been fed with the parameters of friction and stiffness corresponding the mean value of the all possible tyre-road conditions explored.

  It is worth highlighting that, as expected, it is not possible to perform the DLC maneuver with the icy road with the above configuration. Indeed, as appears clear in the Figure 16, the rear axle achieves the maximum slip ratio, not allowing to complete the simulation in safety.

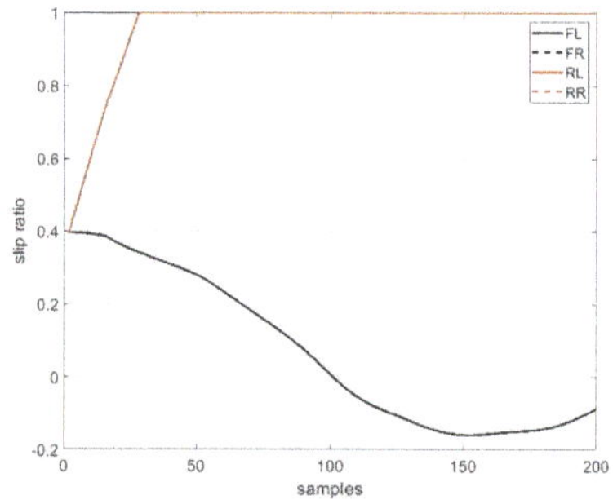

Figure 16. Slip ratio achieved for the four tyres.

For this reason, in the following figures, only dry, wet, and snowy road conditions are reported. In the Figure 17a,b it is possible to observe how the difference between the three pavement surfaces are less pronounced towards the results discussed in Scenario B. Moreover, the vehicle in contact with the wet road achieves a maximum velocity, even higher than with the dry surface, completing the maneuver in less time Figure 18c).

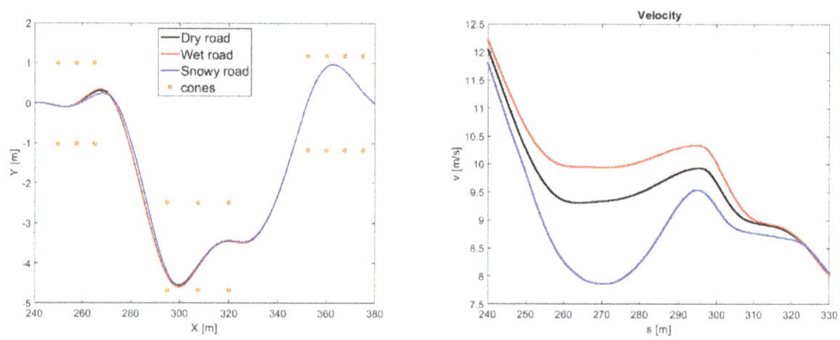

(a) (b)

Figure 17. (**a**) Vehicle trajectory performed in the DLC maneuvers in a dry, wet, and snow road, with new tyre in optimal range temperature. (**b**) Vehicle velocity

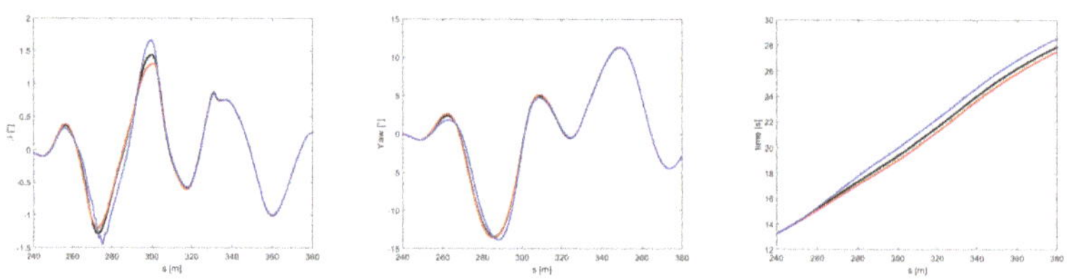

(a) (b) (c)

Figure 18. (**a**) Side slip angle. (**b**) Yaw angle. (**c**) Time.

The reason for such behavior can be conducted to the conservative control action, particularly visible in dry boundary condition, since the absolute difference in terms of the friction limit is particularly high between the plant and the controller models in this scenario. Indeed, in the Figure 18a the the side slip angle is similar for three conditions explored. Furthermore, even the maneuver in snow conditions is achieved in a comparable time period, since the friction limit of the average controller model is similar to the one of the plant model working in snowy boundary conditions Figure 18b.

- *Scenario C2*
  Remarking that an accurate online friction coefficient estimation becomes absolutely necessary to allow exploiting the vehicle dynamics in maximum performance conditions within a combined DLC maneuver, in this paragraph the aim of the authors is to argue another possible query: *In a real scenario, where an onboard tyre-road friction estimator able to estimate (among others) the grip parameter with a certain degree of accuracy and to update the control model parameters in run-time, is available, how a controller model with a percentage error concerning the vehicle instantaneous conditions could perform within the same maneuver?*
  With this purpose, the parameter concerning the tyres' friction of the controller model has been considered with an intrinsic error with a supposed standard deviation of $\pm 15\%$ respect to the actual grip value of the vehicle plant model.

It is worth noting that in a scenario where the grip factor is overestimated, the controller with the global configuration of the cost function computes more aggressive control actions leading to out-of cones trajectories and undesirable sliding effect. To avoid this issue, a robust global configuration has been introduced in order to let the controller being effective in managing the vehicle behavior in overestimated grip-linked scenarios. The above new configuration leads to more conservative actions and, consequently, to a considerable loss of performance in terms of velocity. In particular, the loss in performance in terms of the average speed (in percentage) in the four cases analyzed has been objectively quantified in Table 5. In Figures 19 and 20, the performance obtained by the two configurations in terms of trajectories, speed, side-slips, and yaw angles are also compared. Notably, the side-slip in Figure 5 reaches peaks of 5 degrees, confirming that the configurations obtained controls the vehicle at the limit of handling.

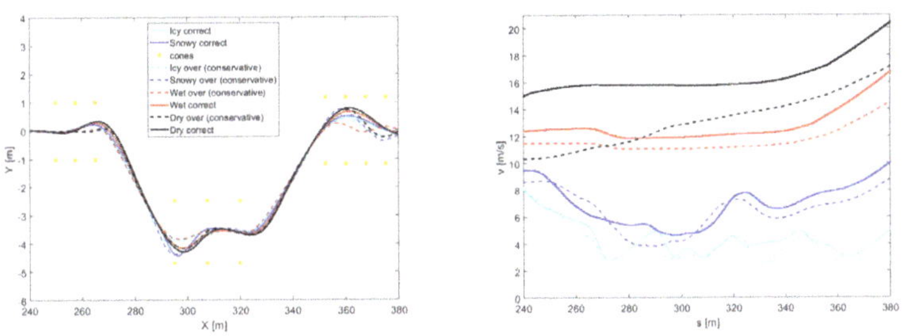

(a) (b)

Figure 19. (**a**) Vehicle trajectory performed in the DLC maneuvers in conservative vs global configuration. (**b**) Vehicle velocities

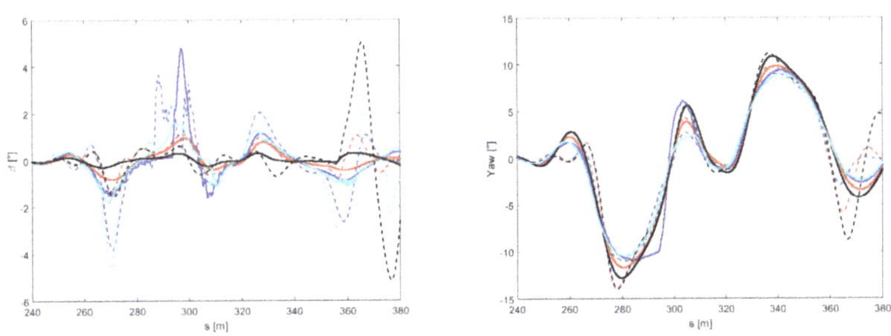

(a) (b)

Figure 20. (**a**) Side slip angles. (**b**) Yaw angles.

Table 5. Summary of the difference in velocity mean values (%) and lateral error assumed for each road scenario.

Road Surface [−]	Friction Estimation [−]	Longitudinal Velocity v [%]
Icy	correct	–
Icy	overestimation	−16.02
Snowy	correct	–
Snowy	overestimation	−8.36
Wet	correct	–
Wet	overestimation	−8.30
Dry	correct	–
Dry	overestimation	−21.00

5. Conclusions

The global autonomous mobility industry, growing at a rapid pace, is an intrinsically multidisciplinary field that aims at designing advanced onboard control logics by integrating principles from different disciplines including mechanical, control, and computer science engineering, legal, social, and economic fields. The multidisciplinary proposal will provide a paradigm shift in the development of strategies and a solid breakthrough towards enhanced development of the driving automatization systems.

In the proposed work, the authors have investigated how accurate information regarding the state of the real system of the parameters concerning the controller model could affect the behavior of the real system, represented in the form of the high-fidelity validated plant model. The influence of the tyre thermal dynamics, the impact of the possible ageing effects and the contact with different road pavements have been examined. Wrong parameters in the definition of the internal model of the NMPC might compromise the control performance, especially when the vehicle is supposed to drive at the limit of handling conditions. Specifically, a controller characterized by an overestimation of the grip conditions is forced to compute too aggressive control actions that might bring the vehicle in unstable and unsafe conditions, that are very difficult to handle for the controller itself. On the contrary, an underestimation of the grip might reduce the performance of the controller, which is forced to compute too conservative control actions. Moreover, the parameters of the cost function play an important role in defining the level of performance that the controller is required to achieve. A high weight on the travel time forces the vehicle to drive fast along the path, hence requiring effective proper internal model parameters to describe the vehicle behavior at the limit of handling. Instead, a more conservative tuning (i.e., high weights on side-slip, lateral error, orientation error) can be effective with also less precise coefficients, as the vehicle is not supposed to travel at the limit of handling. Due to these statements, in future work, the effects of including a real-time estimate of the tyre and the environment states, along with an adapting strategy for the weights of the cost function will be included in the whole analysis.

The investigation constitutes a part of the broader panorama of studies on novel model-based ADAS systems, which could become adaptive to the system state and boundary conditions, by means of a real-time physical model-based estimator of adherence, sensitive to environmental conditions and to the instantaneous state of tyres. In this way, the physical limits of the real system could become totally exploited by the control logic, minimizing the safety-related risks, and unleashing the potential of physical modeling to the next level of vehicle control.

Author Contributions: Conceptualization, A.S.; methodology, A.S., M.B. and E.M.; software, V.M.A., E.M. and E.P.; validation, A.S. and E.M.; formal analysis, A.S., E.M. and A.C.; investigation, V.M.A., E.M. and E.P.; resources, A.S., M.B., A.C. and E.F.; data curation, V.M.A. and E.M.; writing, A.S., V.M.A. and E.M.; visualization, M.B. and A.C.; supervision, A.S., M.B. and E.F.; project administration, A.S. M.B. and E.F.; funding acquisition, A.S. and E.F. All authors have read and agreed to the published version of the manuscript.

Funding: This research received no external funding.

Institutional Review Board Statement: Not applicable.

Informed Consent Statement: Not applicable.

Data Availability Statement: Data available on request due to industrial confidential agreements.

Conflicts of Interest: The authors declare no conflicts of interest.

References

1. West, W.; Limebeer, D. Optimal tyre management for a high-performance race car. *Veh. Syst. Dyn.* **2020**, 1–19. [CrossRef]
2. da Cunha Rodovalho, E.; de Tomi, G. Reducing environmental impacts via improved tyre wear management. *J. Clean. Prod.* **2017**, *141*, 1419–1427. [CrossRef]
3. Singh, K.B.; Sivaramakrishnan, S. An adaptive tire model for enhanced vehicle control systems. *SAE Int. J. Passeng. Cars-Mech. Syst.* **2015**, *8*, 128–145. [CrossRef]
4. Frezza, R.; Saccon, A.; Bacchet, D. SmartDriver: A sensor based model of a car driver for virtual product development. *Proc. IEEE/Asme Int. Conf. Adv. Intell. Mechatron.* **2003**, *1*, 366–370.
5. Bellet, T.; Deniel, J.; Bornard, J.C.; Richard, B. Driver modeling and simulation to support the virtual human centered design of future driving aids. In Proceeding of the INCOSE International Conference on Human-Systems Integration (HSI2019), Biarritz, France, 11–13 September 2019; pp. 11–13.
6. Marino, R.; Scalzi, S.; Netto, M. Nested PID steering control for lane keeping in autonomous vehicles. *Control Eng. Pract.* **2011**, *19*, 1459–1467. [CrossRef]
7. Wang, R.; Li, Y.; Fan, J.; Wang, T.; Chen, X. A Novel Pure Pursuit Algorithm for Autonomous Vehicles Based on Salp Swarm Algorithm and Velocity Controller. *IEEE Access* **2020**, *8*, 166525–166540. [CrossRef]
8. Choi, C.; Kang, Y. Simultaneous braking and steering control method based on nonlinear model predictive control for emergency driving support. *Int. J. Control Autom. Syst.* **2017**, *15*, 345–353. [CrossRef]
9. Bruschetta, M.; Picotti, E.; De Simoi, A.; Chen, Y.; Beghi, A.; Nishimura, M.; Tezuka, Y.; Ambrogi, F. Real-Time Nonlinear Model Predictive Control of a Virtual Motorcycle. *IEEE Trans. Control Syst. Technol.* **2021**, *29*, 2214–2222 [CrossRef]
10. Chen, Y.; Bruschetta, M.; Cuccato, D.; Beghi, A. An adaptive partial sensitivity updating scheme for fast nonlinear model predictive control. *IEEE Trans. Autom. Control* **2018**, *64*, 2712–2726. [CrossRef]
11. Frasch, J.V.; Gray, A.; Zanon, M.; Ferreau, H.J.; Sager, S.; Borrelli, F.; Diehl, M. An auto-generated nonlinear MPC algorithm for real-time obstacle avoidance of ground vehicles. In Proceedings of the 2013 European Control Conference (ECC), Zurich, Switzerland, 17–19 July 2013; pp. 4136–4141.
12. Houska, B.; Ferreau, H.; Diehl, M. ACADO Toolkit—An Open Source Framework for Automatic Control and Dynamic Optimization. *Optim. Control Appl. Methods* **2011**, *32*, 298–312. [CrossRef]
13. Carvalho, A.; Gao, Y.; Gray, A.; Tseng, H.E.; Borrelli, F. Predictive control of an autonomous ground vehicle using an iterative linearization approach. In Proceedings of the 16th International IEEE conference on intelligent transportation systems (ITSC 2013), The Hague, The Netherlands, 6–9 October 2013; pp. 2335–2340.
14. Zhang, V.; Thornton, S.M.; Gerdes, J.C. Tire modeling to enable model predictive control of automated vehicles from standstill to the limits of handling. In Proceedings of the 14th International Symposium on Advanced Vehicle Control, Nagoya, Japan, 14–18 September 2018.
15. Pacejka, H. *Tire and Vehicle Dynamics*; Butterworth-Heinemann Elsevier Ltd.: Oxford, UK, 2005.
16. Bruschetta, M.; Picotti, E.; Mion, E.; Chen, Y.; Beghi, A.; Minen, D. A Nonlinear Model Predictive Control based Virtual Driver for high performance driving. In Proceedings of the 2019 IEEE Conference on Control Technology and Applications (CCTA), Hong Kong, China, 19–21 August 2019; pp. 9–14. [CrossRef]
17. Liniger, A.; Domahidi, A.; Morari, M. Optimization-based autonomous racing of 1: 43 scale RC cars. *Optim. Control Appl. Methods* **2015**, *36*, 628–647. [CrossRef]
18. Kabzan, J.; Valls, M.I.; Reijgwart, V.J.; Hendrikx, H.F.; Ehmke, C.; Prajapat, M.; Bühler, A.; Gosala, N.; Gupta, M.; Sivanesan, R.; et al. Amz driverless: The full autonomous racing system. *J. Field Robot.* **2020**, *37*, 1267–1294. [CrossRef]
19. Liu, H.; Wei, H.; Zuo, T.; Li, Z.; Yang, Y.J. Fine-tuning ADAS algorithm parameters for optimizing traffic safety and mobility in connected vehicle environment. *Transp. Res. Part C Emerg. Technol.* **2017**, *76*, 132–149. [CrossRef] [PubMed]

20. Levinson, J.; Askeland, J.; Becker, J.; Dolson, J.; Held, D.; Kammel, S.; Kolter, J.Z.; Langer, D.; Pink, O.; Pratt, V.; et al. Towards fully autonomous driving: Systems and algorithms. In Proceedings of the 2011 IEEE Intelligent Vehicles Symposium (IV), Baden-Baden, Germany, 5–9 June 2011; pp. 163–168.
21. Persson, B.N. Contact mechanics for randomly rough surfaces. *Surf. Sci. Rep.* **2006**, *61*, 201–227. [CrossRef]
22. Farroni, F.; Sakhnevych, A.; Timpone, F. Physical modelling of tire wear for the analysis of the influence of thermal and frictional effects on vehicle performance. *Proc. Inst. Mech. Eng. Part L J. Mater. Des. Appl.* **2017**, *231*, 151–161. [CrossRef]
23. Sippl, C.; Bock, F.; Wittmann, D.; Altinger, H.; German, R. From simulation data to test cases for fully automated driving and ADAS. In *IFIP International Conference on Testing Software and Systems*; Springer: Berlin/Heidelberg, Germany, 2016; pp. 191–206.
24. Chen, Y.; Bruschetta, M.; Picotti, E.; Beghi, A. Matmpc-a matlab based toolbox for real-time nonlinear model predictive control. In Proceedings of the 2019 18th European Control Conference (ECC), Naples, Italy, 25–28 June 2019; pp. 3365–3370.
25. Berntorp, K.; Quirynen, R.; Di Cairano, S. Friction Adaptive Vehicle Control. U.S. Patent Application No. 16/299,285, 17 September 2020.
26. Laurense, V.A.; Goh, J.Y.; Gerdes, J.C. Path-tracking for autonomous vehicles at the limit of friction. In Proceedings of the 2017 American Control Conference (ACC), Seattle, WC, USA, 24–26 May 2017; pp. 5586–5591.
27. Zhao, J.; Zhang, J.; Zhu, B. Development and verification of the tire/road friction estimation algorithm for antilock braking system. *Math. Probl. Eng.* **2014**, *2014*, 786492. [CrossRef]
28. Santini, S.; Albarella, N.; Arricale, V.M.; Brancati, R.; Sakhnevych, A. On-Board Road Friction Estimation Technique for Autonomous Driving Vehicle-Following Maneuvers. *Appl. Sci.* **2021**, *11*, 2197. [CrossRef]
29. Papelis, Y.E.; Watson, G.S.; Brown, T.L. An empirical study of the effectiveness of electronic stability control system in reducing loss of vehicle control. *Accid. Anal. Prev.* **2010**, *42*, 929–934. [CrossRef] [PubMed]
30. Punzo, V.; Zheng, Z.; Montanino, M. About calibration of car-following dynamics of automated and human-driven vehicles: Methodology, guidelines and codes. *Transp. Res. Part C Emerg. Technol.* **2021**, *128*, 103165. [CrossRef]
31. Kim, A.; Otani, T.; Leung, V. Model-Based Design for the development and System-Level testing of ADAS. In *Energy Consumption and Autonomous Driving*; Springer: Berlin/Heidelberg, Germany, 2016; pp. 39–48.
32. Shakouri, P.; Czeczot, J.; Ordys, A. Simulation validation of three nonlinear model-based controllers in the adaptive cruise control system. *J. Intell. Robot. Syst.* **2015**, *80*, 207–229. [CrossRef]
33. Popov, V.L. *Contact Mechanics and Friction*; Springer: Berlin/Heidelberg, Germany, 2010.
34. Wang, H.; Al-Qadi, I.L.; Stanciulescu, I. Simulation of tyre–pavement interaction for predicting contact stresses at static and various rolling conditions. *Int. J. Pavement Eng.* **2012**, *13*, 310–321. [CrossRef]
35. Persson, B.N. Rubber friction: Role of the flash temperature. *J. Phys. Condens. Matter* **2006**, *18*, 7789. [CrossRef] [PubMed]
36. Farroni, F. TRICK-Tire/Road Interaction Characterization & Knowledge-A tool for the evaluation of tire and vehicle performances in outdoor test sessions. *Mech. Syst. Signal Process.* **2016**, *72*, 808–831.
37. Farroni, F.; Lamberti, R.; Mancinelli, N.; Timpone, F. TRIP-ID: A tool for a smart and interactive identification of Magic Formula tyre model parameters from experimental data acquired on track or test rig. *Mech. Syst. Signal Process.* **2018**, *102*, 1–22. [CrossRef]
38. Shim, T.; Ghike, C. Understanding the limitations of different vehicle models for roll dynamics studies. *Veh. Syst. Dyn.* **2007**, *45*, 191–216. [CrossRef]
39. Pacejka, H.B. *Tyre and Vehicle Dynamics*, 2nd ed.; Butterworth-Heinemann Elsevier Ltd.: Oxford, UK, 2006.
40. Damiano, C.; Flavio, F.; Aleksandr, S.; Gianluca, S.; Antonio, S.; Francesco, T. On the Implementation of an Innovative Temperature-Sensitive Version of Pacejka's MF in Vehicle Dynamics Simulations. In *Conference of the Italian Association of Theoretical and Applied Mechanics*; Springer: Berlin/Heidelberg, Germany, 2019; pp. 1084–1092.
41. Cossalter, V.; Da Lio, M.; Lot, R.; Fabbri, L. A general method for the evaluation of vehicle manoeuvrability with special emphasis on motorcycles. *Veh. Syst. Dyn.* **1999**, *31*, 113–135. [CrossRef]
42. Gao, Y.; Gray, A.; Frasch, J.V.; Lin, T.; Tseng, E.; Hedrick, J.K.; Borrelli, F. Spatial predictive control for agile semi-autonomous ground vehicles. In Proceedings of the 11th International Symposium on Advanced Vehicle Control, Seoul, Korea, 9–12 September 2012.
43. Gillespie, T.D. *Vehicle Dynamics*; Society of Automotive Engineers: Warrendale, PA, USA, 1997.
44. Guiggiani, M. *The Science of Vehicle Dynamics*; Springer: Pisa, Italy, 2014.
45. Chen, Y.; Cuccato, D.; Bruschetta, M.; Beghi, A. An inexact sensitivity updating scheme for fast nonlinear model predictive control based on a curvature-like measure of nonlinearity. In Proceedings of the 2017 IEEE 56th Annual Conference on Decision and Control (CDC), Melbourne, Australia, 12–14 December 2017; pp. 4382–4387.
46. Bock, H.G.; Plitt, K.J. A multiple shooting algorithm for direct solution of optimal control problems. In *Proceedings of the IFAC World Congress*; Elsevier: Budapest, Hungary, 1984.
47. Diehl, M.; Bock, H.G.; Schlöder, J.P.; Findeisen, R.; Nagy, Z.; Allgöwer, F. Real-time optimization and nonlinear model predictive control of processes governed by differential-algebraic equations. *J. Process. Control* **2002**, *12*, 577–585. [CrossRef]
48. Chen, Y.; Scarabottolo, N.; Bruschetta, M.; Beghi, A. Efficient move blocking strategy for multiple shooting-based non-linear model predictive control. *IET Control Theory Appl.* **2019**, *14*, 343–351. [CrossRef]
49. Zeilinger, M.N.; Jones, C.N.; Morari, M. Robust stability properties of soft constrained MPC. In Proceedings of the 49th IEEE Conference on Decision and Control (CDC), Atlanta, GA, USA, 15–17 December 2010; pp. 5276–5282. [CrossRef]

50. Brach, R.; Brach, M. *The Tire-Force Ellipse (Friction Ellipse) and Tire Characteristics*; SAE Technical Paper 2011-01-0094. 2011. Available online: https://doi.org/10.4271/2011-01-0094 (accessed on 12 April 2011).
51. Arefnezhad, S.; Ghaffari, A.; Khodayari, A.; Nosoudi, S. Modeling of double lane change maneuver of vehicles. *Int. J. Automot. Technol.* **2018**, *19*, 271–279. [CrossRef]
52. Nilsson, J.; Brännström, M.; Coelingh, E.; Fredriksson, J. Lane change maneuvers for automated vehicles. *IEEE Trans. Intell. Transp. Syst.* **2016**, *18*, 1087–1096. [CrossRef]

sensors

MDPI

Letter

Nonlinear Ride Height Control of Active Air Suspension System with Output Constraints and Time-Varying Disturbances

Rongchen Zhao [1], Wei Xie [2,*], Jin Zhao [3], Pak Kin Wong [4] and Carlos Silvestre [2,†]

1 School of Mechanical and Electrical Engineering, Guizhou Normal University, Guizhou 550001, China; zhaorongchen@outlook.com
2 Department of Electrical and Computer Engineering, University of Macau, Macau 999078, China; csilvestre@um.edu.mo
3 Key Laboratory of Advanced Manufacturing Technology of the Ministry of Education, Guizhou University, Guizhou 550025, China; zhaoj@gzu.edu.cn
4 Department of Electromechanical Engineering, University of Macau, Macau 999078, China; fstpkw@um.edu.mo
* Correspondence: weixie@um.edu.mo
† Carlos Silvestre is on leave from the Instituto Superior Técnico of the Universidade de Lisboa, Portugal.

Abstract: This paper addresses the problem of nonlinear height tracking control of an automobile active air suspension with the output state constraints and time-varying disturbances. The proposed control strategy guarantees that the ride height stays within a predefined range, and converges closely to an arbitrarily small neighborhood of the desired height, ensuring uniform ultimate boundedness. The designed nonlinear observer is able to compensate for the time-varying disturbances caused by external random road excitation and perturbations, achieving robust performance. Simulation results obtained from the co-simulation (AMESim-Matlab/Simulink) are given and analyzed, demonstrating the efficiency of the proposed control methodology.

Keywords: nonlinear height control; active air suspension; output constraints; random road excitation; disturbance observer design

Citation: Zhao, R.; Xie, W.; Zhao, J.; Wong, P.K.; Silvestre, C. Nonlinear Ride Height Control of Active Air Suspension System with Output Constraints and Time-Varying Disturbances. *Sensors* **2021**, *21*, 1539. https://doi.org/10.3390/s21041539

Academic Editors: Juan A. Cabrera and Maria Gabriella Xibilia

Received: 7 December 2020
Accepted: 19 February 2021
Published: 23 February 2021

1. Introduction

Active vehicle suspensions are effective ways to isolate or dissipate the vibration energies transferred from irregular road excitation to vehicle body [1–3]. With the development of automobile industry, the active suspension has demonstrated its capability in (1) improving ride comfort, i.e., reducing vehicle body acceleration, and (2) the safety performance constraint, such as suspension dynamic displacement, tire dynamic payload, and actuator input saturation [2,4–6]. As it is convenient to employ electronically-controlled active air suspension (AAS) systems to adjust the ride height by inflating and deflating the air spring, they have drawn attention from automobile manufactories (e.g., Tesla) and have been extensively utilized in commercial vehicles [7–9].

However, there are still many challenges in regulating the ride height motion of the vehicle body (with the AAS system) robustly and accurately under random road excitation. Moreover, the adjustment of ride height usually changes the stiffness and hysteresis, and generates perturbations in the AAS system [10,11]. In addition, because of the mechanical structure and travel limitations of the AAS, the ride height movement should always be constrained in a reliable range for safety performance [12–14]. Therefore, an appropriate ride height controller should be designed for the AAS systems in the presence of perturbations and output constraints.

Aiming to deal with the aforesaid problems, many results have been reported, such as robust H_∞ control [15,16], sliding mode control (SMC) [11,17,18], fuzzy logic [19], neural network-based [20], and backstepping control techniques [14,21]. In [16], a robust H_∞ controller for AAS systems was proposed, where the ride comfort and time domain hard constraints were considered. However, the model uncertainties are linearized by transformation of their utilized dynamic model [16]. As this model cannot capture the actual behavior of the AAS system, it could deteriorate the height tracking performance. Nonlinear controllers were proposed by employing SMC technique to handle the external random road excitation and perturbations in the AAS system [11,17,18]. However, the authors of [11] dealt with the time-varying disturbances by choosing high control gains for robustness resulting in unwanted oscillations. To cope with this drawback, an adaptive SMC scheme was proposed by using neural networks to increase SMC properties in [18]. Simulation results and Lyapunov-based stability proof were presented, demonstrating the proposed control method can stabilize displacement and speed of the suspension systems. Similarly, the backstepping control has been extensively investigated by employing fuzzy logic and neural networks for enhancing the control performance [19,20]. However, in order to make the approximation error arbitrarily small, the numbers of FLS rules or neurons should be increased, resulting in a heavy computational burden. Meanwhile, to our best knowledge, the numbers of the needed rules or neurons are difficult to be determined for keeping the estimation error bounded in a specific range. In [14,21], nonlinear backstepping-based height tracking controllers were designed, where some conservativeness was adopted in the control law to reduce the effects of time-varying disturbances.

In addition to the challenge raised from developing control strategy for handling disturbances, the output height constraint is also considered as a critical issue due to the mechanical structure limitation of the AAS system. Although the nonlinear ride height controllers based on the classic Quadratic Lyapunov function are presented to track predefined trajectories in the presence of perturbations and the height constraint are neglected for the ride height control applications with the AAS system [14,21]. By using the backstepping control approaches, the Barrier Lyapunov Functions (BLFs) have been developed and defined as control Lyapunov candidates for achieving the constrained objectives control [13,22–24]. In [22], the BLFs are employed in the controller design. Moreover, an asymmetric BLF is presented and employed in the constrained controller design to handle the external disturbances without violation of the output constraints [23]. In addition, the author of [13] proposed the constrained adaptive controller for damping force control by using the BLFs, improving ride comfort, and satisfying the performance constraints. However, the height motion control of AAS systems in the presence of output constraint has not been addressed yet.

Inspired by the above discussion, this paper presents a novel solution to address the height tracking control problem of nonlinear AAS system with output constraint and external time-varying disturbances. The novelties and contributions of this paper are summarized as follows.

- A nonlinear height tracking controller for the nonlinear AAS system is proposed, guaranteeing that (i) the output height always stays in a predefined range and (ii) uniform ultimate boundedness is achieved.
- A nonlinear disturbance observer is designed to compensate the time-varying disturbances caused by external random road excitation and perturbations in the AAS system.

With respect to the state-of-the art approaches, the main merits of the proposed constrained control strategy are as follows. In this research, unlike the linearized models used in [16], the mathematical model with the time-varying disturbances is employed to describe the perturbations in the AAS system. Unlike the disturbance rejection methods presented in [14,19–21], a time-varying disturbance observer is designed in this paper, guaranteeing that the estimation error is bounded by certain value. The designed disturbance observer can guarantee the estimate converges closely to zero. Moreover, we take the output constraints into consideration by using the BLFs in the backstepping controller design. By contrast, the output constraint was neglected in [11,17,19,20].

The remainder of this paper is organized as follows. In Section 2, the notation used throughout this paper is introduced. Section 3 presents a mathematical model of the AAS system and formulates the control problem. In Section 4, a nonlinear disturbance observer is designed, and a constrained ride height controller is proposed based on the backstepping control technique with BLFs. To validate the efficiency of proposed control strategy, co-simulation results are presented and analyzed in Section 5. At last, Section 6 summarizes the contents of this paper and describes the future work.

2. Notation

In this paper, $\mathbb{R}^n$ denotes the $n-$dimensional euclidean space. A function f is of class C^n if the derivatives $f', f'', ..., f^n$ exist and are continuous. For a vector $\mathbf{x} \in \mathbb{R}^n$, its norm is defined as $\|\mathbf{x}\| = \sqrt{\mathbf{x}^\top\mathbf{x}}$. $\int_0^t \mathbf{x}\, d\tau$ denotes the integral of $\mathbf{x}$, and τ denotes the integration variable. For the reader's reference, Table 1 summarizes the main symbols and descriptions for the model, controller, and parameter estimators in the paper.

Table 1. Symbols and their descriptions.

Symbol	Description	Symbol	Description
h_s (m)	height of vehicle sprung mass	m (kg)	sprung mass of quarter vehicle
h_u (m)	unsprung mass displacement	m_u (kg)	unsprung mass of quarter vehicle
h_d (m)	desired height	$\dot{m}_{des}$ (g)	desired change of air mass for air spring
h_r (m)	road excitation	$\bar{A}_s$ (mm^2)	area of adjustable air spring
h_0 (m)	initial height of sprung mass	$\bar{h}_d$ (m)	maximum value of desired height
b (N $\cdot$ s $\cdot$ m^{-1})	damping coefficient of damper	h_{max} (m)	maximum value of sprung height
p_s (Pa)	air spring pressure	$d(t)$	time-varying disturbances
p_0 (Pa)	initial air pressure	d_{max}	maximum value of disturbances
v_s (m^3)	air spring volume	k_t (N $\cdot$ m^{-1})	tire stiffness
$\dot{Q}$ (J $\cdot$ s^{-1})	heat transfer rate	u	control input
F_s	air spring force	F_d	damping force

3. Problem Formulation

The objective of this section is to formulate the problem of nonlinear ride height tracking control with application to the AAS system in the presence of output constraints and the time-varying disturbances. We start by presenting the mathematical model of a quarter vehicle with AAS system. Then, the problem of constrained ride height tracking control is formulated with the maximum boundary value of vehicle ride height and the time-varying disturbances.

3.1. ASS Modeling

In order to describe the dynamic characteristics, a mathematical model of a quarter vehicle with AAS is employed as a part of model-based height controller design for realizing the control objective successfully. The schematic diagram of a quarter vehicle with AAS is shown in Figure 1, and the dynamic equations of the sprung mass and unsprung mass are given by

$$\begin{aligned} m_s\ddot{h}_s &= F_s - F_d - F_g \\ m_u\ddot{h}_u &= -F_s + F_d - F_t - D_t \end{aligned} \tag{1}$$

where $F_g = m_s g$ denotes the gravitational force; m_s is the sprung mass of a quarter vehicle; m_u is the unsprung mass, which denotes mass of the wheel assembly; and F_s and F_d represent the forces produced by the air spring and damper, respectively. F_t and D_t are the elasticity force and damping force of the tire, respectively. Forces produced by the nonlinear air spring, the linear damper, and the tire yield,

$$\begin{aligned} F_s &= \bar{A}_s(p_s - p_a) \\ F_d &= b(\dot{h}_s - \dot{h}_u) \\ F_t &= k_t(h_u - h_r) \\ D_t &= b_t(\dot{h}_u - \dot{h}_r) \end{aligned} \quad (2)$$

where p_a is the atmospheric pressure, $s_p = 10^5$, and $\bar{A}_s = A_s s_p$ is the effective area of the adjustable air spring. h_r denotes the random road excitation, and h_u and h_s are the displacements of the unsprung mass and the sprung mass of the quarter vehicle, respectively. b is the damping coefficient, k_t and b_t are the stiffness and damping coefficient of the tire, respectively.

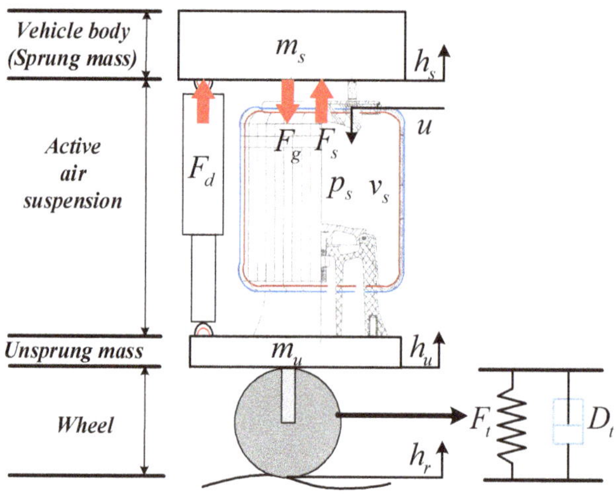

Figure 1. Schematic diagram of quarter vehicle with active air suspension (AAS).

Due to the perturbations in the AAS system, the time-varying disturbances should be considered in the employed model for ride height control. From (1), the quarter vehicle model with the AAS can be then expressed in a compact state-space form as

$$\begin{aligned} \dot{h}_1 &= h_2 \\ \dot{h}_2 &= m_s^{-1}\left(\bar{A}_s(h_3 - p_{atm}) - b(h_2 - \dot{z}_u) - m_s g\right) \\ \dot{h}_3 &= u - \gamma A_s h_2 h_3 v_s^{-1} + \dot{Q}(\gamma - 1)(s_p v_s)^{-1} + d(t) \end{aligned} \quad (3)$$

where $h_1 = h_s$, $h_2 = \dot{h}_s$, $h_3 = p_s s_p^{-1}$. v_s represents the air volume; $\dot{Q} = h_t A_{heat}(T_e - T_{as})$ is the heat transfer rate between the inner and the outer sides of the control volume, in which h_t is the heat transfer coefficient; A_{heat} represents the area of the heat transfer; T_e and T_{as} are temperatures of the outer and the inner sides of the control volume, respectively; and u is viewed as the control input for inflating and deflating the adjustable air spring. The time-varying disturbances are denoted by $d(t)$ with the following assumption:

Assumption 1. *Disturbances $d(t)$ are unknown, time-varying and satisfy*

$$|d(t)| \le d_{max}, \quad |\dot{d}(t)| \le \bar{d}$$

where d_{max} and $\bar{d}$ are known positive numbers.

3.2. Output Constraint and Barrier Lyapunov Function

In practice, because of the structure limitation, the output height of the AAS system should be constrained by $h_{\max}$, which denotes the maximum ride height of the AAS. Inspired by the work in [24], the following BLF is used.

$$V_b = \frac{1}{2}Y^2(z_1) \tag{4}$$

where

$$Y(z_1) = \frac{\mu^2 z_1}{\mu^2 - z_1^2} \tag{5}$$

where $z_1 = h_1 - h_d$ is the velocity tracking error, h_d denotes the desired height under the assumption that $|h_d| < \bar{h} < h_{\max}$, and $\mu = h_{\max} - \bar{h}$. To facilitate the analysis, we formulate a simple lemma, given as

Lemma 1. *For any two nonzero scalars $x \in \mathbb{R}$, $y \in \mathbb{R}$, if $|x| < x_{\max}$, $|y| < y_{\max}$, $x_{\max} > y_{\max} > 0$, then we have*

$$|x| - |y| \leq |x - y|. \tag{6}$$

Furthermore, based on Equation (6), if we have $|x - y| < x_{\max} - y_{\max}$, $|y| < y_{\max}, x_{\max} > y_{\max} > 0$, we can obtain $|x| < x_{\max}$.

Remark 1. *The BLF V_b is positive definite and C^1 continuous for $|z_1| < \mu$.*

Remark 2. *If there is no constraint on h_1, that is, $h_{\max} \to +\infty$, the BLF becomes*

$$V_b = \frac{1}{2}z_1^2 \tag{7}$$

which is a quadratic Lyapunov function.

3.3. Problem Statement

For AAS systems, the following control objectives should be considered in the ride height controller design.

- The proposed ride height controller can guarantee the accurate trajectory tracking performance in the presence of time-varying disturbances.
- Due to the mechanical structure and travel limitation of the AAS, the dynamic height should be restrained within its allowable maximum value, which is expressed as $|h_s| < h_{\max}$.

4. Nonlinear Backstepping Controller Synthesis

In this section, the control objective is to design the virtual control input u for the ASS that ensures convergence of the ride height to an arbitrarily small neighborhood of the desired height without violating the requirement of output constraint $|h_1| < h_{\max}$. A disturbance observer $\hat{d}(t)$ is first designed to estimates $d(t)$, and then a constrained controller is designed based on the backstepping technique by using the BLF. Details are given in the sequel.

4.1. Disturbance Observer Design

To design the disturbance observer for $d(t)$, we take a clue from the work in [25] and define two auxiliary terms as

$$\begin{aligned} \xi &= d(t) - \phi(h_3), \\ \hat{\xi} &= \hat{d}(t) - \phi(h_3), \end{aligned} \quad (8)$$

where $\phi(h_3) = \lambda_d h_3$, λ_d is a positive estimation gain, and $\hat{d}(t)$ is the estimation of $d(t)$. From (8), we have

$$\hat{d}(t) = \hat{\xi} + \phi(h_3) \quad (9)$$

and the estimation error is

$$d_e = \xi - \hat{\xi}. \quad (10)$$

Computing the time derivative of ξ, we obtain

$$\dot{\xi} = \dot{d}(t) - \frac{\partial\phi(h_3)}{\partial h_3}\Big[u - \gamma A_s h_2 h_3 v_s^{-1} + \dot{Q}(\gamma - 1)(s_p v_s)^{-1} + d(t)\Big]. \quad (11)$$

Then, we introduce the time derivative of the estimated $\hat{\xi}$, given as

$$\dot{\hat{\xi}} = -\frac{\partial\phi(h_3)}{\partial h_3}\Big[u - \gamma A_s h_2 h_3 v_s^{-1} + \dot{Q}(\gamma - 1)(s_p v_s)^{-1} + \hat{d}(t)\Big], \quad (12)$$

leading to

$$\begin{aligned} \dot{d}_e &= \dot{\xi} - \dot{\hat{\xi}} \\ &= \dot{d}(t) - \lambda_d d_e. \end{aligned} \quad (13)$$

The main result is summarized in the following Lemma.

Lemma 2. *Through the use of designed disturbance observer (9), the estimate $|d_e(t)|$ exponentially converges to the circle centered at the origin with radius $d_e(t)(4\varepsilon(\lambda_d - \varepsilon))^{-\frac{1}{2}}$, which can be made arbitrarily small by increasing the estimation gain λ_d, where ε is a positive constant.*

Proof. We start the proof by defining a Lyapunov candidate function, given as

$$V_d(t) = \frac{1}{2}d_e(t)^2. \quad (14)$$

Computing the time derivative of $V_d(t)$, we have

$$\begin{aligned} \dot{V}_d(t) &= -\lambda_d d_e(t)^2 + d_e(t)\dot{d}(t) \\ &\leq -2(\lambda_d - \varepsilon)V_d(t) + \bar{d}^2(4\varepsilon)^{-1}, \end{aligned} \quad (15)$$

where $\lambda_d > \varepsilon$. Solving (15), we obtain

$$V_d(t) \leq e^{-2(\lambda_d - \varepsilon)}V_d(0) + \bar{d}^2(8\varepsilon(\lambda_d - \varepsilon))^{-1}. \quad (16)$$

From here we can conclude that V_d converges to a circle of radius $\bar{d}^2(8\varepsilon(\lambda_d - \varepsilon))^{-1}$. It follows that $|d_e(t)|$ converges to a circle of radius $\bar{d}(4\varepsilon(\lambda_d - \varepsilon))^{-\frac{1}{2}}$, which can be made arbitrarily small by increasing λ_d. □

4.2. Constrained Controller Design

Let the desired height h_d be a curve of class at least C^3, with all its time derivatives bounded. In order to address the constrained height tracking problem, we consider (4) as an initial Lyapunov function candidate given by

$$V_1 = V_b = \frac{1}{2}\Upsilon(z_1)^2, \tag{17}$$

whose time derivative yields

$$\dot{V}_1 = \Upsilon(z_1)\dot{\Upsilon}(z_1), \tag{18}$$

where

$$\dot{\Upsilon}(z_1) = \frac{2\mu^2 z_1^2}{(\mu^2 - z_1^2)^2}\dot{z}_1 + \frac{\mu^2}{\mu^2 - z_1^2}\dot{z}_1, \tag{19}$$

For the sake of simplicity, we define δ_1 and δ_2 as

$$\delta_1 = \frac{2\mu^2 z_1^2}{(\mu^2 - z_1^2)^2}, \delta_2 = \frac{\mu^2}{\mu^2 - z_1^2}, \tag{20}$$

then, Equation (18) can be rewritten as

$$\dot{V}_1 = \Upsilon(z_1)(\delta_1\dot{z}_1 + \delta_2\dot{z}_1). \tag{21}$$

Isolating a negative definite term in $\Upsilon(z_1)$ and rearranging the terms of $\dot{V}_1$, we get

$$\dot{V}_1 = -W_1(z_1) + \Upsilon(z_1)(\delta_1\dot{z}_1 + \delta_2\dot{z}_1 + k_1\Upsilon(z_1)). \tag{22}$$

where $W_1(z_1) = k_1\Upsilon(z_1)^2$, and k_1 is a positive number. Following the backstepping technique, we define the new error z_2 as

$$z_2 = \delta_1\dot{z}_1 + \delta_2\dot{z}_1 + k_1\Upsilon(z_1), \tag{23}$$

and rewriting (22), we have

$$\dot{V}_1 = -W_1(z_1) + \Upsilon(z_1)z_2. \tag{24}$$

Constructing a new Lyapunov function candidate by incorporating z_2, we obtain

$$V_2 = \frac{1}{2}\Upsilon(z_1)^2 + \frac{1}{2}z_2^2, \tag{25}$$

with time derivative

$$\begin{aligned}\dot{V}_2 = -W_2(z_1, z_2) + z_2\Big((m_s^{-1}\bar{A}_s(h_3 - p_{atm}) - m_s^{-1}b(h_2 - \dot{z}_u) - g - \ddot{h}_d)(\delta_1 + \delta_2) \\ + \Upsilon(z_1) + (\dot{\delta}_1 + \dot{\delta}_2)\dot{z}_1 + k_1\dot{\Upsilon}(z_1) + k_2 z_2\Big),\end{aligned} \tag{26}$$

where $W_2(z_1, z_2) = W_1(z_1) + k_2 z_2^2$, k_2 is a positive number,

$$\dot{\delta}_1 = \frac{4\mu^2 z_1\dot{z}_1}{(\mu^2 - z_1^2)^2} + \frac{8\mu^2 z_1^3\dot{z}_1}{(\mu^2 - z_1^2)^3}, \dot{\delta}_2 = \frac{2\mu^2 z_1\dot{z}_1}{(\mu^2 - z_1^2)^2}. \tag{27}$$

Furthermore, we can rewrite (26) as

$$\dot{V}_2 = -W_2(z_1, z_1) + z_2(\delta_1 + \delta_2)\Big[m_s^{-1}\bar{A}_s(h_3 - p_{atm}) - m_s^{-1}b(h_2 - \dot{z}_u) - g - \ddot{h}_d + \frac{\Upsilon(z_1)}{\delta_1 + \delta_2} + \Big(\frac{\dot{\delta}_1 + \dot{\delta}_2}{\delta_1 + \delta_2} + k_1\Big)\dot{z}_1 + \frac{k_2 z_2}{\delta_1 + \delta_2}\Big]. \tag{28}$$

Continuing with the backstepping procedure, we define the last error term as

$$\begin{aligned} z_3 = {} & m_s^{-1}\bar{A}_s(h_3 - p_{atm}) - m_s^{-1}b(h_2 - \dot{z}_u) - g - \ddot{h}_d + \frac{\Upsilon(z_1)}{\delta_1 + \delta_2} \\ & + \Big(\frac{\dot{\delta}_1 + \dot{\delta}_2}{\delta_1 + \delta_2} + k_1\Big)\dot{z}_1 + \frac{k_2 z_2}{\delta_1 + \delta_2}, \end{aligned} \tag{29}$$

and augment the Lyapunov function candidate as

$$V_3 = V_2 + \frac{1}{2}z_3^2. \tag{30}$$

The closed-loop time derivative is then

$$\begin{aligned} \dot{V}_3 = {} & -W_3(z_1, z_2, z_3) + z_3\Big[m_s^{-1}\bar{A}_s\dot{h}_3 - m_s^{-1}b(h_2 - \dot{z}_u) - h_d^{(3)} \\ & + \Big(\frac{\dot{\delta}_1 + \dot{\delta}_2}{\delta_1 + \delta_2} + k_1 + k_2 + 1\Big)\ddot{z}_1 + \Big(\frac{\ddot{\delta}_1 + \ddot{\delta}_2}{\delta_1 + \delta_2} - \frac{\dot{\delta}_1 + \dot{\delta}_2}{(\delta_1 + \delta_2)^2} \\ & + \frac{k_1 k_2}{\delta_1 + \delta_2}\Big)\dot{z}_1 + \frac{\dot{\delta}_1 + \dot{\delta}_2}{(\delta_1 + \delta_2)^2}k_1 k_2 \Upsilon(z_1) \\ & + (\delta_1 + \delta_2)z_2 + k_3 z_3\Big]. \end{aligned} \tag{31}$$

where

$$\begin{aligned} \ddot{\delta}_1 &= \frac{4\mu^2\dot{z}_1^2}{(\mu^2 - z_1^2)^2} + \frac{40\mu^2 z_1^2\dot{z}_1^2}{(\mu^2 - z_1^2)^3} + \frac{48\mu^2 z_1^4\dot{z}_1^2}{(\mu^2 - z_1^2)^4} + \frac{4\mu^2 z_1\ddot{z}_1}{(\mu^2 - z_1^2)^2} + \frac{8\mu^2 z_1^3\ddot{z}_1}{(\mu^2 - z_1^2)^3}, \\ \ddot{\delta}_1 &= \frac{2\mu^2\dot{z}_1^2}{(\mu^2 - z_1^2)^2} + \frac{8\mu^2 z_1^2\dot{z}_1^2}{(\mu^2 - z_1^2)^3} + \frac{2\mu^2 z_1\ddot{z}_1}{(\mu^2 - z_1^2)^2}, \end{aligned} \tag{32}$$

and $W_3(z_1, z_2, z_3) = W_2(z_1, z_2) + k_3 z_3^2$, and k_3 is a positive number

Here, note that the time derivative of $\dot{V}_3$ is dependent on the disturbances $d(t)$ through the dependency of $\dot{h}_3$ in these quantities. In order to exploit the dependency of $\dot{h}_3$ in the uncertain quantities, the time derivative $\dot{h}_3$ can be expressed as

$$\dot{h}_3 = u - \gamma A_s h_2 h_3 v_s^{-1} + \dot{Q}(\gamma - 1)(s_p v_s)^{-1} + \hat{d}(t) + d_e(t), \tag{33}$$

where $d_e(t)$ is the estimation error. We now establish the final Lyapunov function candidate by adding the terms of disturbance estimate error to V_3 as follows,

$$V_{3b} = \dot{V}_3 + \frac{1}{2}d_e(t)^2. \tag{34}$$

Computing the time derivative of V_{3b}, we obtain

$$\begin{aligned} \dot{V}_{3b} = {} & -W_3(z_1, z_2, z_3) + z_3\Big[M + m_s^{-1}\bar{A}_s\Big(u - \gamma A_s h_2 h_3 v_s^{-1} + \dot{Q}(\gamma - 1)(s_p v_s)^{-1} \\ & + \hat{d}(t)\Big) - m_s^{-1}b(h_2 - \dot{z}_u) - h_d^{(3)}\Big] + z_3 m_s^{-1}\bar{A}_s d_e(t) + d_e(t)(\dot{d}(t) - d_e(t)). \end{aligned} \tag{35}$$

where

$$M = \Big(\frac{\dot{\delta}_1 + \dot{\delta}_2}{\delta_1 + \delta_2} + k_1 + k_2 + 1\Big)\ddot{z}_1 + \Big(\frac{\ddot{\delta}_1 + \ddot{\delta}_2}{\delta_1 + \delta_2} - \frac{\dot{\delta}_1 + \dot{\delta}_2}{(\delta_1 + \delta_2)^2} + \frac{k_1 k_2}{\delta_1 + \delta_2}\Big)\dot{z}_1 + \frac{\dot{\delta}_1 + \dot{\delta}_2}{(\delta_1 + \delta_2)^2} k_1 k_2 Y(z_1) + (\delta_1 + \delta_2) z_2 + k_3 z_3. \tag{36}$$

Here, we notice that apart from the time derivative of disturbances $\dot{d}(t)$ and estimated error $d_e(t)$, $\dot{V}_{3b}$ is also dependent on the z_1, z_2, z_3. To cancel the dependency of $\dot{V}_{3b}$ on z_1, z_2, z_3 in (35), the virtual control law u is chosen as

$$u = \bar{A}_s^{-1} m_s(-M + h_d^{(3)}) + \bar{A}_s^{-1} b(\dot{h}_2 + \ddot{z}_u) + \gamma A_s h_2 h_3 v_s^{-1} - \dot{Q}(\gamma - 1)(s_p v_s)^{-1} - \hat{d}(t), \tag{37}$$

Substituting (33) and (37) into (35), in closed-loop, we have

$$\dot{V}_{3b} = -k_1 Y(z_1)^2 - k_2 z_2^2 - k_3 z_3^2 - \lambda_d d_e(t)^2 + z_3 m_s^{-1} \bar{A}_s d_e(t) + \dot{d}(t) d_e(t). \tag{38}$$

The main result of this paper is summarized in the following theorem.

Theorem 1. *Let $h_d \in C^3$ in (5) be the desired height whose time derivatives are bounded, and $|z_1(0)| < \mu$. By considering the designed time-varying disturbance observer (9) and input (37), the errors $\mathbf{z} = [Y(z_1), z_2, z_3, d_e]^\top$ converge to an arbitrarily small neighborhood of zero, achieving uniform ultimate boundedness without violating the output constraint.*

Proof. Let us go back to (38) and apply Young's inequality, we have

$$\dot{V}_{3b} \leq -k_1 Y(z_1)^2 - k_2 z_2^2 - \Big(k_3 - \frac{m_s^{-1}\bar{A}_s}{4}\Big) z_3^2 - \Big(\lambda_d - m_s^{-1}\bar{A}_s - 1\Big) d_e(t)^2 + \frac{d_{max}^2}{4} \tag{39}$$

where k_3 are chosen such that $k_3 > m_s^{-1}\bar{A}_s/4$, $\lambda_d > 1 + m_s^{-1}\bar{A}_s$. Setting $\mathbf{z} = [Y(z_1), z_2, z_3, d_e]^\top$, $\dot{V}_{3b}$ can be further written as

$$\begin{aligned} \dot{V}_{3b} &\leq -k_{\min}||\mathbf{z}||^2 + \frac{d_{max}^2}{4} \\ &= -k_{\min}\Big(||\mathbf{z}||^2 - \frac{d_{max}^2}{4k_{\min}}\Big) \end{aligned} \tag{40}$$

which is negative definite for

$$||\mathbf{z}|| > \sqrt{\frac{d_{max}^2}{4k_{\min}}} + \epsilon$$

where ϵ is an arbitrarily small positive constant. It follows that $||\mathbf{z}||$ is ultimately bounded by

$$z_{\max} = \sqrt{\frac{d_{max}^2}{4k_{\min}}} + \epsilon \tag{41}$$

which can be made arbitrarily small by increasing the control gains, k_1, k_2, k_3 and λ_d. Consequently, global uniform ultimate boundedness is achieved. Notice that $|Y(z_1)| > |z_1|$, therefore, bounded $|Y(z_1)|$ leads to bounded $|z_1|$. Moreover, it is important to point out that if the output constraint is violated, $|Y(z_1)|$ will be infinity. However, as we established above, for $|z_1(0)| < \mu$, the error $\|\mathbf{z}\|$ will converge to a bounded value instead of infinity, from which we can conclude that the output constraint is guaranteed. □

Remark 3. *From Theorem 1, we know that larger k_1, k_2, k_3, λ_d would lead to smaller ultimate error. However, larger gains could also cause unwanted oscillation. Consequently, we cannot choose* *them arbitrarily large. In summary, we need to find a trade-off between the amplitude of oscillation and tracking accuracy.*

5. Simulation Verification

In order to verify the performance of proposed controller, a co-simulation is conducted in this section by combining the virtual plant of quarter vehicle with the AAS system in AMEsim software with the proposed controller in Matlab/Simulink to regulate the sprung height by inflating and deflating the air spring. The control block diagram of co-simulation is displayed in Figure 2. Unlike the mathematical model of the controller implemented in Matlab/Simulink, the AMESim-based quarter vehicle plant is established based on the actual pneumatic system configuration so that it is closer to the actual pneumatic system.

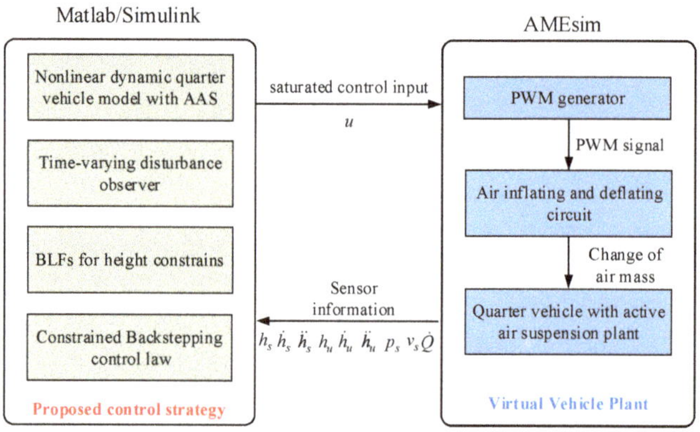

Figure 2. Co-simulation block diagram.

5.1. Simulation Conditions

The desired height h_d is a sine trajectory, given by

$$h_d = 0.015\sin(\omega t), \tag{42}$$

where $\omega = 0.5$ (rad/s) . The time-varying disturbances are chosen as

$$d(t) = \sin(\pi/2t)\sin(\pi t) + 10\sin(\pi t)\cos(2\pi t) + \vartheta(t), \tag{43}$$

where $\vartheta(t)$ is a class of band-limited white noise. The road excitation h_u is set as the class C of ISO profile with a driving speed of 50 $(\text{km}\cdot\text{h}^{-1})$, whose graphic representation is shown in Figure 2. The main parameters used in the co-simulation are given in Table 2, where the control gains are chosen through a trial–error process.

Table 2. Parameters for co-simulation.

Parameter	Value	Parameter	Value
$\bar{A}$	178 (mm^2)	k_1	9
b	1140 $(\text{N}\cdot\text{s}\cdot\text{m}^{-1})$	k_2	40
h_0	0.2047 (m)	k_3	4
$\bar{h}$	0.4047 (m)	λ_d	100
m_s	300 (kg)	γ	1.4
m_u	30 (kg)	p_0	5.11 (Bar)
d_{max}	1.8	p_a	1.01 (Bar)
k_t	7.5×10^6 $(\text{N}\cdot\text{m}^{-1})$	b_t	300 $(\text{N}\cdot\text{s}\cdot\text{m}^{-1})$

5.2. Simulation Results and Analysis

The co-simulation results of the proposed control strategy for ride height control with the AAS system are displayed in Figures 3–7. As shown in Figure 3, the height of vehicle sprung mass with the proposed controller can track the desired height within 1 s. Moreover, the tracking errors, with $\mu = 10$ mm and $\mu = 5$ mm, all stay within the predefined range of ± 10 mm and ± 5 mm as shown in Figure 4. Moreover, compared with the proposed controller without considering the output state constraint (i.e., choosing $\mu_{\max}$ very large), although the height of vehicle sprung mass is able to track the desired value as displayed in Figure 5, the height tracking error exceeds its allowable maximum value that results probably in a poor performance or even insecurity as illustrated in Figure 6. Furthermore, the time-varying disturbances $d(t)$ could be estimated by the developed nonlinear disturbance observer and $\hat{d}(t)$ can also be kept within the range of ± 0.1 as depicted in Figure 7. It means that the designed disturbance observer is effective. Additionally, in order to simulate the real operating conditions, white noise is considered during the height measurement procedure. As Figure 8 displayed, the proposed controller still can track the desired height under the presence of measurement noise.

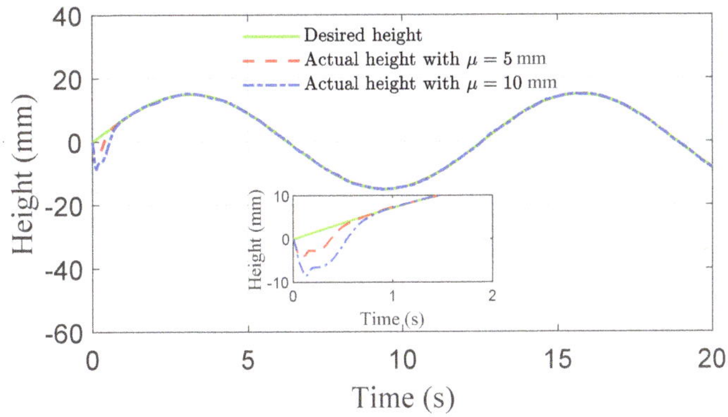

Figure 3. Height tracking performance of quarter vehicle with AAS in co-simulation.

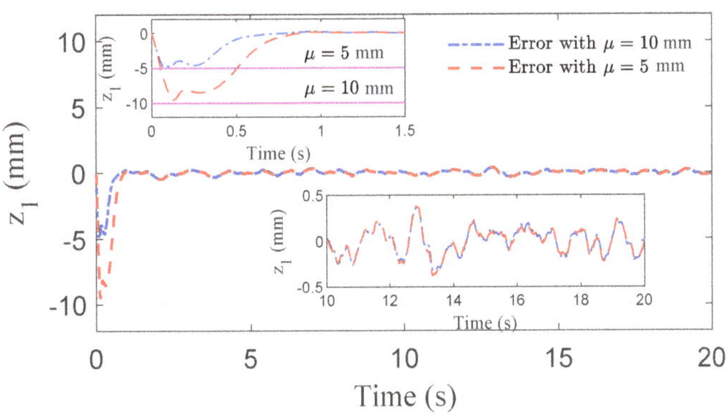

Figure 4. Height tracking error of quarter vehicle with AAS in co-simulation, where $|z_1|$ always stays within its corresponding bound.

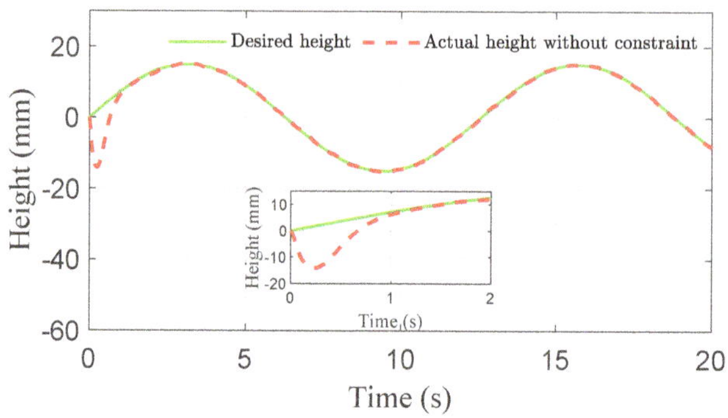

Figure 5. Height comparison of quarter vehicle with AAS in co-simulation.

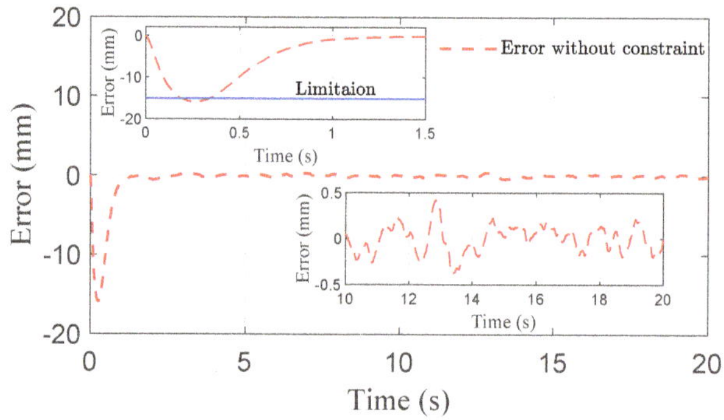

Figure 6. Error comparison of quarter vehicle with AAS in co-simulation, where $|z_1|$ exceeds its allowable maximum value.

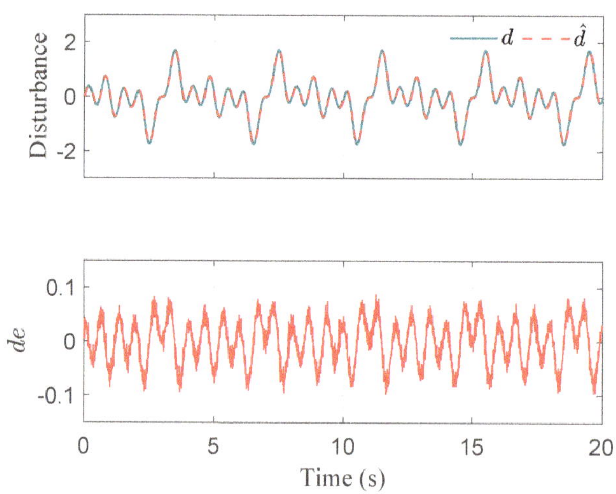

Figure 7. Disturbance estimation of quarter vehicle with AAS in co-simulation.

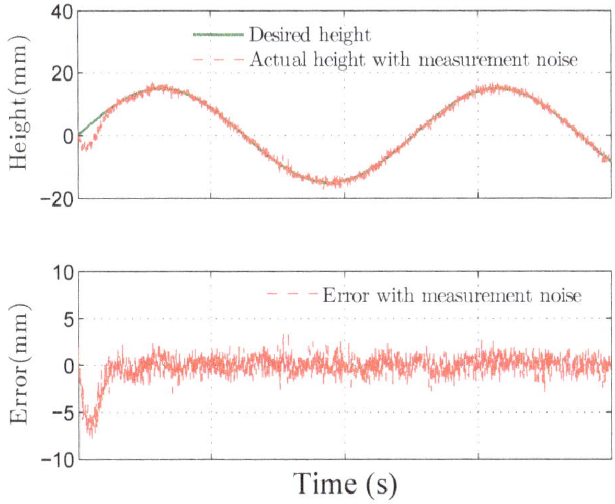

Figure 8. Tracking height and error of quarter vehicle with measurement noise in co-simulation.

5.3. Comparison of Simulation Results

In order to further demonstrate the benefits of the considerations of output state constraint and time-varying disturbances, Figures 9–11 also show the simulation results of ride height with nonlinear robust controller [14] and hybrid model predictive controller (HMPC) [7] under the same simulation parameters, disturbances, and road excitation. As demonstrated in Figure 9, the height of vehicle sprung mass reaches the target value within 1 s, which is much shorter than 4 s obtained for the controller presented in [14]. Meanwhile, during the time from 10 s to 20 s, the steady-state error achieved by the robust controller presented in [14] is bounded by 0.8 mm, which is larger than the bound of 0.5 mm obtained with the proposed controller, as depicted in Figure 10. Moreover, the proposed controller can track the desired height during both leveling up and lowering down processes so that the height of vehicle sprung mass reaches the target height as illustrated in Figure 11. The desired height used in the test is presented in [7]. The simulation results in Figures 10 and 11, and the performance comparison in Tables 3 and 4 indicates that the proposed control technique is more effective than the robust controller presented in [14] and the HMPC presented in [7].

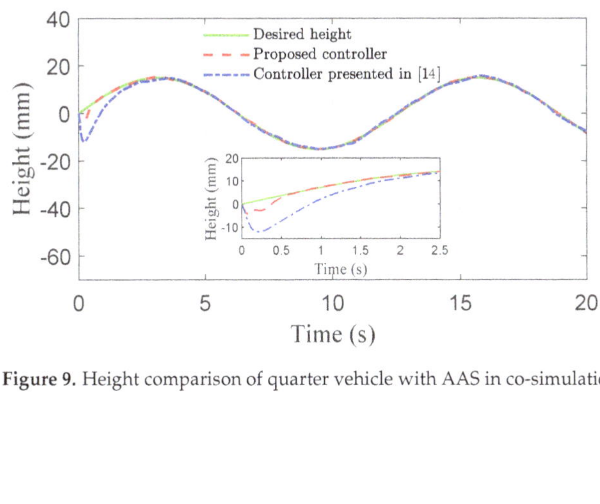

Figure 9. Height comparison of quarter vehicle with AAS in co-simulation.

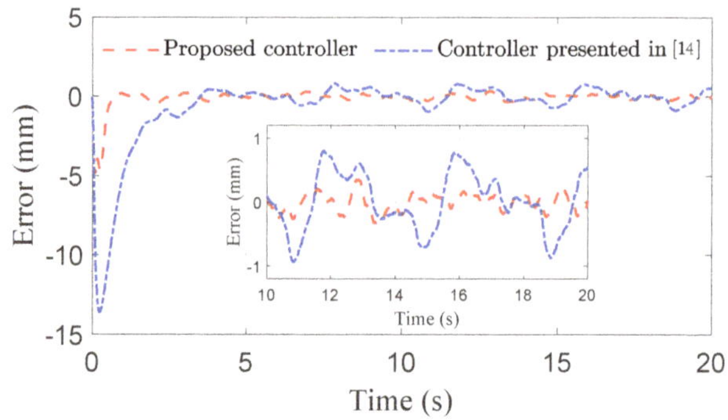

Figure 10. Error comparison of quarter vehicle with AAS in co-simulation.

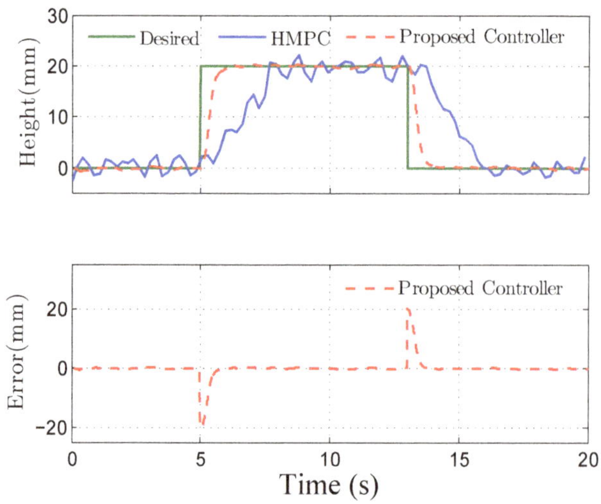

Figure 11. Tracking performance comparison of quarter vehicle with hybrid model predictive controller (HMPC) presented in [7].

Table 3. Performance index comparison of co-simulation.

Performance Index	Robust Controller in [14]	Proposed Controller	Improvement *
RMS of tracking error	4.6824×10^{-1} (mm)	1.6230×10^{-1} (mm)	65.3%
SD of tracking error	4.6809×10^{-1} (mm)	1.6012×10^{-1} (mm)	65.6%
Adjusting time	4 (s)	1 (s)	75%

* denotes relative to robust controller presented in [14].

Table 4. Performance index comparison of co-simulation.

Performance Index	HMPC in [7]	Proposed Controller	Improvement *
RMS of tracking error	7.7785 (mm)	4.3781 (mm)	43.71%
SD of tracking error	6.0746 (mm)	3.2452 (mm)	46.58%
Adjusting time	3 (s)	1 (s)	66.67%

* denotes relative to HMPC presented in [7].

6. Conclusions

This paper presents a solution to the task of vehicle height tracking for an electronically controlled AAS system. By employing the BLF-based backstepping technique, a novel constrained control strategy is proposed to drive the vehicle height to the neighborhood of preset desired values in the presence of output state constrains and perturbations in the AAS system, achieving uniform ultimate boundedness. To realize the robust performance, a nonlinear disturbance observer is introduced in the adaptive control law to compensate for the time-varying disturbances caused the external random road excitation and perturbations, achieving robust performance. Co-simulation results illustrate that the proposed control strategy is effective, robust, and superior to other recent techniques. With respect to our future research, it includes (i) designing a robust height tracking controller for a full-car model with the AAS system, and (ii) developing a noise filter and delay compensator for the system so as to improve the closed-loop performance in real applications.

Author Contributions: Conceptualization, R.Z. and W.X.; methodology, R.Z. and W.X.; validation, R.Z. and W.X.; formal analysis, R.Z. and W.X.; investigation, R.Z. and W.X.; resources, R.Z. and W.X.; writing—original draft preparation, R.Z. and W.X.; writing—review and editing, R.Z. and W.X.; supervision, J.Z., P.K.W. and C.S.; project administration, R.Z.; funding acquisition, R.Z. All authors have read and agreed to the published version of the manuscript.

Funding: This work was funded by the Ph.D. Research Project of Guizhou Normal University (Grant Number GZNUD [2019] 22), the Cooperation Project of Guizhou Education Department (Grant Number KY [2021] 297), the National Science Foundation (Grant Number 51965008), the Key Science Project of Guizhou (Grant Number ZNWLQC [2019] 3012), the University of Macau Research Grant (Grant Numbers MYRG-2019-00028-FST and MYRG-2018-00198-FST), Guangdong Basic and Applied Basic Research Foundation (Grant number 2019A1515011602), and the Portuguese Fundação para a Ciência e a Tecnologia (FCT) through ISR (Grant Number LARSyS UID/EEA/50009/2019).

Institutional Review Board Statement: Not applicable.

Informed Consent Statement: Not applicable.

Data Availability Statement: The data presented in this study are available on request from the corresponding author.

Conflicts of Interest: The authors declare no conflict of interest.

References

1. Cao, D.; Song, X.; Ahmadian, M. Editors' perspectives: Road vehicle suspension design, dynamics, and control. *Veh. Syst. Dyn.* **2011**, *49*, 3–28. [CrossRef]
2. Zhao, J.; Wong, P.K.; Ma, X.B.; Xie, Z.C. Chassis integrated control for active suspension, active front steering and direct yaw moment systems using hierarchical strategy. *Veh. Syst. Dyn.* **2017**, *55*, 72–103. [CrossRef]
3. Riofrio, A.; Sanz, S.; Boada, M.J.L.; Boada, B.L. A LQR-Based Controller with Estimation of Road Bank for Improving Vehicle Lateral and Rollover Stability via Active Suspension. *Sensors* **2017**, *17*, 2318. [CrossRef] [PubMed]
4. Aly, A.A.; Salem, F.A. Vehicle suspension systems control: A review. *Int. J. Control Autom. Syst.* **2013**, *2*, 46–54.
5. Huang, Y.B.; Na, J.;Wu, X.; Liu, X.Q.; Guo, Y. Adaptive control of nonlinear uncertain active suspension systems with prescribed performance. *ISA Trans.* **2015**, *54*, 145–155. [CrossRef] [PubMed]
6. Zhao, J.; Wong, P.K.; Xie, Z.C.; Wei, C.Y.; Zhao, R.C. Design and evaluation of a rode comfort of based suspension system using an optimal stiffness-determination method. *Trans. Can. Soc. Mech. Eng.* **2016**, *40*, 773–785. [CrossRef]
7. Sun, X.Q.; Cai, Y.F.; Chen, L.; Liu, Y.L.; Wang, S.H. Vehicle height and posture control of the electronic air suspension system using the hybrid system approach. *Veh. Syst. Dyn.* **2016**, *54*, 328–352. [CrossRef]
8. Sun, X.Q.; Cai, Y.F.; Wang, S.H.; Liu, Y.L.; Chen, L. Design of a hybrid model predictive controller for the vehicle height adjustment system of an electronic air suspension. *Proc. IMechE Part D J. Automob. Eng.* **2016**, *230*, 1504–1520. [CrossRef]
9. Gao, Z.; Chen, S.; Zhao, Y.; Nan, J. Height Adjustment of Vehicles Based on a Static Equilibrium Position State Observation Algorithm. *Energies* **2016**,*11*, 455. [CrossRef]
10. Lee, S.J. Development and analysis of an air spring model. *Int. J. Automot. Technol.* **2010**, *11*, 471–479. [CrossRef]
11. Kim, H.; Lee, H. Height and leveling control of automotive air suspension system using sliding mode approach. *IEEE Trans. Veh. Technol.* **2011**, *60*, 2027–2041.
12. Sun, W.; Pan, H.; Zhang, Y.F.; Gao, H.J. Multi-objective control for uncertain nonlinear active suspension systems. *Mechatronics* **2014**, *24*, 318–327. [CrossRef]
13. Pang, H.; Zhang, X.; Xu, Z. Adaptive backstepping-based tracking control design for nonlinear active suspension system with parameter uncertainties and safety constraints. *ISA Trans.* **2019**, *88*, 23–36. [CrossRef]

14. Zhao, R.C.; Xie, W.; Wong, P.K.; Cabecinhas, D.; Silvestre, C. Robust ride height control for active air suspension systems with multiple unmodeled dynamics and parametric uncertainties. *IEEE Access* **2019**, *7*, 59185–59199. [CrossRef]
15. Sun, W.; Gao, H.; Kaynak, O. Finite frequency control for active vehicle suspension systems. *IEEE Trans. Control Syst. Technol.* **2011**, *19*, 416–422. [CrossRef]
16. Kong, Y.S.; Zhao, D.X.; Yang, B.; Han, C.H.; Han, K. Robust non-fragile $H_\infty/L_2 - L_\infty$ control of uncertain linear system with time-delay and application to vehicle active suspension. *Int. J. Robust Nonlinear Control* **2015**, *25*, 2122–2141. [CrossRef]
17. Rath J.J.; Defoort M.; Karimi, H.R.; Veluvolu, K.C. Output Feedback Active Suspension Control With Higher Order Terminal Sliding Mode. *IEEE Trans. Ind. Electron.* **2017**, *64*, 1392–1403. [CrossRef]
18. Liu Y.; Chen H. Adaptive Sliding Mode Control for Uncertain Active Suspension Systems With Prescribed Performance. *IEEE Trans. Syst. Man. Cybern* **2020**, 1–9 . [CrossRef]
19. Na, J.; Huang, Y.; Wu, X.; Su, S.F.; Li, G. Adaptive finite-time fuzzy control of nonlinear active suspension systems with input delay. *IEEE Trans. Cybern.* **2019**, *50*, 2639–2650. [CrossRef]
20. Liu, Y.J.; Zeng, Q.; Tong, S.; Chen, P.; Liu, L. Adaptive neural network control for active suspension systems with time-varying vertical displacement and speed constraints. *IEEE Trans. Ind. Electron.* **2019**, *66*, 9458–9466. [CrossRef]
21. Zhao, R.C.; Xie, W.; Wong P.K.; Cabecinhas, D.; Silvestre, C. Adaptive vehicle posture and height synchronization control of active air suspension systems with multiple uncertainties. *Nonlinear Dyn.* **2020**, *99*, 2109–2127. [CrossRef]
22. Tee, K.P.; Ge, S.S.; Tay, E.H. Barrier Lyapunov functions for the control of output-constrained nonlinear systems. *Automatica* **2009**, *45*, 918–927.
23. Tee, K.P.; Ren, B.; Ge, S.S. Control of nonlinear systems with time-varying output constraints. *Automatica* **2011**, *47*, 2511–2516. [CrossRef]
24. Jin, X. Adaptive Fixed-Time Control for MIMO Nonlinear Systems With Asymmetric Output Constraints Using Universal Barrier Functions. *IEEE Trans. Autom. Control* **2018**, *64*, 3046–3053. [CrossRef]
25. Xie, W.; Reis, J.; Cabecinhas, D.; Silvestre, C. Design and experimental validation of a nonlinear controller for underactuated surface vessels. *Nonlinear Dyn.* **2020**, *102*, 2563–2581. [CrossRef]

sensors

MDPI

Article

Roll Angle Estimation of a Motorcycle through Inertial Measurements

Diego Maceira, Alberto Luaces, Urbano Lugrís, Miguel Á. Naya and Emilio Sanjurjo *

Laboratorio de Ingeniería Mecánica, University of A Coruña, 15403 A Coruña, Spain; diego.maceira@udc.es (D.M.); alberto.luaces@udc.es (A.L.); urbano.lugris@udc.es (U.L.); miguel.naya@udc.es (M.Á.N.)

* Correspondence: emilio.sanjurjo@udc.es

Abstract: Currently, the interest in creating autonomous driving vehicles and progressively more sophisticated active safety systems is growing enormously, being a prevailing importance factor for the end user when choosing between either one or another commercial vehicle model. While four-wheelers are ahead in the adoption of these systems, the development for two-wheelers is beginning to gain importance within the sector. This makes sense, since the vulnerability for the driver is much higher in these vehicles compared to traditional four-wheelers. The particular dynamics and stability that govern the behavior of single-track vehicles (STVs) make the task of designing active control systems, such as Anti-lock Braking System (ABS) systems or active or semi-active suspension systems, particularly challenging. The roll angle can achieve high values, which greatly affects the general behavior of the vehicle. Therefore, it is a magnitude of the utmost importance; however, its accurate measurement or estimation is far from trivial. This work is based on a previous paper, in which a roll angle estimator based on the Kalman filter was presented and tested on an instrumented bicycle. In this work, a further refinement of the method is proposed, and it is tested in more challenging situations using the multibody model of a motorcycle. Moreover, an extension of the method is also presented to improve the way noise is modeled within this Kalman filter.

Keywords: roll angle estimator; Kalman filter; LQR controller; inertial sensors; motorcycle lean angle

Citation: Maceira, D.; Luaces, A.; Lugrís, U.; Naya, M.Á.; Sanjurjo, E. Roll Angle Estimation of a Motorcycle through Inertial Measurements. *Sensors* **2021**, *21*, 6626. https://doi.org/10.3390/s21196626

Academic Editor: Juan A. Cabrera

Received: 17 August 2021
Accepted: 28 September 2021
Published: 5 October 2021

1. Introduction

Single track vehicles (STVs) present some intrinsic advantages as a mobility solution: they are lighter, use less space, and have better fuel economy, particularly at low speeds. These advantages make them great candidates for urban mobility. However, due to their lack of static stability, they also present some intrinsic challenges regarding safety and autonomous driving.

One magnitude that is of particular interest is the roll angle with respect to gravity (i.e., with respect to the vertical), since it is the magnitude used to keep the balance of the vehicle when it is in motion. Therefore, any control system that aims at controlling the balance of a STV, or any advanced driving aid for human-driven vehicles, has to take the roll angle into account. The roll angle determination has other potential applications.

For example, it is required in scientific experiments, such as those performed for the rider control parameter identification [1] or to evaluate the maneuverability of a motorcycle [2]. Unfortunately, there is not any means of measuring the roll angle of a STV in an economic and reliable way. Usually, the practical approach is measuring this magnitude indirectly through state observers, which combines the information provided by some sensors with a mathematical model to estimate the variable of interest. Consequently, this problem has been studied for several years, trying to improve the estimation algorithms.

There are some works using video-based approaches or distance sensors. The roll angle measurements using these types of devices are not reliable enough to be installed on commercial vehicles, since they are sensitive to the dirt. Moreover, they provide roll angle estimations with reference to the ground (video-based approaches use the horizon line (e.g., [3]), and distance sensors use the distance to the ground).

However, in order to keep the balance of a STV, the roll angle with respect to gravity is needed. Therefore, we will focus on works relying on inertial and odometric sensors, since they are already widespread in the automotive field for safety applications. The reason is that they are affordable and reliable because they do not depend on the lighting (as do the video-based approaches), the ground properties, and geometry (as do the distance sensors), nor the sky visibility (as does Global Positioning System (GPS)).

In [4], an extended Kalman filter (EKF) based on an analytical dynamics model was presented. This model is quite complex, and hence it can provide a lot of information of the vehicle; however, at the same time, the estimations rely on a correct characterization of the multiple parameters of the model. Moreover, the parameter values have to be adapted if applied to a different model of motorcycle. Similarly, in [5], a dynamics model of a motorcycle is used into an observer, although the latter is an unknown-input high-order sliding-mode observer instead of an EKF. The fuzzy logic approach has also been applied to the estimation of the states of motorcycles using dynamical models [6].

However, whenever possible, simpler models are preferred, since they are easier to adjust, and serve a broader range of vehicles with only minor or no tuning at all. The following works belong to this category. In [7,8], an observer based on the frequency separation is presented and validated. In [9], an EKF and an unscented Kalman filter (UKF) were compared, achieving a similar level of accuracy. In [10], another EKF based on inertial measurement was presented. It was applied to a racing motorcycle and validated on racing tracks. All these methods, and some other similar works not cited here, are compared with different tracks, sensors, vehicles, and riding conditions. Therefore, it is difficult to make an objective comparison of their effectiveness.

The present work is based on the roll angle estimation method presented in [11], where a roll angle estimator based on the EKF and angular rate measurements was validated at low speed on a flat floor with a bicycle. Due to limitations of the experimental setup, only low speeds were considered in leveled roads, with thin tires and no suspensions and without knowledge of the rider motions. In order to fill the gaps left by [11], in this work, a multibody model of a motorcycle is developed.

This multibody model is covered in Section 2, including the force models, such as toroidal wheels, suspensions, and a rider that can move laterally to displace its torso inwards or outwards during the turns, which are covered in Section 2.2. In order to complete the desired maneuvers, some controllers are needed, where the drive and braking forces are controlled by longitudinal controller, and the steer torque is governed by a lateral controller, which is in charge of keeping the motorcycle balance while following the predefined trajectories. These controllers are covered in Section 2.3.

Then, Section 2.4 describes the sensor models obtained from the multibody simulations. A sequence of pseudorandom noise is added to the signal of every sensor to provide realistic noise properties. The multibody model is run in several scenarios, which are described in Section 2.5. Finally, two roll angle observers based on the one presented in [11] are developed and tested in Section 3 using the noisy signals previously obtained from the multibody simulations. The roll angle estimations are compared in Section 4 to the roll angle obtained by the multibody simulation, which is used as the reference in this work.

2. Methods

This work is focused on the estimation of the roll angle of a motorcycle. In order to achieve this, a multibody motorcycle model is developed. Sensor models are implemented on the multibody model, and these data are used to run a state observer that estimates the roll angle of the multibody model. Since the sensors of the multibody model are "perfect", a sequence of pseudorandom numbers is added to mimic the noise of a realistic signal.

This section deals with the description of the multibody model, including its kinematics and the main force models, namely, tire forces, suspensions, drive and brake forces, etc.

Since the motorcycle is an intrinsically unstable system, a controller is required to maintain the balance and follow the prescribed trajectory, which is also described in this section. There are two more controllers: a longitudinal controller, which adapts the speed of the motorcycle depending on the curvature of the upcoming path, and a rider controller, which is used to control the position of the rider's torso, which can be in a neutral position, tilting inside the curve, or tilting outside the curve. Finally, this section describes the sensor data obtained from the multibody model and the designed tracks where it is simulated.

2.1. Multibody Model

The model used is a seven-element assembly without closed kinematic loops, where six elements belong to motorcycle parts, and one of them represents the driver's torso. As a result, a 12 Degrees of Freedom (DOF) model is obtained: six DOF from the chassis rigid body condition, five revolute joints from the two wheels, swingarm, steer and torso roll movement, and one prismatic joint between the fork bars and the fork bottles. All the motorcycle elements and DOFs are represented in Figure 1.

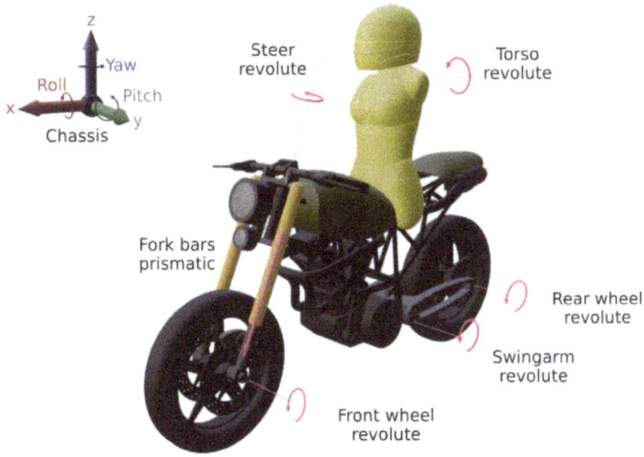

Figure 1. Multibody model elements and DOFs.

Mass and inertia properties for each element of the model are shown in Table 1. Some of them are taken from the bibliography [12] and other are obtained from a CAD tool.

Table 1. Mass and inertial properties of all the model elements.

Element	Mass (kg)	Inertia Tensor [Ix Iy Iz Ixy Ixz Iyz] (kg m²)					
Chassis	165.13	6.14	12.51	8.55	0.01	−2.43	0.03
Swingarm	8	0.10	0.23	0.31	0	−0.01	0
Rear wheel	14.7	0.43	0.81	0.43	0	0	0
Front wheel	11.9	0.32	0.62	0.32	0	0	0
Steer & Fork bottles	10.21	0.23	0.11	0.16	0	−0.01	0
Fork bars	3.13	0.05	0.02	0.03	0	0	0
Torso	42	2.32	2.13	0.49	0	−0.23	0

A relative coordinates formulation was used, which is a well suited option to take advantage of the topology of the mechanism. Each solid in the chain is defined relative to the previous one, using algorithms to calculate kinematics and dynamics terms by means of recursive methods. An in-depth description of this formulations is covered in [13]. The trapezoidal rule was chosen as the integrator, with a 1 millisecond fixed time step to solve the dynamics.

2.2. Force Models

Some forces are taken into account to achieve a realistic behavior of the motorcycle and its dynamics, such as gravity influence, forces from the motorcycle subsystems, such as brakes, tires and suspensions, and different torques applied to rear wheel, to steer mechanism and to the driver's torso. Rear wheel torque represents power transmission as an in-wheel hub motor would produce, steer torque represents the rider input to control the vehicle, and torso torque aims to represent the influence of the rider's movements in the roll angle estimation.

2.2.1. Brake Models

The brake model employed in this research is based on the tangential force model developed in [14], which takes into account sliding and stiction phenomena, with original contributions from [15]. In essence, a dry friction model is used in series with a spring-damper model. When the spring force exceeds the maximum achievable braking torque for a given braking pressure, one of the ends of the spring is allowed to slide so that the maximum braking force is not exceeded.

The front wheel has more braking capabilities than the rear one, as in a conventional motorcycle, but both brakes act together when braking. The maximum brake torque applied to each wheel is obtained by estimating the longitudinal force that the tires can perform according to Equation (1).

$$F_{long} = \mu N, \tag{1}$$

where μ is the tire-friction road considered, and N is the vertical force supported by the wheel. In case of the front wheel, this force is considered to be the whole weight of the vehicle and the driver, since in an emergency braking situation, almost all the weight rests on the front wheel. The normal force considered for the rear wheel is the weight that it supports when going straight and without any longitudinal acceleration.

2.2.2. Tire Models

Tire behavior and properties play a crucial role in the evaluation of the motorcycle dynamics. In this work, toroidal tires are used, since it is a good approximation to the behavior of motorcycle tires. They are defined with an outer radius, R, which represents the undeformed tire outer radius, and the torus tube radius, r, which should be selected to represent, in the most accurate way, the tire curvature near the contact patch.

Tire force models are divided between the normal force model and tangential force model. Normal forces calculation is closely related with the contact routine used to detect the intersection between the tire and the floor. In this research, a triangle mesh was used to characterize the floor, and thus the first step to calculate normal forces is to solve the contact problem between the torus and the triangle mesh. The contact algorithm between analytic torus and the triangle mesh used in this work is described in [16]. After a wheel contact is detected, the normal force is modeled as a spring-damper force. However, the force of the spring-damper element is limited such that it can produce compression forces over the road, but traction forces are not allowed.

Related to the tangential force model, part of the TMeasy tire model was applied [17], using the same simplifications as in [18]. This is an empirical and physical tire model, in the sense that first, curve fitting using few parameters is necessary to adjust the tire characteristic curves, and then dynamic behaviors of the tire are considered. The model takes into account both the longitudinal and lateral forces experienced by the tires, and these forces are slip-dependent in both directions.

As a consequence, the effects of tire deflection affect the tire behavior when the forces vary. This phenomenon is specially important when the vehicle moves at very low speed, because small displacements can produce high slips. If tire deflection is not considered, these high slips introduce unrealistic high forces. What actually happens is that the forces acting in the contact patch deflect longitudinally and laterally the tire as shown in Figure 2.

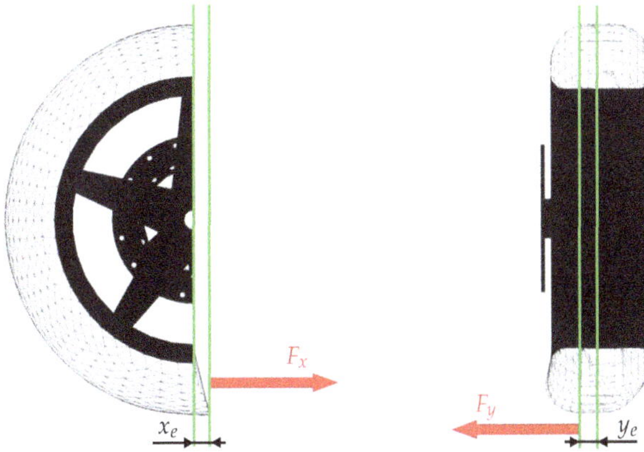

Figure 2. Longitudinal an lateral deflections in a toroidal tire due to tangential forces

2.2.3. Suspension Models

The motorcycle has traditional suspension systems for this kind of vehicles. Thus, for the front assembly, an inverted telescopic fork was chosen. For the rear frame, a monoshock absorber links the swingarm and the chassis. In both cases, a linear spring-damper model is used.

2.2.4. Drive Torque

The model simulates power transmission from an electric motor by introducing a torque in the rear wheel. Torque values have a dependency with angular rate, hence tabulated values with realistic properties are used. This torque value will be calculated by the longitudinal motorcycle controller, which will be described in depth in Section 2.3.2.

2.2.5. Steer Torque

Steer torque is the only input used to control the lateral dynamics of the motorcycle. It is used to control both the lateral equilibrium and to follow the prescribed trajectory. It highly affects the forces experienced by the tires, modifying completely the whole vehicle dynamics. This variable is managed by the lateral motorcycle controller, which will be covered in Section 2.3.3.

2.2.6. Torso Torque

This work aims at analyzing the rider's influence on the estimation of the vehicle roll angle. For this purpose, a torque between the torso and chassis elements is introduced in order to verify if the estimation is influenced by these movements. There is a specific controller managing this torque value, and it will be explained in Section 2.3.4. In this work, the torso controller is not used as a tool for motorcycle stabilization.

2.3. Controller Models

Since the aim of this work is to test the roll angle estimation in different conditions and scenarios, the motorcycle model has to be able to perform in a wide variety of circumstances. To achieve that, several controllers were implemented. First, longitudinal control manages accelerating and braking phases. Secondly, lateral control guarantees the lateral dynamic equilibrium while following the specified trajectories. The longitudinal controller has to be able to keep the velocity, which allows the lateral controller to perform the desired maneuver.

Lastly, a torso controller is used to perform maneuvers with a lean relative angle between torso and motorcycle, inside and outside the turns, as an option.

2.3.1. Path Tracking

In order to follow a predetermined path, it is necessary to use a curve definition system that can encode the trajectory that the motorcycle has to follow. In this work, splines are employed. Each scenario has a finite number of splines. They are drawn using the Blender software (Blender Foundation, Amsterdam, The Netherlands), and then their parameters are exported to a text file to be read by the simulation program.

Then, a point referred to the chassis frame is defined, and its position is compared to the path previously defined. This point is not necessarily coincident with the chassis. In this work, it is ahead of the chassis position to evaluate the deviation with respect to the trajectory in advance, as a human driver would do. This reference point position is speed-dependent: the faster the motorcycle goes, the farther and higher the point is found, as shown in Equations (2) and (3), where $v_{ch,x}$ and $v_{ch,y}$ stand for chassis velocity components, and rp_x and rp_z are the coordinates of the reference point expressed in the local reference system of the chassis (in SI units). In this work, the position of this reference point has an important effect on the overall behavior of the controller.

$$rp_x = 0.75\sqrt{v_{ch,x}^2 + v_{ch,y}^2} \tag{2}$$

$$rp_z = 0.1\, rp_x \tag{3}$$

The constants were adjusted by trial and error. The height of the reference point is also variable with speed. This height is relevant in tilting vehicles, since it produces a lateral displacement of the reference point when the vehicle is tilted. Once this point is defined, it is necessary to calculate the distance between the point and the nearest spline of the track. In this work, we calculate the distance to the spline perpendicular to the longitudinal axis of the motorcycle model.

With this information, the position, velocity, and orientation errors can be calculated, in order to correct the motorcycle position with respect to the desired path. These error values are the data that the lateral controller needs to calculate the roll angle target required to correct the motorcycle trajectory respect to the desired path. This will be addressed in Section 2.3.3.

2.3.2. Longitudinal Controller

This controller manages the acceleration and braking phases. The longitudinal control is designed to perform the maneuver as fast as possible, taking into account the curvature of the trajectory and the power limits of the motorcycle. The controller evaluates the path ahead the motorcycle 10 s in advance with respect to the reference point described on the previous section, and thus there is enough time to start braking before arriving too fast at a turn. This value is translated into a distance variable by means the Equation (4), which means the position in advance is calculated taking into account the current speed of the motorcycle.

$$x_{i+t_{preview}} = t_{preview}\, v_i \tag{4}$$

Once the position in the spline is known, the maximum speed is obtained in relation to the curvature of the path (κ), by means of the steady-state cornering equilibrium equation [19], expressed in Equation (5), where ϕ_{max} stands for the maximum roll angle the motorcycle can achieve, which is a configuration parameter for the controller.

$$V_{max} = \sqrt{\frac{g \tan \phi_{max}}{\kappa}} \tag{5}$$

The value of V_{max} cannot achieve infinite values and is limited to a max value, which is a configuration parameter. Once known, the current speed is evaluated against it, and the controller will accelerate if the current speed value is lower than V_{max}, or will brake if the motorcycle speed is too high. Both of the actions are managed by proportional controllers, as shown in Equations (6) and (7).

$$accel = K_{acel}\,(V_{max} - V_i) \tag{6}$$
$$brak = K_{br}\,(V_i - V_{max}) \tag{7}$$

When accelerating, the value of *accel* is translated into a rear wheel torque, taking into account tabulated torque values and the wheel angular rate. When braking, the value of *brak* is transformed into brake torque in both wheels, but with different values, since the brake power capabilities are different, as mentioned before.

2.3.3. Lateral Controller

In order to obtain full stability during maneuvers, a lateral controller was implemented. For this purpose, a Linear Quadratic Regulator (LQR) controller [20] was used, and a good behavior was achieved. Since the LQR controller requires a dynamic model in state-space form, the motorcycle multibody model was adapted into a Whipple model, and thus the initial seven-solid model is transformed into a four-solid model, as described in [21]. Model adaptation is shown in Figure 3. From the Whipple model, only the lateral dynamics are used.

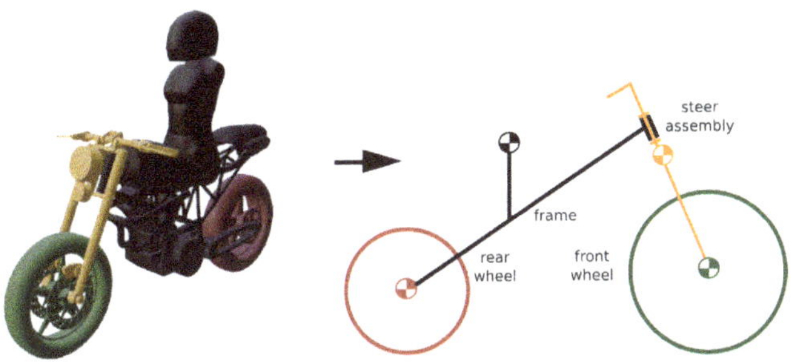

Figure 3. Multibody model transform to Whipple's model.

Linearized dynamic equations of the model are expressed by Equation (8), where $\mathbf{M}$, $\mathbf{C}_1$, $\mathbf{K}_0$ and $\mathbf{K}_2$ are obtained from a set of 25 parameters of the motorcycle (see Table 2), $\mathbf{q} = [\phi\ \delta]^\top$ is a vector that contains roll (ϕ) and steer (δ) angles, $\mathbf{f} = [T_\phi\ T_\delta]^\top$ is a vector that contains roll and steer torques (the roll torque is considered to be null in this work), and g and v stand for gravity acceleration and forward velocity, respectively. The employed values are shown in Table 2.

$$\mathbf{M}\ddot{\mathbf{q}} + v\mathbf{C}_1\dot{\mathbf{q}} + [g\mathbf{K}_0 + v^2\mathbf{K}_2]\,\mathbf{q} = \mathbf{f} \tag{8}$$

Equation (8) can be expressed in state-space form as shown in Equations (9) and (10), where u is the input vector, $\mathbf{x}$ is the state vector and y is the system output.

$$\dot{\mathbf{x}} = \mathbf{A}\mathbf{x} + \mathbf{B}u \tag{9}$$
$$y = \mathbf{C}\mathbf{x} + \mathbf{D}u \tag{10}$$

Table 2. Whipple's model parameters employed.

Parameter	Symbol	Value, Unit
wheelbase	w	1.3295 m
trail	c	0.0696 m
steer axis tilt	λ	0.4363 rad
gravity	g	9.81 m/s²
forward velocity	v	20 m/s
rear wheel		
radio	r_R	0.3069 m
mass	m_R	14.7 kg
mass moments of inertia	(I_{Rxx}, I_{Ryy})	(0.4332, 0.8134) kg m²
front wheel		
radio	r_F	0.2819 m
mass	m_F	11.9 kg
mass moments of inertia	(I_{Fxx}, I_{Fyy})	(0.3265, 0.6209) kg m²
rear body, chassis and torso		
position center of mass	(x_B, z_B)	(0.6344, −0.4741) m
mass	m_B	215.13 kg
mass moments of inertia	$\begin{bmatrix} I_{Bxx} & 0 & I_{Bxz} \\ 0 & I_{Byy} & 0 \\ I_{Bxz} & 0 & I_{Bzz} \end{bmatrix}$	$\begin{bmatrix} 22.3589 & 0.0197 & -2.0664 \\ 0.0197 & 29.6528 & 0.0316 \\ -2.0664 & 0.0316 & 10.3619 \end{bmatrix}$ kg m²
front handlebar and fork assembly		
position center of mass	(x_H, z_H)	(1.0669, −0.7446) m
mass	m_H	13.3516 kg
mass moments of inertia	$\begin{bmatrix} I_{Hxx} & 0 & I_{Hxz} \\ 0 & I_{Hyy} & 0 \\ I_{Hxz} & 0 & I_{Hzz} \end{bmatrix}$	$\begin{bmatrix} 0.4221 & -0.0002 & -0.1112 \\ -0.0002 & 0.3457 & 0.0002 \\ -0.1112 & 0.0002 & 0.2722 \end{bmatrix}$ kg m²

As Equation (8) is second order with respect to time and Equation (9) is first order, some changes are necessary. Taking $\mathbf{f} = [0\ T_\delta]^\top$, $\mathbf{x} = [\mathbf{q}\ \dot{\mathbf{q}}]^\top$ and its derivative as $\dot{\mathbf{x}} = [\dot{\mathbf{q}}\ \ddot{\mathbf{q}}]^\top$, Equation (11) is obtained.

$$\begin{bmatrix} \dot{\mathbf{q}} \\ \ddot{\mathbf{q}} \end{bmatrix} = \begin{bmatrix} 0_{2\times2} & I_{2\times2} \\ -\mathbf{M}^{-1}[v^2\,\mathbf{K}_2 + g\,\mathbf{K}_0] & -\mathbf{M}^{-1}\,v\,\mathbf{C}_1 \end{bmatrix} \begin{bmatrix} \mathbf{q} \\ \dot{\mathbf{q}} \end{bmatrix} + \begin{bmatrix} 0 & 0 & 0 & (-\mathbf{M}^{-1})_{2,2} \end{bmatrix}^\top T_\delta \tag{11}$$

Once the parameters of Table 2 are calculated, $\mathbf{M}$, $\mathbf{C}_1$, $\mathbf{K}_0$ and $\mathbf{K}_2$ can be obtained. The following values are the ones employed on this work.

$$\mathbf{M} = \begin{bmatrix} 81.6343 & 4.2211 \\ 4.2211 & 1.0320 \end{bmatrix} \qquad \mathbf{K}_0 = \begin{bmatrix} -119.8071 & -8.7515 \\ -8.7515 & -3.6985 \end{bmatrix} \tag{12}$$

$$\mathbf{K}_2 = \begin{bmatrix} 0 & 84.9797 \\ 0 & 6.6004 \end{bmatrix} \qquad \mathbf{C}_1 = \begin{bmatrix} 0 & 60.8120 \\ -2.2266 & 5.8313 \end{bmatrix} \tag{13}$$

To ensure system stability, state feedback is used, defining the system input as a negative feedback of the state, as in Equation (14), finding a $\mathbf{K}$ matrix that stabilizes the system.

$$u = -\mathbf{K}\mathbf{x} \tag{14}$$

Stabilization is achieved if the real parts of all eigenvalues of the system matrix are negative. Thus, Equation (9) can be combined with Equation (14), obtaining:

$$\dot{\mathbf{x}} = \mathbf{A}\mathbf{x} - \mathbf{B}\mathbf{K}\mathbf{x} = (\mathbf{A} - \mathbf{B}\mathbf{K})\mathbf{x} \tag{15}$$

Now, system stability is determined by the eigenvalues of $\mathbf{A} - \mathbf{BK}$, and thus a $\mathbf{K}$ matrix can be calculated with the values that ensure stability, since $\mathbf{A}$ and $\mathbf{B}$ are constants. In LQR controllers, the value of $\mathbf{K}$ is the one that minimizes the following cost function:

$$J = \int_0^{\infty} \left(\mathbf{x}^T \mathbf{Q} \mathbf{x} + R u^2 \right) dt \tag{16}$$

The function combines the quadratic values, integrated over time, of the magnitudes that should be minimized: the states and the control inputs. Each of them is weighted by a term, $\mathbf{Q}$ for the states and R for the input. The values of this terms can be adjusted in order to assign more weight to the control effort (increasing R value), or penalizing more the state errors by increasing the values of $\mathbf{Q}$.

The trajectory of the motorcycle can be controlled through its roll angle ϕ, and thus we need to set it to a certain value at every time step. However, the LQR controller defined thus far is only a regulator, i.e., it drives the states to zero. Since the resulting system is of type 0 [20], we need to transform it into a type 1 system, in order to make it capable to track a reference value of the roll angle. This can be achieved by adding an integrator at the input. Therefore, the system size increases, as a new state is added, ξ, which is the integral of the tracking error, which means the integral of the difference between the controller target (r) and its output (ϕ). The input u changes its value to:

$$u = -\mathbf{K}\mathbf{x} + k_I \, \xi \tag{17}$$

and the value of ξ should be calculated during the runtime as shown in Equation (18), where dt is the simulation time step.

$$\xi_i = \xi_{i-1} + dt(r - \phi) \tag{18}$$

Hence, the system turns into Equation (19):

$$\dot{\mathbf{x}}' = \begin{bmatrix} \dot{\mathbf{x}} \\ \dot{\xi} \end{bmatrix} = \begin{bmatrix} \mathbf{A} - \mathbf{BK} & \mathbf{B}k_I \\ \mathbf{DK} - \mathbf{C} & -\mathbf{D}k_I \end{bmatrix} \begin{bmatrix} \mathbf{x} \\ \xi \end{bmatrix} + \begin{bmatrix} 0 \\ 1 \end{bmatrix} r \tag{19}$$

where $\mathbf{x}'$ is the augmented state vector, and $\dot{\mathbf{x}}'$ is its derivative:

$$\mathbf{x}' = \begin{bmatrix} \phi & \delta & \dot{\phi} & \dot{\delta} & \xi \end{bmatrix}^{\top} \tag{20}$$

We can now define $\mathbf{K}' = \begin{bmatrix} \mathbf{K} & -k_I \end{bmatrix}$. Therefore, we can rewrite Equation (17) as follows:

$$u = -\mathbf{K}'\mathbf{x}' \tag{21}$$

The values of $\mathbf{Q}$ and R used in this work for Equation (16) are:

$$\mathbf{Q} = \begin{bmatrix} 0 & 0 & 1 \times 10^{-1} & 0 & 500 \end{bmatrix}^{\top}, R = 1 \times 10^{-4} \tag{22}$$

These values were adjusted by trial and error to obtain a realistic behavior.

The $\mathbf{K}'$ values obtained for a 20 m/s forward velocity can be seen in Equation (23). It was not necessary to calculate $\mathbf{K}'$ values for different velocities, as could be expected, since the controller works fine with the values obtained for the forementioned velocity for all the speed range used in this work.

$$\mathbf{K}' = 1 \times 10^3 \begin{bmatrix} -1.1768 & 0.1063 & -0.2079 & 0.0189 & 2.2361 \end{bmatrix} \tag{23}$$

Now, the controller target, which represents the control input (u), should be assigned. Hence, some terms are calculated, such as position (u_i), velocity (v_i), and angular errors (α_i) between the motorcycle position and the spline curve:

$$u_i = \cos\psi\,(y_{sp} - y_{rp}) - \sin\psi\,(x_{sp} - x_{rp}) \tag{24}$$

$$v_i = \mathbf{sp}_n \times \mathbf{rp}_{vel} \tag{25}$$

$$\alpha_i = \alpha - \psi \tag{26}$$

where ψ stands for the yaw motorcycle angle, x_{sp} and y_{sp} are the spline coordinates, x_{sp} and y_{sp} are the reference point coordinates mentioned in Equations (2) and (3), $\mathbf{sp}_n$ is the spline normal vector, $\mathbf{rp}_{vel}$ is the reference point velocity, and α is the angle of the spline tangent vector. With these values,

$$K_1 = -k_1\,u_i \tag{27}$$

$$K_2 = -k_2\,v_i \tag{28}$$

$$K_3 = -k_3\,\alpha_i \tag{29}$$

Finally, the roll target expression is obtained:

$$\phi_{target} = K_1 + K_2 + K_3 \tag{30}$$

The steer torque expression becomes:

$$\tau_{steer} = -\mathbf{K}'\,\mathbf{x}' + k_d\,v_{steer} \tag{31}$$

The second term in Equation (31) acts as a steer damper, in order to minimize all the small instabilities coming from different sources (contact forces, tire forces, controller, etc.), achieving a better performance without becoming slow on response. The variable v_{steer} refers to the steer velocity, while k_d is a damping coefficient, which, in our case, takes a value of 10 Ns/rad.

2.3.4. Torso Controller

Torso roll movement has a capital influence on the motorcycle dynamics, since it changes the center of mass of the vehicle. If the torso goes inside a turn, the motorcycle lean angle can be reduced, whereas going outside instead forces the motorcycle to increase its roll angle to tackle the same turn at the same speed. In order to analyze this effect in the estimator, a torso controller was implemented. The controller allows to configure three different positions as seen in Figure 4: inside position, neutral position, and outside position.

Figure 4. Inside, neutral, and outside configurations in a torso controller.

In the neutral position, the torso has no influence in the maneuver, since it stays aligned with the motorcycle. On the other hand, the inside and outside positions modify the relative angle between the torso and the motorcycle, changing the center of mass position. The input angle for the controller is defined by Equation (32).

$$\alpha_{obj} = 0.5\, F\, \phi \tag{32}$$

where ϕ is the vehicle roll angle, F is a factor that can take three values: 0 for a neutral position, 1 for an inside position, and -1 for an outside position. The 0.5 was used in this work, but other values could be used if more or less rider lean is desired. Once obtained, a PD controller can be defined as:

$$\tau_{torso} = K\,\epsilon - C\,\dot{\alpha}_{torso} \tag{33}$$

$$\epsilon = \alpha_{obj} - \alpha_{torso} \tag{34}$$

Proportional (K) and derivative (C) term values used in this work are 1.5×10^3 and 1×10^3.

2.4. Sensor Models

As described in [11], it is assumed that wheel speed sensor and angular rate measurements are available. These sensors are already available in many commercial motorcycles. Sensor data is built from the values of the multibody simulation.

In order to obtain vehicle forward speed, a wheel speed sensor is employed. Through it, longitudinal speed can be easily estimated by Equation (35). Values from any wheel can be taken, since there is no significant difference between them, except during aggressive accelerating or braking phases. When the wheel slides, the speed estimation will not be correct. Moreover, when the vehicle is tilted, the effective radius of the wheel is reduced. The multibody simulation used in this work can represent these two situations.

$$v_{long} = \omega_{wheel}\, r_{wheel} \tag{35}$$

An Inertial Measurement Unit (IMU) sensor was modeled in order to obtain angular rate measurements from the chassis element. The sensor is attached to the chassis reference frame, but in order to obtain the correct measurements, the IMU longitudinal axis should be aligned with the roll axis of the vehicle. This is particularly important since if this condition is not fulfilled, angular rate values will not be the correct ones. A rotation matrix can be applied to the sensor system reference in order to fix the possible wrong alignment between the IMU axis and the longitudinal axis of the motorcycle. This can be seen in Figure 5. Through Equation (36), the right angular rate values can be obtained.

$$\omega_{\mathrm{IMU}} = \mathbf{R}_{\mathrm{IMU}}^{\top}\, \mathbf{R}_{\mathrm{ch}}^{\top}\, \omega_{\mathrm{ch}} \tag{36}$$

where $\mathbf{R}_{\mathrm{IMU}}$ is the rotation matrix of the IMU relative to the chassis reference system, and $\mathbf{R}_{\mathrm{ch}}$ is the rotation matrix of the chassis reference system with respect to the global axes. Even if the IMU is perfectly aligned when the motorcycle goes at constant speed in straight line, there will be some situations in which the axis coincidence condition will not be fulfilled.

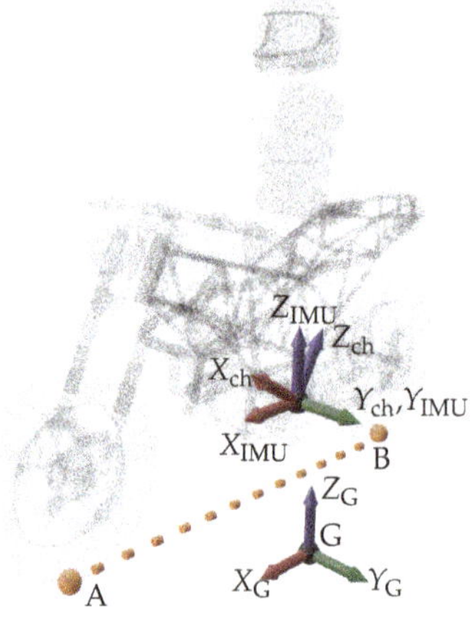

Figure 5. IMU position adjust.

For example, during a turn, when the motorcycle leans, the wheel contact points change their position due to the width difference between the two tires (in this case, as usual, the rear tire is wider than the front one). When the steering is turned, the front wheel contact point changes due to the fork trail, which also leads to a misalignment. Suspension movements can also produce similar effects. In all of these situations, an error between the IMU longitudinal axis and the motorcycle roll axis is expected, but it is an unavoidable fact.

In the case of both sensors, in order to achieve a more realistic situation, some additive random noise is added to the data in order to mimic real sensor characteristics. The noise is modeled as additive Gaussian white noise. However, neither the plant nor the measurement white noise. Therefore, for both sensors, values employed in the estimator (obs) are obtained through the simulation ones (mb) as:

$$\omega_{\mathrm{IMU}}^{\mathrm{obs}} = \omega_{\mathrm{IMU}}^{\mathrm{mb}} + N_{\mathrm{IMU}} \tag{37}$$

$$\omega_{\mathrm{ws}}^{\mathrm{obs}} = \omega_{\mathrm{ws}}^{\mathrm{mb}} + N_{\mathrm{ws}} \tag{38}$$

where N_{IMU} and N_{ws} are pseudorandom numbers following a normal distribution with 0 mean, which simulate the noise of the IMU and the wheel speed sensor, respectively. In the case of the IMU, the standard deviation used for the simulated noise is 9.839×10^{-4} rad/s (this value was obtained from a measurement taken with a low-cost IMU). For the wheel speed sensor, since no data was available, a conservative estimation of 1 m/s was taken for the standard deviation of the sensor noise.

2.5. Scenarios and Maneuvers

The results from the previous work shown in [11] were obtained for a bicycle with no suspensions. Therefore, the results were only verified for low speed, thin tires, and without suspensions. Moreover, due to the experimental setup, the motion of the rider was unknown, and the measurement of the roll angle for verification purposes could only be performed on flat surfaces. Since this work is based on a simulation, both the scenario slope and bank angle can be controlled, and every magnitude can be measured, including the motion of the rider, which is part of the model. Some scenarios were built in order to test the estimations. The goal is to check the behavior of the estimator in all kind of situations, such as level grounds, slopes or bends. With this in mind, six scenarios are proposed, as shown in Figure 6: three with level ground and another three with slopes and/or bank angles.

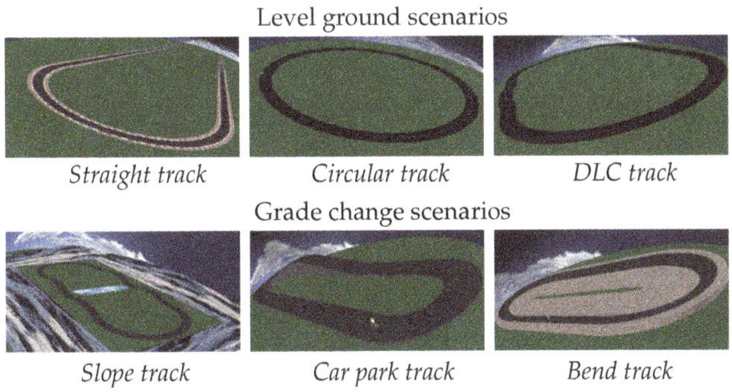

Figure 6. Scenarios created to test the estimator behavior.

The level ground scenarios are the *Straight track*, *Circular track*, and *DLC track* scenarios, while the *Slope track*, the *Car park track*, and the *Bend track* are non-planar tracks.

The *Straight track* is actually an oval track with two 2 km long straights, intended to study high-speed maneuvers and the transition from straight to turn. The *Circular track* is almost circular (it was approximated by four splines, and thus it is not a perfect circumference) to verify the steady-state cornering behavior of the observer. In this case, it has a radius of 50 m. The *DLC track* is also an oval track, but it features a double lane change in one of its straights.

When considering the non-planar tracks, all of them are ovals. The *Slope track* has two large bumps on each straight. Each one reaches a height of 6.5 m over a length of 38.4 m, which gives an average slope of 17%. After they reach their maximum height, they go back to the base level. The *Car park track* has an ascending slope combined with a turn, with a trajectory resembling a screw thread. The slope of this part is 22%. After that, there is a straight descent with a slope of 18% until the initial level is achieved. Finally, the *Bend track* is a flat circuit, but with a bank angle of 15°. The range of speeds used for the maneuvers reproduced in every scenario are shown in Table 3. The full data set obtained from the simulations is provided as supplementary material with this paper.

Table 3. Speeds used for the maneuvers in every test track (in m/s).

Track	Minimum Speed	Average Speed	Maximum Speed
Straight	29.00	38.15	50.00
Circular	15.00	15.88	16.21
DLC	10.00	19.38	23.00
Slope	5.79	12.62	16.41
Car park	3.84	8.41	15.00
Bend	8.78	11.58	15.00

3. Roll Angle Estimator

The tool described in this section aims at estimating the roll angle of a STV with respect to gravity. To achieve that, it is assumed that the data from previously described sensors are available. Those sensors are affordable and easy to install on a vehicle, but they do not provide any direct measurement of the roll angle. To resolve this issue, information from both sensors and knowledge from the system has to be blended. In this work, a Kalman filter algorithm is employed to develop two variants of the roll angle estimator.

3.1. The Kalman Filter

The Kalman filter is a stochastic estimator that combines predictions from a model with measurements coming from sensors. The equations of the discrete version of Kalman filter are reproduced here. The reader interested in a deeper understanding of the Kalman filter is referred to any of the books on the topic, such as [22,23].

The filter runs in two stages: prediction and correction. During the prediction phase, the state $\mathbf{x}$ and the covariance matrix of its estimation error $\mathbf{P}$ are propagated by means of the model:

$$\hat{\mathbf{x}}_k^- = \mathbf{F}\hat{\mathbf{x}}_{k-1}^+ + \mathbf{G}\mathbf{u}_{k-1} \tag{39}$$

$$\mathbf{P}_k^- = \mathbf{F}_{k-1}\mathbf{P}_{k-1}^+\mathbf{F}_{k-1}^\top + \mathbf{\Sigma}^P \tag{40}$$

where $\mathbf{F}$ stands for the transition model of the system, $\mathbf{G}$ is the input matrix, $\hat{\mathbf{x}}_k^-$ is the estimation of the state vector in time step k before the measurement is applied, $\hat{\mathbf{x}}_{k-1}^+$ is the estimation of the state vector of the $k-1$ time step after the corresponding measurements were applied, and $\mathbf{u}_k$ is the input of the system. If measurements from sensors are available, they are used at the correction phase to improve the estimation from the prediction stage.

First, innovation $\tilde{\mathbf{y}}$ is calculated as the difference between the measurements from sensors ($\mathbf{o}_k$) and the expected sensor readings according to the model ($\mathbf{H}\hat{\mathbf{x}}_k^-$). Kalman gain ($\mathbf{K}$) and the innovation covariance matrix ($\mathbf{S}$):

$$\tilde{\mathbf{y}}_k = \mathbf{o}_k - \mathbf{H}\hat{\mathbf{x}}_k^- \tag{41}$$

$$\mathbf{S}_k = \mathbf{H}\mathbf{P}_k^-\mathbf{H}^\top + \mathbf{\Sigma}^S \tag{42}$$

$$\mathbf{K}_k = \mathbf{P}_k^-\mathbf{H}^\top\mathbf{S}_k^{-1} \tag{43}$$

The value of $\mathbf{S}$ represents the uncertainty in the system state projected via the sensor function ($\mathbf{H}\mathbf{P}_K^-\mathbf{H}^\top$) plus an additional uncertainty, $\mathbf{\Sigma}^S$, due to the sensor noise. Finally, the estimation of the state and its covariance are updated by means of the Kalman gain:

$$\mathbf{x}_k^+ = \mathbf{x}_k^- + \mathbf{K}_k\tilde{\mathbf{y}}_k \tag{44}$$

$$\mathbf{P}_k^+ = (\mathbf{I}_8 - \mathbf{K}_k\mathbf{H})\mathbf{P}_k^- \tag{45}$$

If measurements from sensors are not available, the correction stage is omitted.

3.2. Dynamical Model of the Filter

The model employed by the filter has two states: the roll angle and the bias of the angular rate sensor along the body fixed x-axis. The relationship among the angular rates measured by the body-mounted angular rate sensors ($\omega_b^B, \omega_y^B, \omega_z^B$) and the time derivative of the roll ($\dot{\phi}$), pitch ($\dot{\theta}$), and yaw ($\dot{\psi}$) angles of the vehicle body follows from:

$$\mathbf{R}^\top\dot{\mathbf{R}} = \tilde{\omega}^B = \begin{bmatrix} 0 & -\omega_z^B & \omega_y^B \\ \omega_z^B & 0 & -\omega_x^B \\ -\omega_y^B & \omega_x^B & 0 \end{bmatrix} \tag{46}$$

and can be expressed as:

$$\dot{\phi} = (\omega_y^B \sin\phi + \omega_z^B \cos\phi)\tan\theta + \omega_x^B \tag{47}$$

$$\dot{\theta} = \omega_y^B \cos\phi - \omega_z^B \sin\phi \tag{48}$$

$$\dot{\psi} = \frac{\omega_y^B \sin\phi + \omega_z^B \cos\phi}{\cos\theta} \tag{49}$$

Assuming a small pitch angle, $|\theta| \approx 0$, Equation (47) becomes:

$$\dot{\phi} \approx \omega_x^B \tag{50}$$

The bias (b_x) of the x angular rate sensor (ω_x^B) can be modeled as a random walk, i.e., assuming that it is constant and that the variations are produced by the plant noise. Once

b_x is known, ω_x^B can be corrected. Therefore, after applying the forward Euler integration method, the dynamic model of the filter becomes:

$$\begin{bmatrix} \hat{\phi} \\ \hat{b}_x \end{bmatrix}_k^- = \begin{bmatrix} 1 & -dt \\ 0 & 1 \end{bmatrix} \begin{bmatrix} \hat{\phi} \\ \hat{b}_x \end{bmatrix}_{k-1}^+ + \begin{bmatrix} dt \\ 0 \end{bmatrix} \omega_{x,k-1}^B \tag{51}$$

where the states at the present time step k are expressed as a function of the states and the inputs of the previous time step $k-1$, being dt the integration time step. In the previous equation we can identify the system and input matrices (**F** and **G**) as follows:

$$\mathbf{F} = \begin{bmatrix} 1 & -dt \\ 0 & 1 \end{bmatrix} \tag{52}$$

$$\mathbf{G} = \begin{bmatrix} dt \\ 0 \end{bmatrix} \tag{53}$$

3.3. Absolute Measurements of the Roll Angle

In the correction stage of this Kalman filter, absolute measurements of the roll angle are needed. None of the sensors considered in this work provides a roll angle measurement, therefore, a model was built from the sensor measurements, and employed as absolute roll angle measurement. This is the measurement employed at the Kalman filter correction stage, denoted as $\mathbf{o}_k$ in Equation (41). The model is the same as described in [11], and is created by combining two ways of obtaining a roll angle estimation, coming from two different assumptions. The first one is obtaining from the steady-state cornering equilibrium, which will be referenced as ϕ_d on this work:

$$\phi_d = \arctan\left(\frac{\dot{\psi} v}{g}\right) \tag{54}$$

where v is the forward vehicle speed (obtaining from wheel speed sensor), g is the gravity acceleration, and $\dot{\psi}$ is the yaw velocity, which is assumed that ω_z^B (obtaining from IMU) is an enough accurate estimation. This method works well for small roll angles on level roads, but it tends to underestimate the roll angle in more realistic conditions due to the gyroscopic effect and the thickness of the tires, which are not considered in this model.

On the other hand, roll angle can be obtained under a null pitch rate condition, which means the Equation (48) becomes:

$$0 = \omega_y^B \cos(\phi) - \omega_z^B \sin(\phi) \tag{55}$$

and then, the another estimation of the roll angle, referenced as ϕ_ω in this case, can be calculated as:

$$\phi_\omega = \arctan\left(\frac{\omega_y^B}{\omega_z^B}\right) = sgn(\omega_z^B) \arcsin\left(\frac{\omega_y^B}{\sqrt{(\omega_y^B)^2 + (\omega_z^B)^2}}\right) \tag{56}$$

where $sgn(\omega_z^B)$ is the sign of the z angular rate. This method is more convenient for greater roll angles because the predicted roll tends to oscillate around the true value, although is noisier than the steady-state based one. In order to take the best from both estimations, they are combined using a weighted mean. The weighing function changes its value depending on the last available estimation provided by the equations of the steady-state cornering equilibrium, ϕ_d.

This is the main difference with respect to the method employed in [11], where the last estimation provided by the Kalman filter ($\hat{\phi}$) was used instead. The change was motivated because it was seen that, in some cases, an instantaneous wrong angle estimation could lead to a inadequate weighting value, which could eventually make the estimation diverge definitively from the true value. In this work, the weighting function, which can be seen in Figure 7, is defined by:

$$W = \exp\left(\frac{-\phi_d^2}{\bar{\phi}^2}\right) \tag{57}$$

where $\bar{\phi}^2$ is a constant value that can be used to adjust the behavior of the weighting function. In this work, a value of $\bar{\phi}^2 = 0.04$ was used, with the dynamically estimated roll angle ϕ_d^2 expressed in radians.

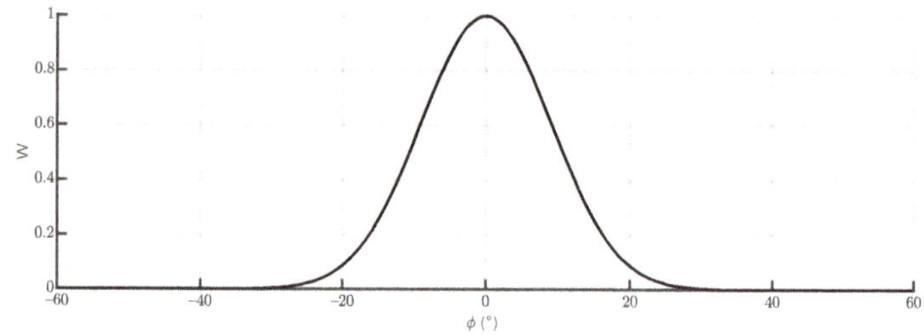

Figure 7. Shape of the weight function used to combine the two estimated measurements of the roll angle.

Therefore, the roll angle measurement $\mathbf{o}_k$ to be employed in Equation (41) is ϕ_m, built as weighted combination of ϕ_d and ϕ_ω:

$$\phi_m = W\phi_d + (1 - W)\,\phi_\omega \tag{58}$$

Since this measurement is the same magnitude than the first component of the state vector, the output matrix **H** results as follows:

$$\mathbf{H} = \begin{bmatrix} 1 & 0 \end{bmatrix} \tag{59}$$

3.4. Adaptation of the Kalman Filter for Colored Noise

The Kalman filter was formulated assuming that the noise affecting both the plant and the measurements is additive white Gaussian noise. However, neither the plant nor the measurement noises are white for the models used in this work. Instead, the lower frequencies are predominant in both noises (for instance, if the measurement is overestimating the roll angle at a given moment, it is more probable than the next time step it will be also overestimating it). This kind of behavior, usually known as colored noise, can be modeled as follows [22]:

$$Cn_{i+1} = W \times Cn_i + Wn_i \tag{60}$$

where the subindex i represents the time step, Cn is the colored noise, Wn is the white noise, and W is a parameter that can take any value from 0 to 1 to modulate the influence of the previous value of the noise. If $W = 0$, the model would produce white noise. If $W = 1$, the noise would behave as a Markov model. If we introduce this noise model for the noise of the roll angle of the dynamic model of the Kalman filter, and also for the noise of the sensor, the resulting filter matrices will be as follows:

$$\mathbf{F}' = \begin{bmatrix} 1 & -dt & 0 & 1 \\ 0 & 1 & 0 & 0 \\ 0 & 0 & W_1 & 0 \\ 0 & 0 & 0 & W_2 \end{bmatrix} \tag{61}$$

$$\mathbf{G}' = \begin{bmatrix} dt \\ 0 \\ 0 \\ 0 \end{bmatrix} \tag{62}$$

$$\mathbf{H}' = \begin{bmatrix} 1 & 0 & 1 & 0 \end{bmatrix} \tag{63}$$

where $\mathbf{F}'$, $\mathbf{G}'$, and $\mathbf{H}'$ are the system, input, and output matrices of this augmented filter, respectively, W_1 is the weighting parameter of the measurement noise, and W_2 the weighting parameter of the plant noise (applied only to the first state of the filter). When this model is applied, the measurement noise is incorporated into the colored noise, and, therefore, the covariance matrix of the measurement noise should be set to zero. Other than that, the method is identical to the previous one.

4. Results and Discussion

One of the key aspects of the performance of a Kalman filter is a proper adjustment of the values of the plant and the measurement noise covariance matrices. In the algorithm presented in this work, the weighting parameter for the measurement estimation also has to be set. Moreover, in the filter adapted to deal with colored noise there are two additional parameters, i.e., the weights used to model low frequency component of the noise. In this work, this tuning process was made by trial and error, seeking the minimization of the Root Mean Square Estimation Error (RMSEE) of the worst maneuver. The parameters used in this work are shown in Table 4.

Table 4. Values of the parameters used in this work (in SI units). In this work, both the simple Kalman filter presented here, and the previous version in [11] were used with the same parameters for a fair comparison. Note that the $\mathbf{\Sigma}^S$ for the Kalman filter for colored noise is null. The reason is that the measurement noise is contained in the third row of the plant noise.

Parameter	$\mathbf{\Sigma}^P$	$\mathbf{\Sigma}^S$	$\bar{\phi}^2$	Noise Weights
Kalman filter	$\begin{bmatrix} 5\times10^{-7} & 0 \\ 0 & 1\times10^{-8} \end{bmatrix}$	1.5	0.04	-
Kalman filter for colored noise	$\begin{bmatrix} 1\times10^{-6} & 0 & 0 & 0 \\ 0 & 1\times10^{-8} & 0 & 0 \\ 0 & 0 & 0.5 & 0 \\ 0 & 0 & 0 & 1\times10^{-6} \end{bmatrix}$	0	0.04	$W_1 = 0.8$ $W_2 = 0.5$

The results obtained for all the six tests scenarios are shown in Figure 8. The numeric results of the RMSEE for both methods are shown in Table 5. The method as described in [11] is also added to the comparison. In all these tests, the rider kept a neutral position. The results are very good for all the maneuvers with flat floor, with a RMSEE below 1° for all estimation methods. The maneuvers on inclined floors are a harder challenge for the estimator, mainly the *Car park ramp track* and the *Bend track*. In these maneuvers the RMSEE are around 2.5°, with the Kalman filter for colored noise having slightly better estimations for both maneuvers.

With different parameters, the results of both maneuvers can be improved, but improving the results in one of the maneuvers would produce deterioration of the performance on the other. The estimation at the *Car park ramp track* is, in general, good, but the accuracy degrades after the turns. This might be because the bumps of this track start after the turns and the motorcycle goes over them when it is still leaned, thus degrading the performance of all the indicators, because neither the pitch nor the pitch rate are null.

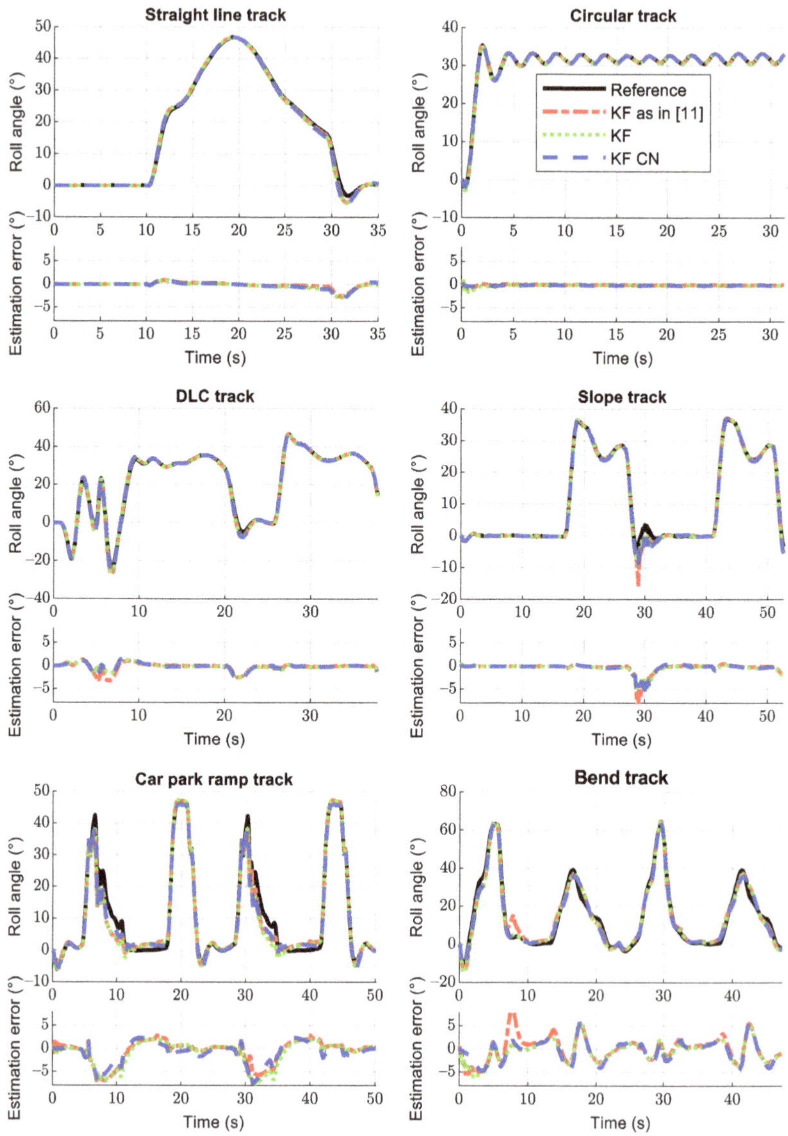

Figure 8. Estimated roll angle and roll angle error for the six different scenarios. All of these tests were performed with the rider in a neutral position. The reference is the black solid line, the red dash dotted line is the Kalman filter as presented in [11], the green dotted line is the variation of that Kalman filter presented here, and the blue dashed line represents the Kalman filter adapted for colored noise.

Table 5. Root mean square estimation errors for the three estimation algorithms on the six scenarios (in degrees).

Maneuver	Straight	Circular	DLC	Slope	Car Park	Bend
Kalman filter as in [11]	0.60	0.12	0.91	1.21	2.50	2.71
Kalman filter	0.69	0.27	0.71	0.90	2.81	2.28
Kalman filter for colored noise	0.78	0.11	0.82	1.05	2.34	2.35

The estimation method as presented in [11] provides good estimations in general, but it seems to be less robust than the new variations, as can be seen in the plots for the maneuvers in the *Slope track* and *Bend track*, where some short but important deviations can be appreciated. The risk with this method is that a temporal wrong estimation can lead to an inadequate weighting value to combine the two roll angle estimations used at the correction phase of the filter. This can potentially turn into a positive feedback that could make the estimation diverge from the true value.

The Kalman filter adapted for colored noise produces the best worst-case scenario estimations, but only by a narrow margin, and it is not the best in all the maneuvers. Moreover, it is more difficult to adjust because it has more parameters to tune, and thus it does not have any practical advantage in its current form.

Nevertheless, there is a potential benefit from using the colored noise adaptation for the filter, although it is not explored in this work. Since the characterization of the noise is improved, the statistical characteristics of the innovation sequence would be improved as well, thus allowing to use innovation-based adaptive Kalman filters, which should be able to improve the performance of the method for all the maneuvers that have now the worst estimation results. It was recently shown that innovation-based Kalman filters can work reasonably well even if the statistical properties of the noise are not perfect, and even with nonlinear models [24].

Another point that we wanted to address with this work was the influence of the rider position on the roll angle estimation error. For the sake of conciseness, only the *Circular track* was studied. Therefore, three simulations were run on this track, the only difference being the driver keeping its neutral position, leaning inwards, or leaning outwards. The results are shown in Figure 9, and the numerical results of the RMSEE are shown in Table 6. As expected, when the rider tilts inwards, the roll angle to perform a given maneuver is reduced, and the opposite happens when the rider leans outwards. Regarding the roll angle estimation accuracy, the difference between the different rider positions does not produce any significant perturbation.

Table 6. Root mean square estimation errors for the three estimation algorithms on the *Circular track* with the rider in neutral position, leaning inwards, and leaning outwards (in degrees).

Maneuver	Neutral	Inwards	Outwards
Kalman filter as in [11]	0.12	0.13	0.12
Kalman filter	0.27	0.28	0.28
Kalman filter for colored noise	0.11	0.10	0.17

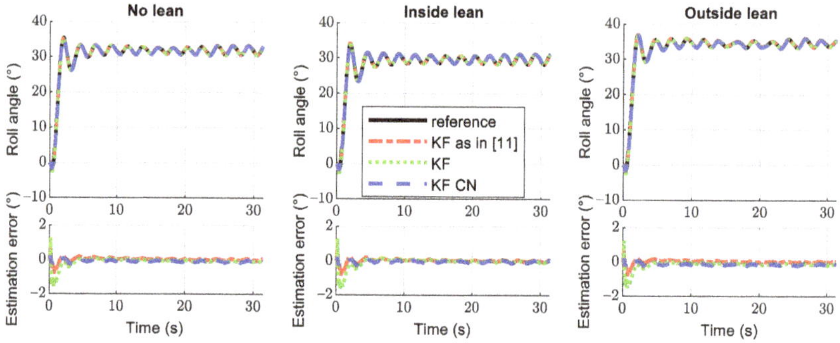

Figure 9. Estimated roll angle and roll angle error for the *Circular test track*. The left plot represents the maneuver with the rider in a neutral position, the central plot represents the rider tilting inwards during the turn, and the right plot represents the rider tilting outwards during the turn.

5. Conclusions

This paper is based on the roll angle estimation algorithm presented in [11], where it was first presented and tested at low speed with an instrumented bicycle. However, due to technical limitations of the experimental setup, part of the validation could not be performed. This work aims at verifying the performance of the roll angle observer in more challenging conditions. In order to achieve this goal, a multibody model of a motorcycle was developed, including toroidal-shaped wheels to consider the displacement of the contact point with wide tires. The multibody model also considers the suspensions, whose movement produces misalignments that could potentially degrade the performance of the roll angle observer. In addition, non planar scenarios modeled as triangle meshes, and rider motion were also considered. Finally, higher speeds than those considered in [11] were tested here.

The multibody model, governed by a longitudinal controller in charge of keeping an adequate speed and by a lateral controller used to keep the balance of the motorcycle and follow the prescribed trajectories, performed maneuvers in six different scenarios. From these maneuvers, some measurements were obtained, mimicking the properties of actual sensors by adding some white Gaussian noise. These measurements were used to verify the performance of the state observer, which was slightly modified to improve the robustness.

Moreover, an augmented version of the observer was devised to deal with the colored noise present at the plant and measurements used at the Kalman filter. Although this last observer provided slightly better results than the simpler version assuming white noise, the improvement is so subtle that in its current state, it would not be justified its practical use, given the more difficult tuning process and the increase in the size of the problem.

However, this new observer should provide (with the proper parameter tuning) statistical properties more consistent with the theoretical assumptions made to develop the Kalman filter. Therefore, the innovation sequence of the Kalman filter could be used to implement an innovation-based adaptive Kalman filter, which, in turn, should provide better estimation results, specially for the maneuvers that have the worst results in the present paper.

Supplementary Materials: Data from the simulations are provided as supplementary material at https://www.mdpi.com/article/10.3390/s21196626/s1. The data are organized in .csv files, and a text file (readme.md) provides the description of the data.

Author Contributions: Conceptualization, E.S. and M.Á.N.; methodology, D.M, A.L., U.L., M.Á.N. and E.S.; software, D.M., A.L., U.L. and E.S.; validation D.M, A.L., U.L., M.Á.N. and E.S.; formal analysis, D.M, A.L., U.L., M.Á.N. and E.S.; investigation, D.M, A.L., U.L., M.Á.N. and E.S.; resources, D.M, A.L., U.L., M.Á.N. and E.S.; data curation, D.M., A.L. and E.S.; writing—original draft preparation, D.M. and E.S.; writing—review and editing, D.M, A.L., U.L., M.Á.N. and E.S.; visualization, D.M.; supervision, E.S.; project administration, E.S.; funding acquisition, M.Á.N. All authors have read and agreed to the published version of the manuscript.

Funding: This research was partially financed by the Spanish Ministry of Science, Innovation and Universities and EU-EFRD funds under the project "Técnicas de co-simulación en tiempo real para bancos de ensayo en automoción" (TRA2017-86488-R).

Institutional Review Board Statement: Not applicable.

Informed Consent Statement: Not applicable.

Data Availability Statement: Data from the simulations are provided as supplementary material.

Conflicts of Interest: The authors declare no conflict of interest.

References

1. Schwab, A.L.; De Lange, P.D.; Happee, R.; Moore, J.K. Rider control identification in bicycling using lateral force perturbation tests. *Proc. Inst. Mech. Eng. Part K J. Multi-Body Dyn.* **2013**, *227*, 390–406. [CrossRef]
2. Cossalter, V.; Sadauckas, J. Elaboration and quantitative assessment of manoeuvrability for motorcycle lane change. *Veh. Syst. Dyn.* **2006**, *44*, 903–920. [CrossRef]
3. Schlipsing, M.; Schepankek, J.; Salmen, J. Video-Based Roll Angle Estimation for Two-Wheeled Vehicles. In Proceedings of the 2011 IEEE Intelligent Vehicles Symposium (IV), Baden-Baden, Germany, 5–9 June 2011. [CrossRef]
4. Teerhuis, a.P.; Jansen, S.T. Motorcycle state estimation for lateral dynamics. *Veh. Syst. Dyn.* **2012**, *50*, 1261–1276. [CrossRef]
5. Nehaoua, L.; Ichalal, D.; Arioui, H.; Davila, J.; Mammar, S.; Fridman, L.M. An Unknown-Input HOSM Approach to Estimate Lean and Steering Motorcycle Dynamics. *IEEE Trans. Veh. Technol.* **2014**, *63*, 3116–3127. [CrossRef]
6. Dabladji, M.E.H.; Ichalal, D.; Arioui, H.; Mammar, S. Unknown-input observer design for motorcycle lateral dynamics: TS approach. *Control Eng. Pract.* **2016**, *54*, 12–26. [CrossRef]
7. Boniolo, I.; Tanelli, M.; Savaresi, S.M. Roll angle estimation in two-wheeled vehicles. In Proceedings of the 2008 IEEE International Conference on Control Applications, San Antonio, TX, USA, 3–5 September 2008; pp. 31–36. [CrossRef]
8. Boniolo, I.; Tanelli, M.; Savaresi, S. Roll angle estimation in two-wheeled vehicles. *IET Control Theory Appl.* **2009**, *3*, 20–32. [CrossRef]
9. Corbetta, S.; Boniolo, I.; Savaresi, S.M. Attitude estimation of a motorcycle via Unscented Kalman Filter. In Proceedings of the 5th IFAC Symposium on Mechatronic Systems, Cambridge, MA, USA, 13–15 September 2010; pp. 511–516. [CrossRef]
10. Romualdi, L.; Mancinelli, N.; De, F.; Sorrentino, S. A new application of the Extended Kalman Filter to the estimation of roll angles of a motorcycle with Inertial Measurement Unit. *FME Trans.* **2020**, *48*, 255–265. [CrossRef]
11. Sanjurjo, E.; Naya, M.Á.; Cuadrado, J.; Schwab, A.L. Roll angle estimator based on angular rate measurements for bicycles. *Veh. Syst. Dyn.* **2018**, *57*, 1705–1719. [CrossRef]
12. Sharp, R.; Evangelou, S.; Limebeer, D. Advances in the Modelling of Motorcycle Dynamics. *Multibody Syst. Dyn.* **2004**, *12*, 251–283. [CrossRef]
13. Cuadrado, J.; Dopico, D.; Naya, M.Á.; Gonzalez, M. Real-Time Multibody Dynamics and Applications. In *Simulation Techniques for Applied Dynamics*; Springer: Vienna, Austria, 2008; pp. 247–311. [CrossRef]
14. Dopico, D.; Luaces, A.; Gonzalez, M.; Cuadrado, J. Dealing with multiple contacts in a human-in-the-loop application. *Multibody Syst. Dyn.* **2011**, *25*, 167–183. [CrossRef]
15. Sanjurjo, E. State Observers Based on Detailed Multibody Models Applied to an Automobile. Ph.D. Thesis, Universidade da Coruña, A Coruna, Spain, 2016.
16. Luaces Fernández, A. Contact and HiL Interaction in Multibody Based Machinery Simulators. Ph.D. Thesis, Universidade da Coruña, A Coruna, Spain, 2013.
17. Hirschberg, W.; Rill, G.; Weinfurter, H. Tire model TMeasy. *Veh. Syst. Dyn.* **2007**, *45*, 101–119. [CrossRef]
18. Pastorino, R.; Sanjurjo, E.; Luaces, A.; Naya, M.Á.; Desmet, W.; Cuadrado, J. Validation of a Real-Time Multibody Model for an X-by-Wire Vehicle Prototype Through Field Testing. *J. Comput. Nonlinear Dyn.* **2015**, *10*, 031006. [CrossRef]
19. Cossalter, V. *Motorcycle Dynamics*; Lulu: Morrisville, NC, USA, 2006.
20. Ogata, K. *Modern Control Engineering*, 5th ed.; Prentice Hall: New Jersey, NJ, USA, 2010.
21. Meijaard, J.P.; Papadopoulos, J.M.; Ruina, A.; Schwab, A.L. Linearized dynamics equations for the balance and steer of a bicycle: A benchmark and review. *Proc. R. Soc. A Math. Phys. Eng. Sci.* **2007**, *463*, 1955–1982. [CrossRef]
22. Simon, D. *Optimal State Estimation: Kalman, H Infinity, and Nonlinear Approaches*; Wiley: Hoboken, NJ, USA, 2006.
23. Grewal, M.; Andrews, A. *Kalman Filtering: Theory and Practice Using MATLAB®*; Wiley: Hoboken, NJ, USA, 2008.
24. Rodríguez, A.J.; Sanjurjo, E.; Pastorino, R.; Naya, M.Á. Multibody-Based Input and State Observers Using Adaptive Extended Kalman Filter. *Sensors* **2021**, *21*, 5241. [CrossRef] [PubMed]

sensors

MDPI

Article

Semantics Aware Dynamic SLAM Based on 3D MODT

Muhammad Sualeh and Gon-Woo Kim *

Intelligent Robotics Laboratory, Control and Robotics Engineering Department, Chungbuk National University, Cheongju 28644, Chungbuk, Korea; er.sualeh@gmail.com
* Correspondence: gwkim@cbnu.ac.kr

Abstract: The idea of SLAM (Simultaneous Localization and Mapping) being a solved problem revolves around the static world assumption, even though autonomous systems are gaining environmental perception capabilities by exploiting the advances in computer vision and data-driven approaches. The computational demands and time complexities remain the main impediment in the effective fusion of the paradigms. In this paper, a framework to solve the dynamic SLAM problem is proposed. The dynamic regions of the scene are handled by making use of Visual-LiDAR based MODT (Multiple Object Detection and Tracking). Furthermore, minimal computational demands and real-time performance are ensured. The framework is tested on the KITTI Datasets and evaluated against the publicly available evaluation tools for a fair comparison with state-of-the-art SLAM algorithms. The results suggest that the proposed dynamic SLAM framework can perform in real-time with budgeted computational resources. In addition, the fused MODT provides rich semantic information that can be readily integrated into SLAM.

Keywords: semantics; 3D multiple object detection; multiple object tracking; dynamic SLAM

Citation: Sualeh, M.; Kim, G.-W. Semantics Aware Dynamic SLAM Based on 3D MODT. *Sensors* **2021**, *21*, 6355. https://doi.org/10.3390/s21196355

Academic Editor: Juan A. Cabrera

Received: 9 August 2021
Accepted: 21 September 2021
Published: 23 September 2021

1. Introduction

In the realm of robotics, the SLAM paradigm is a well-established research area. Even though there are several efficient solutions to the problem, most of the approaches rely on the static world assumption [1]. The use of current algorithms in a real-world setting, where a dynamic and unstructured environment is a given, is hampered by this assumption. Although the advances in data driven approaches have enabled near-real-time environmental perception, and have shown promising performances [2,3], real-time requirements become a hurdle in resource-constrained computing platforms. Incorporating the semantics information from the environment into a geometric SLAM formulation also necessitates a 3D-pose of the detected objects, as well as semantic segmentation [4–6], which further adds to the complexity [4–6]. The learning-based techniques inherently necessitate bulky computing resources to meet the real-time requirements. The SLAM techniques that utilize such approaches require a lot of computing power or cannot guarantee real-time operation.

The objective of this work is to effectively merge the two broad paradigms of SLAM and 3D MODT, such that both complement the individual findings while being capable of performing independently. The feature based visual SLAM approaches rely on feature tracking to estimate the camera pose over time. Unless the region of visual feature is highly dynamic, it is hard to categorize and filter the features pertaining to the dynamic regions. Thus, if left untreated, the information pertaining to dynamic regions of the scene gets incorporated in the pose estimation process, which leads to an inaccurate SLAM system. Traditionally, semantic segmentation masks are provided alongside the image to enable SLAM for an informed selection of visual features at the early stage. This process is however computationally demanding and real-time constraint is often compromised. Furthermore, the semantic segmentation only provides class labels and contour of the detected objects. Hence, there is a need for an alternate approach that can effectively provide equivalent semantic information of the environment without demanding extraordinary computational demands and that can guarantee real-time operation.

The capability of detecting and avoiding information related to the dynamic objects in the environment for mapping and localization purposes is the concept underlying the term dynamic SLAM. The dynamic objects are often considered as outliers, and the outlier rejection techniques such as RANSAC are employed to filter such data. In the literature, several attempts are made to deal with the presence of dynamic objects under feature-based SLAM [7–10] and direct SLAM approaches [11–14]. The further categorization of dynamic SLAM is based on the type of semantics integration adopted. The loosely coupled approaches [8,15–18] perform environmental perception and SLAM separately. On the other hand, tightly coupled approaches solve both problems in an integrated fashion [19–25]. Both approaches have their pros and cons, but the core issue boils down to the computational resource requirements of perception and real-time capability of the entire framework, which is largely left unattended.

The proposed framework operates on a Visual-LiDAR setup that provides calibrated and time synchronized inputs. The core of the framework is a lightweight neural network model that takes the image sequence from camera and produces classification and localization of objects in the image frame. LiDAR data, on the other hand, and treated by the MODT module, which clusters and tracks the potentially trackable objects. The sensor fusion module utilizes the tracking information to associate visual classifications with tracked clusters. The classified point pertaining to the tracked dynamic objects is projected onto the image frame and up sampled to form a dynamic region mask. The mask of the dynamic region coupled with the image is provided to visual the SLAM module. This enables the SLAM module to make an informed selection of visual features by avoiding the features pertaining to the dynamic regions of the scene.

The key contributions in this work are the semantic mask generation, point cloud classification, visual-LiDAR based MODT, visual features selection in SLAM, and the integration of SLAM and MODT. The mask is generated by fusing the information from the visual detector and LiDAR based tracker. Instead of frame-wise classification of the LiDAR point cloud, the tracked clusters are temporally classified using visual detection information. The SLAM module can make effective use of masks by avoiding the features pertaining to dynamic regions of the scene.

To evaluate and analyze the effect of the proposed dynamic SLAM framework, a comparison is made between basic ORB-SLAM2 [26] and the proposed MODT based dynamic SLAM framework. The Tracking KITTI Datasets [27] are used for fair evaluation against the provided ground truths over well-established evaluation metrics. The results suggest that the proposed approach is an effective solution for the dynamic environments. Furthermore, the framework is evaluated over selected sequences of Raw KITTI datasets [28] are for comparison with state-of-the-art dynamic SLAM approaches.

The remainder of the paper is structured as follows: in Section 2, related works are briefly described. In Section 3, an overview of the framework is presented. The proposed methodology is introduced in Section 4. The evaluation results and comparison with state of the art are laid down in Section 5, followed by conclusions in Section 6.

2. Related Works

In recent years, several attempts have been made to address the dynamic SLAM problem by integrating semantics and environmental perception in SLAM. A feature-based SLAM paradigm [7] projects the map features onto the current frame to classify the dynamic part of the scene. Three-dimensional object tracking is performed in [8] to identify dynamic objects in the scene. Similarly, detection and tracking approaches are proposed in [9] but are restricted to humans being the dynamic objects. The technique proposed in [10] assigns probabilistic weights to the objects, based on the class. The direct SLAM approaches for dynamic environments have also been proposed [11–14]. The stereo camera-based scene flow representation is utilized in [11] to detect moving objects. RGB optical flow-based segmentation is performed in [12]. A method is proposed in [13], that exploits consecutive depth images to differentiate static and dynamic parts of the scene.

Similarly, the work in [14] focuses on the intensity difference of the consecutive RGB images. The above-mentioned approaches consider the assumption that the dynamic object in the scene remains dynamic for the entire view time. Consequently, the dynamic object at rest is still considered static, which eventually becomes a part of the mapping process. The advances in the deep learning paradigm have enabled prior classification of dynamic objects based on the classes. References such as [8,9] can detect and classify the dynamic objects; however, detection of changes produced in the environment by static objects is still challenging. A fusion of multi-view geometry and deep learning techniques caters to such shortcomings [29,30].

The MODT integrated SLAM paradigm can broadly be classified into the categories of Loosely Coupled and Tightly Coupled approaches [25]. The loosely coupled approaches perform MODT and SLAM separately, whereas tightly coupled approaches operate in an integrated fashion [19–25].

Conventionally MODT based SLAM approaches are loosely coupled [8,15–18] and solve MODT and SLAM separately. An implementation in [8] tracks 3D objects and exploits the information provided by the SLAM, and the tracker allows the features pertaining to the objects. A derivation of the Bayes formula of SLAM with tracking of moving objects is devised in [15]. A graph-based approach in [16] makes use of Expectation Maximization, allowing the landmarks to be dynamic. A dense mapping algorithm is proposed in [17] that reconstructs the static background and the dynamic objects in the scene using stereo vision. A dense mesh tracking approach is proposed in [18] that utilizes visual-inertial SLAM in conjunction. The approach is focused on humans, tested on only a simulated environment, and largely relies on camera pose estimation. That is, if the camera pose estimation fails, the MODT also fails. The work in SLAMMOT [31] established a mathematical framework to integrate filtering-based SLAM with dynamic object tracking. Later, an RGB-D camera was employed with the same technique for dense reconstruction of an indoor scene, together with dynamic objects using semantic segmentation [13,14,29,30]. Further, in dense approaches Mask-Fusion [32] and MID-Fusion [33], z techniques were deployed for more accurate semantic segmentation of the dynamic objects in the scene.

Tightly Coupled MODT and SLAM approaches aim to merge information from static and dynamic parts of the scene into a single framework to enhance the estimation accuracy. The work in [19] presented the idea of a tightly coupled approach, however, with limited comparable results. Being end-to-end trained approach estimates 3D pose and dimensions of cars jointly with camera poses. Although the approach provides accurate pose estimations, it suffers with the loss of generality, and thus, huge data would be required to track generic objects. The CubeSLAM [20] is a monocular SLAM based approach that generate 3D bounding box proposals based on 2D bounding boxes and vanishing points. The approach assumes objects to have constant velocity for a fixed time and exploits the constraints pertaining to road structure, non-holonomic wheel motion, and visibility of the cuboid view of objects. The approach in ClusterSLAM [24] proposes a SLAM back end to identify rigid bodies and to estimate the motion. The approach relies on the landmark tracking and association quality. A technique in ClusterVO [21] models the object points for the probability of being dynamic. VDO-SLAM [23] capitalizes on dense optical flow to identify the number of tracked points on the dynamic objects. The bundle adjustment is implemented with cameras, objects, and points; however, at the cost of high computational demand. Similarly, DynaSLAMII [25] proposed a tightly coupled approach for MODT and SLAM that performs the bundle adjustment between camera, points, and objects. The performance relies on high quality semantic segmentation and thus, high computational resources are required.

In contrast to the existing techniques, the proposed framework in this work is unique and hybrid in a way that MODT and SLAM is performed over different sensor modalities. 3D MODT is performed with respect to the vehicle frame of reference and independent of SLAM's estimated pose, whereas SLAM can take full use of semantic MODT information as and when required for feature tracking, mapping, or loop closures. The advantage of such integration is manifold. The object tracking continues even if the pose estimation of SLAM fails. Furthermore, SLAM can make use of masked dynamic regions and provide pose information to MODT to obtain tracking information in a common frame of reference.

3. System Overview

The system pipeline of the proposed framework assumes that the platform is equipped with visual and LiDAR sensors. The visual sensor is primarily used for SLAM and object detection, whereas the LiDAR sensor is used for spatial object tracking. The framework is built on top of ORB-SLAM2 [26] with an additional stream of input for the masked image of dynamic objects. In parallel, an Interactive Multiple Model–Unscented Kalman Filter–Joint Probabilistic Data Association Filter (IMM–UKF–JPDAF) based tracker is operated to track objects in 3D space. To classify the objects, a lightweight visual detector YOLO-v3 [34] is deployed that operates on a reduced image resolution. This increases the inference speed of the visual detector at the cost of frequent missed detections of small and partially occluded objects. However, this shortcoming is leveraged by the object tracker, which preserves the classification history of the tracked objects. The LiDAR point cloud clusters pertaining to the tracked objects are projected onto the image frame and up sampled to generate the dynamic objects mask. The dynamic object mask is exploited by the visual SLAM to choose the visual features for odometry and/or mapping. Figure 1 shows the block diagram of the framework with individual modules.

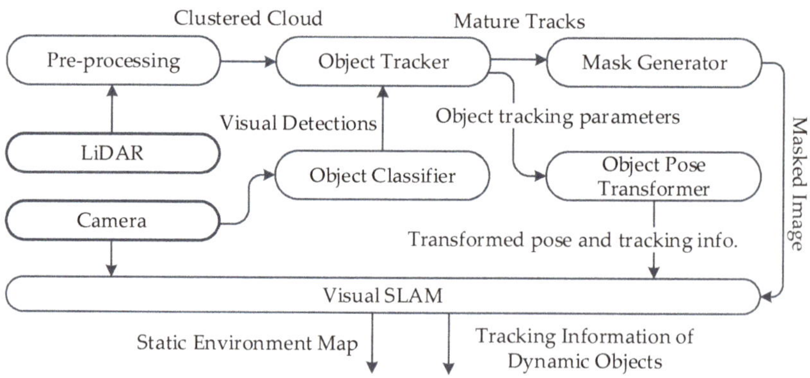

Figure 1. Framework for joint MODT and SLAM.

4. Proposed Framework

The proposed framework is comprised of several modules that interact to solve the dynamic SLAM problem. The operations of individual modules are briefly described in the subsequent subsections and the block diagram is presented in Figure 2. The implementation and evaluation of the framework is based on sensor setup involving stereo camera and 64 channel LiDAR. A stereo camera is used for visual SLAM and visual object detection, whereas a LiDAR sensor is utilized for spatial object detection and tracking. Furthermore, the computational environment used for developing and evaluating the proposed framework is constituted by a desktop computer having an Intel Core i7-7700 CPU with 16 GB of RAM, and Nvidia GTX 1060 GPU. The system runs the Robot Operating System (ROS) "Melodic Morenia" middleware on top of Ubuntu 18.04.5 LTS. The framework is implemented and evaluated in a ROS environment to ensure real-time capability.

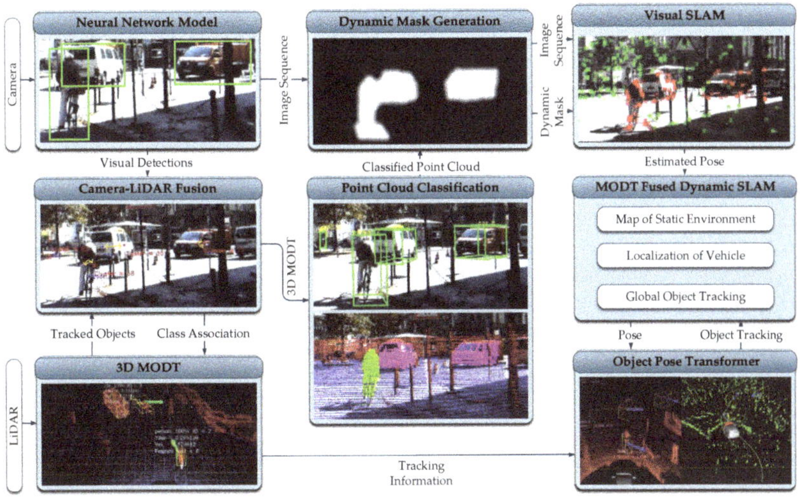

Figure 2. Proposed MODT integrated Dynamic SLAM framework.

4.1. 3D MODT

The 3D MODT module preprocesses the LiDAR point cloud and maintains the tracks of potentially trackable objects. The module has four subcomponents: ground segmentation, clustering, box fitting, and tracking. Each subcomponent is briefly discussed in the subsequent subsections and information flow is described in Figure 3. The LiDAR data after getting treated by the submodules of ground segmentation, clustering, and box fitting are presented to the tracking submodule as measurements for tracking. The tracking submodule maintains a temporal record of tracked objects' dimensions and class associations received from the visual detector. The 3D MODT module provides 3D MODT information with respect to the vehicle reference frame to the camera-LiDAR fusion and pose transformer modules, respectively.

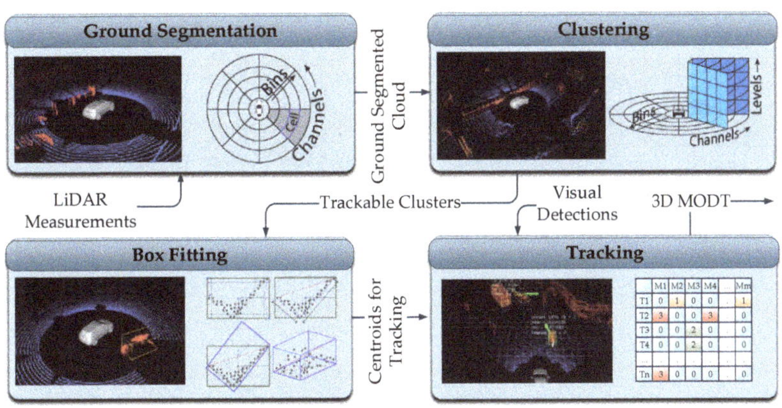

Figure 3. 3D MODT module in the proposed framework.

4.1.1. Ground Segmentation

The ground segmentation is a part of the LiDAR preprocessing step, where LiDAR measurements pertaining to the ground are identified and separated from the point cloud for further processing. Several techniques are proposed in the literature to accomplish ground segmentation, with a grid-based approach being the most relevant and effective [35]. The grid-based approach can accommodate the assumptions of the ground being non-planner, and the point-cloud being a merger of multiple LiDARs measurements. Furthermore, the grid-based approach is computationally efficient because of compact data representation. Moreover, the approach can be easily deployed in any arrangement of the sensor setup without requiring training procedures like data driven approaches. The implementation of ground segmentation in this work follows a grid-based approach that considers a non-planar ground.

Initially invalid and out of rage measurements of the LiDAR point cloud are filtered out and the point cloud is converted into a 2D cylindrical representation. The 2D cylindrical representation is composed of channels and bins, as expressed in Figure 3 from the top view. The channels represent the measurements pertaining to the vertical slices of the LiDAR data, whereas bins are the further divisions of the channels based on the distance from the origin. To segment the ground, bins of each channel are traversed to estimate the local ground level, starting from the origin, where the ground level equals the sensor height. The estimated local ground levels together with vertical distribution of the measurements in a bin and absolute slope between consecutive bins are used to set the ground threshold for each bin. The bin-level ground threshold is applied to label the LiDAR measurements as ground and non-ground measurements. The non-ground labeled point cloud is then fed forward for clustering.

4.1.2. Clustering

The clustering of LiDAR point cloud is referred to as the process of grouping the closely existing LiDAR measurements, the approach adopted in this work falls under the hierarchy-based technique [36]. The 2D cylindrical representation of the non-ground LiDAR measurements from the ground segmentation module is converted to 3D cylindrical representation by vertical distribution of measurements. The 2D grid representation is not effective for clustering in cluttered urban scenarios where elevated structures exist. The 3D cylindrical representation allows the clustering of objects that exist under elevated structures, such as streetlights, traffic signals, overhead bridges, and trees. Furthermore, compact representation of data enables efficient computation of clusters.

In 3D grid representation, each cell of the 2D cylindrical grid is further divided into levels based on the vertical height of LiDAR measurements, as shown in Figure 3. For clustering, each cell with measurements is selected as an index and all neighbor cells are traversed to inspect for a threshold number of measurements. In this fashion, the region for each cluster grows with a unique ID. To address the time complexity, the region of interest is selected based on the effective clustering range of LiDAR measurements. The measurements of LiDARs, for example, with fewer channels get sparse at distances, and the actual shape of the object cannot be identified; thus, clustering of points beyond a reliable range only increases the computation and outliers. With these clustering limits imposed, the clustered point cloud is effectively obtained and is treated with the cluster dimension filter. The dimension filter identifies the clusters pertaining to the potentially trackable objects, considering the size, position, and number of enclosed LiDAR measurements. The clusters that are suitable for tracking are forwarded to the bounding box fitting module.

4.1.3. Box Fitting

The LiDAR measurements experience occlusions, and obtaining the exact 3D shape object is inherently impossible. To estimate the actual shape of the clustered object, box-fitting techniques are devised. In this work, considering the computational limitations and real-time requirements, a feature-based method is deployed [36]. The technique represents the cluster in a minimum rectangular shape in 2D top view, then performs L-shape point cloud fitting as shown in Figure 3. Initially, the farthest pair of points is searched for that exists near a threshold boundary of the minimum rectangle boundary, based on the location of the cluster in the spatial space. The pair of points is used to form a line and a farthest point orthogonal to the line is searched. The three points represent the three corners of the 2D area of the clustered object; the height of the cluster is directly used to define the third dimension of the bounding box. The three dimensions of the estimated bounding box are then used to evaluate the centroid and yaw angle of the cluster relative to the LiDAR sensor. The actual dimensions and centroid of the clustered object at this stage has the effect of occlusions; therefore, the tracker module maintains the history of estimated dimensions of the tracked cluster and performs corrections temporally if required.

4.1.4. Object Tracker

The urban cluttered environment perceived by a LiDAR sensor is affected by numerous uncertainties. Cluttered clusters of trackable objects make data association challenging, whereas dynamic objects tend to follow such motion patterns that further increase the uncertainties. To tackle these uncertainties, an IMM–UKF–JPDAF based tracker is used [36]. The Joint Probabilistic Data Association Filter (JPDAF) based approach is utilized for data association with an assumption of Gaussian distribution to address the uncertainties due to clutter. Furthermore, the Interacting Multiple Model (IMM) approach is adopted for the state estimation of objects to deal with the uncertainties due to motion. The non-linearities of the motion models are accommodated by an Unscented Kalman Filter (UKF) through a Gaussian process. The IMM–UKF–JPDAF based approach effectively solves the recursive state estimation and mode probabilities of object clusters, which is described by a non-linear jump Markov system.

The execution times of the tracker module mainly rely on the number of maintained tracks. Furthermore, JPDAF based approaches tend to face a combinatorial explosion problem, which if left untreated, requires exceptional computational resource, as a hypothesis of all possible combinations of tracks and measurements are made. To tackle this shortcoming, a clustering-based scheme is utilized to limit the association pairs of tracks and measurements to only the gated measurements instead of all possible combinations. The tracking block shown in Figure 3 shows a table of tracks T and measurements M in rows and columns, respectively. The track T2 gets gated measurements of M1 and M4 forming an association cluster of two measurements and a single track. Similarly, tracks T3 and T4 share the measurement M3 forming an association cluster of two tracks and a single measurement. Thus, the JPDA problem is reduced to a set of smaller problems that can be solved efficiently. Moreover, an efficient track management mechanism is set in place that prunes out the tracks pertaining to the inconsistent measurements, resulting in a limited number of tracks to maintain. An additional condition of a tracked object being classified by the visual object detector, reducing the number of tracks further, results in decreased computational load. The prime objective of this module is to estimate the tracking parameters of the potentially trackable objects, such as, pose, velocity, and dimensions.

4.2. Camera-LiDAR Fusion

The fusion approach adopted in this work is regarded as late fusion, where tracked clusters are temporally classified instead of frame-wise classification [36]. This is carried out by two components, class association and class management. The class association component assigns a visually detected object's class to the tracked object's clusters, whereas the class management component assesses the class assignment history to select a class for the tracked object cluster.

To perform the class association, eight corners of the 3D bounding box of the tracked object clusters are projected on the 2D image plane using camera-LiDAR extrinsic parameters. The minimum and maximum pairs of x and y coordinates from the set of 2D projected points forms a 2D bounding box in the image frame representing the 3D tracked object. The Munkres association strategy is applied on the 2D centroids of the visual object detector and 2D bounding box representing the tracked object in the 3D space. This association is maintained by a class association vector that maintains the history of class associations and provides a certainty of classes assigned to the tracked object over time.

Let T^k and D^k be the sets representing tracks and visually detected objects, respectively, at time step *k*. The centroids of tracked clusters o_i^k are projected onto the image frame of the corresponding time step, resulting in a 2D pixel location in an image $\overline{o}_i^k$. Similarly, the centroids of visually detected objects are calculated using bounding box dimensions, represented by m_j^k. The 2D centroids of both sources are utilized to populate the Euclidean distance cost matrix $E^k = \left[c_{ij}^k\right]$, where i $=\{1, 2, \ldots, T\}$ and j $=\{1, 2, \ldots, D\}$. The cost matrix

$$c_{ij}^k = \begin{cases} d\left(\overline{o_i^k}, m_j^k\right) & \text{if } \mathrm{IOU}\left(\overline{t_i^k}, d_j^k\right) > 0.3 \\ 1000 & \text{otherwise} \end{cases} \tag{1}$$

is, however, constrained by the criterion that at least 30% of overlap must exist among the corresponding 2D bounding boxes from the two sources. Following the Munkres association, the algorithm for optimized minimum cost is performed and a set of index pairs Y relating to the associated tracks t_i^k and visual detections d_j^k is acquired. Using the set Y class association matrix $\hat{E}^k = \left[\hat{c}_{ij}^k\right]$, it can be formulated such that,

$$\hat{c}_{ij}^k = \begin{cases} v_j & \text{if } < i,j > \subset Y \\ 0 & \text{otherwise} \end{cases}, \tag{2}$$

where v represents the association of the class of visually detected objects and the dimension index of class association vector $A_v^i = \left(a_1^i, \ldots a_n^i\right)$. The matrix $\hat{E}^k$ finally updates the class association vector A^i and increments the associated class dimension,

$$\hat{c}_{ij}^k = \begin{cases} a_{\hat{c}_{ij}}^i & \text{if } \hat{c}_{ij}^k = 0 \\ a_{\hat{c}_{ij}}^i + 1 & \text{otherwise} \end{cases}. \tag{3}$$

The class association vector A^i, along with the age of track t_{age} is exploited to compute the class certainty P_c^i of tracked objects and the ratio P_o^i of the object, that reasons the tracked object to be valid.

$$P_c^i = \frac{\max(a_v)}{t_{age}} \tag{4}$$

$$P_o^i = \frac{\left(t_{age} - \sum_{v=1}^{n} a_v^i\right)}{t_{age}} \tag{5}$$

The camera-LiDAR fusion thread operates along the LiDAR based MODT and provides the class association certainty to the tracker. In Figure 4, visual classification of tracked clusters is demonstrated. The blue and red dots in the image represent the centroids of visually detected objects and centroids of tracked LiDAR clusters projected on to the image, respectively. The qualified associations update the class association vector of a mature track as demonstrated in Figure 4. The class association vector represents the classes assigned to the tracked cluster, including no associations counting temporally in terms of frames. Together with the tracking age, class certainty of 60% is evaluated for the tracked object being a "car".

4.3. Dynamic Mask Generator

The mask generator module takes the LiDAR point cloud cluster and projects it onto the image frame. The visual features generally exist on the edges and corners of the surfaces, and thus, the mask needs to slightly exceed the exact contour of the detected object. To handle the sparsity of the LiDAR measurements, a 2D Gaussian blur kernel is formulated,

$$G(x, y) = \frac{1}{2\pi\sigma^2} e^{\frac{x^2+y^2}{2\sigma^2}}, \tag{6}$$

where x and y are the coordinates of the projected LiDAR measurement pertaining to the cluster of the dynamic object. The standard deviation σ of the blur is set based on LiDAR and image resolution. For lower ratios of image to LiDAR resolution and larger coverage of LiDAR measurements on the image, lower values of σ are sufficient. The dynamic mask generated from the proposed framework in contrast to the mask generated from Mask R-CNN is presented in Figure 5. In the literature, the reported inference time of Mask R-CNN using standard GPU fails to meet the real-time requirements, whereas the proposed framework provides the comparable dynamic masks well within the sensor sampling time, allowing the entire framework to operate within 100 ms.

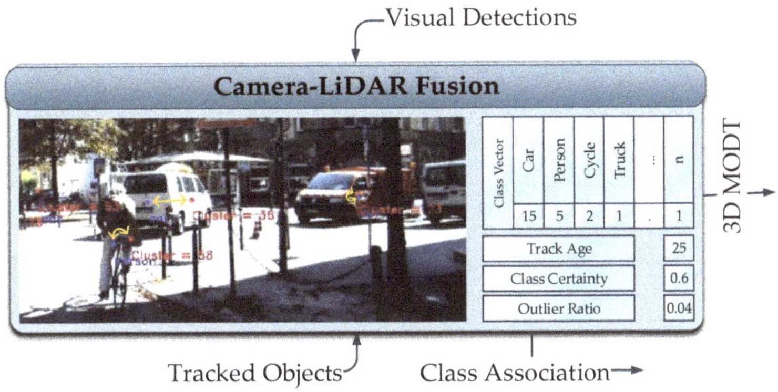

Figure 4. Camera-LiDAR Fusion Module.

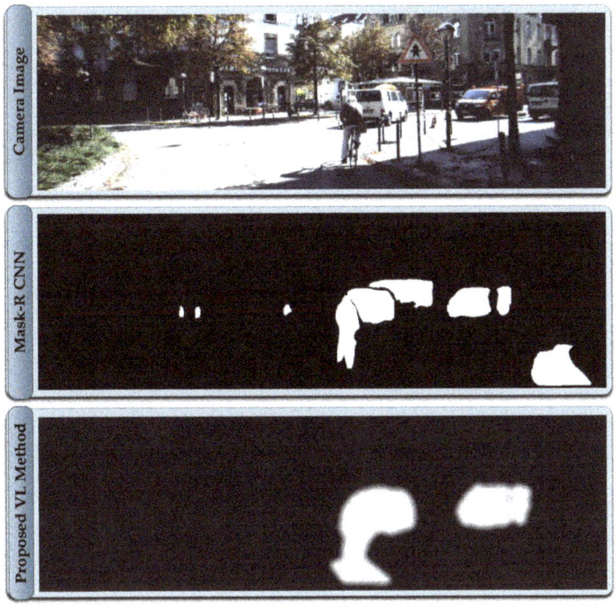

Figure 5. Visual-LiDAR based Mask generated in comparison with Mask R CNN.

4.4. Object Pose Transformer

The primary function of the object pose transformer is to transform the pose and tracking information of objects to the SLAM frame of reference. The module receives the localization information from the SLAM module and tracking information from the

3D MODT module. The object tracking is performed in the vehicle frame, where static–dynamic classification of tracked objects cannot be effectively made without the ego-motion information. For effective utilization of this information, tracking information is transformed into a SLAM frame of reference, where realistic information of the tracked objects can be realized.

4.5. Dynamic SLAM

The core of the proposed framework is an ORB-SLAM2 that gets the capability of making an informed selection of visual features. The module takes images together with the generated mask image as an input. The extracted visual features are labeled as static or dynamic based on the provided mask. The features pertaining to the static part of the scene are validated, whereas features pertaining to dynamic objects in any state are filtered out, as demonstrated in Figure 6. The focus in this implementation is to demonstrate an effective alternative solution to the computationally expensive semantic segmentation algorithms like Mask R CNN [3], widely utilized in the related literature. The proposed framework generates a comparable dynamic mask in real-time without excessive computational resource requirement.

Figure 6. Dynamic SLAM with Visual-LiDAR based dynamic object mask.

The visual features generally exist on the corners and sharp edges of the scene, thus extracting a perfect pixel-level annotation and sharp contours of the dynamic objects are not desired, as features located on the boundary of dynamic objects are dubious. The masks generated in the proposed framework are based on sparse LiDAR data that is treated by a 2D Gaussian blur that ensures that features on the boundary regions of dynamic objects are filtered. Consequently, the generated map is free of information pertaining to dynamic objects, although the tracker module allows identification of the dynamic objects that are currently static such as parked cars and using this information within the SLAM process of pose estimation. This implementation is intentionally left out for future work, as the focus is to demonstrate that a comparable dynamic object's mask to the conventional mask generation approaches can be efficiently achieved. In this work, LiDAR data are only used for spatial object tracking and exploitation of a classified point cloud for dynamic mask generation.

5. Evaluation and Comparison

To evaluate and analyze the effect of the proposed Dynamic SLAM framework, a comparison is made between basic ORB-SLAM2 [26] and the proposed MODT based dynamic SLAM framework. Conventionally, KITTI Datasets [27] are used for fair evaluation against the provided ground truths over well-established evaluation metrics. However, the datasets provided for evaluating SLAM algorithms lack in the presence of dynamic objects, mainly because of static world assumption. For this reason, the datasets provided for object tracking are used with the provided ego-motion data as the ground truth. The true potential of the framework can be fairly tested in the absence of no loop closures, allowing the measurement of maximum drift. Since the datasets are targeted to test object detection and tracking, dynamic objects exist in almost every sequence in abundance.

To perform a detailed and standard analysis, the KITTI Odometry criterion [27] is followed. The Absolute Trajectory Error (ATE) measures the root-mean-square error between predicted camera poses [x, y, z] and ground truth. Relative Pose Error (RPE) measures frame-to-frame relative pose error, since the provided ground truth data are in the GPS/IMU frame, and the trajectories generated by SLAM are in camara coordinates. The generated trajectory is aligned using 7DoF optimization to the ground truth associated poses during evaluation by minimizing the ATE [37]. Table 1. The estimated trajectories evaluation and comparison on KITTI Tracking Datasets

Table 1. The estimated trajectories evaluation and comparison on KITTI Tracking Datasets.

seq	ORB-SLAM2 [26]			Proposed Framework		
	ATE (m)	RPE (m)	RPE (deg)	ATE (m)	RPE (m)	RPE (deg)
00	0.663	0.63	1.345	**0.663**	**0.63**	**1.345**
01	**1.362**	**1.062**	**0.472**	1.704	1.066	0.474
02	**0.127**	**0.704**	0.139	0.148	0.705	**0.139**
03	0.254	**1.699**	0.245	**0.245**	1.703	**0.244**
04	1.73	**1.803**	0.589	**1.275**	1.805	**0.583**
05	0.58	**1.717**	0.117	**0.557**	1.719	**0.116**
06	**0.175**	0.263	**0.481**	0.366	0.254	0.492
07	6.122	0.914	1.107	**3.784**	**0.906**	**1.042**
08	**1.346**	**2.092**	**0.259**	1.369	2.093	0.267
09	8.08	1.289	0.563	**1.677**	**1.26**	**0.521**
10	**0.472**	2.02	0.269	0.55	**2.02**	**0.267**
11	1.324	0.826	0.171	**0.95**	**0.824**	**0.17**
12	0.01	0.003	0.006	**0.01**	**0.002**	**0.005**
13	**0.451**	0.818	**0.383**	0.462	0.818	0.384
14	**0.423**	0.642	**1.436**	0.505	0.634	1.456
15	0.581	0.282	0.04	**0.31**	**0.274**	**0.039**
16	0.023	0.008	0.006	**0.022**	**0.008**	**0.006**
17	0.037	0.006	**0.008**	**0.037**	0.004	0.009
18	**0.866**	**1.088**	**0.118**	1.032	1.096	0.119
19	3.742	0.324	0.26	**1.835**	**0.315**	**0.24**
20	7.891	1.227	0.261	**0.634**	**1.219**	**0.25**
mean	1.812	0.970	0.413	**0.906**	**0.967**	**0.408**

Lower errors are written in bold.

The results of comparison are presented in Table 1. The variation in the scores suggests the presence of dynamic objects. The better performing metrics are presented in bold text. The sequences with dynamic objects having less effect on the SLAM produce near similar metric results. In these sequences the dynamic objects' presence does not cover the entire view for consecutive time frames. In some sequences, the dynamic objects are static, like parked cars, the conventional SLAM takes use of the features belonging to the corresponding regions to produce better results. However, this information becomes the part of the generated maps.

The sequences with dynamic objects are in a motion similar in pattern to the ego-vehicle, such as in the roadway conditions, influencing SLAM the most. The metric scores of sequences 9, 19, and 20 support this argument, where ORB-SLAM2 [26] suffers a lot due to the presence of dynamic objects, and drift creeps into the estimated trajectory. In contrast, the proposed framework efficiently filters the visual features pertaining to the dynamic object regions. Thus, the estimated trajectory is more accurate. The plots of trajectories generated by ORB-SLAM2 [26] and the proposed framework of sequence 20 are presented in Figure 7. The drift in both plots suggest that the informed selection of features can estimate better trajectories, specifically in the scenario where dynamic objects tend to follow a similar motion pattern.

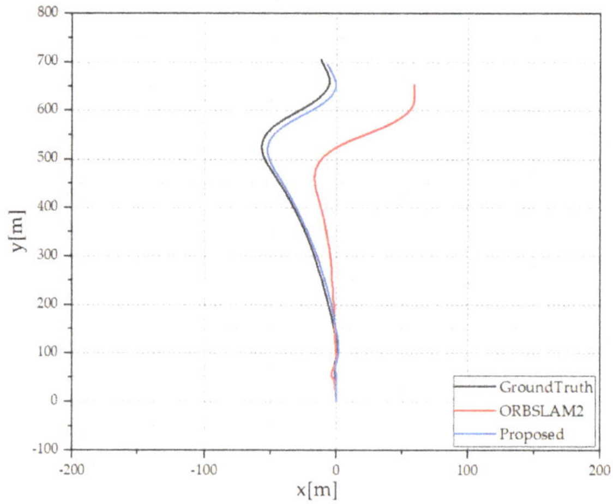

Figure 7. Estimated trajectories of sequence 20 from KITTI Tracking Dataset.

To view the performance of proposed framework in comparison to the state-of-the-art dynamic SLAM frameworks, the metric scores against KITT Raw DATASET [28] sequences are evaluated and presented in Table 2. The evaluation scores of ORB-SLAM2 [26] are reproduced using the similar evaluation tool and computational platform. However, the evaluation scores of related dynamic SLAM frameworks on the same dataset sequences are reported as a reference from the work in [25]. Best mean metric scores suggest that the proposed dynamic SLAM framework can perform at par to the conventional approaches that rely on data driven semantic segmentation approaches to identify dynamic objects in the scene.

Table 2. The estimated trajectories evaluation and comparison on KITTI Raw DATASET.

seq	ORB-SLAM2 [26]			DynaSLAM [29]			ClusterSLAM [24]			ClusterVO [21]			DynaSLAMII [25]			Proposed Framework		
	ATE [m]	**RPE [m]**	**RPE [deg]**	**ATE [m]**	**RPE [m]**	**RPE [deg]**	**ATE [m]**	**RPE [m]**	**RPE [deg]**	**ATE [m]**	**RPE [m]**	**RPE [deg]**	**ATE [m]**	**RPE [m]**	**RPE [deg]**	**ATE [m]**	**RPE [m]**	**RPE [deg]**
0009	1.36	1.06	0.47	0.81	1.80	0.57	0.92	2.34	1.72	**0.79**	2.98	1.72	0.85	1.87	0.57	1.70	**1.06**	**0.47**
0013	0.25	1.69	0.24	0.30	0.99	0.57	2.12	5.50	4.01	0.26	1.16	0.57	0.29	**0.93**	**0.00**	**0.24**	1.70	0.24
0014	1.73	1.80	0.58	0.60	1.62	**0.57**	0.81	2.24	1.72	**0.48**	**1.04**	**0.57**	**0.48**	1.35	**0.57**	1.27	1.80	0.58
0004	0.86	**1.08**	**0.12**	0.56	1.36	0.57	1.12	2.78	1.15	**0.40**	1.77	1.15	0.64	1.41	0.57	1.03	1.09	**0.12**
0047	7.89	1.22	0.26	2.87	5.95	1.15	10.2	8.94	3.44	4.78	6.54	2.86	3.03	6.85	1.15	**0.63**	**1.21**	**0.25**
mean	2.42	1.37	0.33	1.02	2.34	0.68	3.03	4.36	2.40	1.34	2.69	1.37	1.05	2.48	0.57	**0.97**	**1.37**	**0.33**

Better and similar scores are written in bold.

The mean overall scores reflect the promising performance of the proposed algorithm but the individual sequences of 0009, 0014, and 0004 show slightly poor results. The similarity in these sequences is the abundance of dynamic objects being static. ORB-SLAM2 making use of the whole visible region can produce better pose estimation, whereas the dynamic SLAM ignoring the dynamic regions tends to degrade the performance in such scenarios. However, this can be seen as an advantage in terms of SLAM process, where the generated map remains free of dynamic entities, enabling better loop closure and re-localization capabilities. The sequence 0013 is a relatively short time sequence with limited dynamic regions and plentiful visual features; thus, dynamic SLAM tends to produce similar metric scores. Sequence 0047 being a highway scenario with a large trajectory demonstrates the effectiveness of the proposed dynamic SLAM. An interesting find in the reported scores is that the tightly coupled approaches perform more poorly than the loosely coupled approaches. ClusterVO [21], and DynaSLAM II [25] having a tightly coupled approach tend to have lower accuracies on average. On the other hand, DynaSLAM [29] which follows a dynamic masking approach like the proposed framework manages to acquire better metrics in comparison. ClusterSLAM [24] largely relies on the initial camera pose estimations, which further supports the argument that loosely coupled approaches with effective use of redundant information can guarantee better accuracies.

6. Conclusions

In this work, a visual-LiDAR based 3D MODT is integrated with SLAM that caters to the challenges pertaining to the dynamic world. The proposed framework considering the constrained computational resources and real-time requirements performs a temporal classification of tracked objects. An efficient IMM–UKF–JPDAF based tracker spatially tracks the objects while maintaining the class association history, to address the real-time limitations and shortcomings of object detection. The classified LiDAR point cloud is effectively utilized to produce a dynamic object mask, capable of replicating the state-of-the-art semantic segmentation approaches. SLAM exploits the dynamic mask provided by the MODT for informed selection of visual features for tracking and mapping tasks to realize a dynamic SLAM.

The proposed framework is tested on the KITTI Datasets [28] and evaluated against the established metrics. The results and comparison with the related works suggest that the proposed approach is an effective solution for the dynamic SLAM. The key contribution in this work is the efficient generation of dynamic object mask from the Visual-LiDAR based MODT in real-time without the need of exceptional computational resources. Furthermore, a loosely coupled approach for 3D MODT and SLAM is proposed that can exploit the redundant information with a minimalistic interdependence.

In the future, it is intended to incorporate the further static–dynamic classification of characteristically dynamic objects supported by the MODT information. Furthermore, information pertaining to static regions of the scene can be utilized in semantic representation on the maps. Moreover, the tracking and semantic information of the dynamic objects in the environment is at the disposal of SLAM framework, which can be exploited at several stages of SLAM to attain better accuracies and generation of semantic maps.

Author Contributions: Writing—original draft preparation, M.S.; writing—review and editing, G.-W.K. All authors have read and agreed to the published version of the manuscript.

Funding: This research was financially supported in part by the Ministry of Trade, Industry and Energy (MOTIE) and Korea Institute for Advancement of Technology (KIAT) through the International Cooperative R&D program. (Project No. P0004631), and in part by the MSIT (Ministry of Science and ICT), Korea, under the Grand Information Technology Research Center support program (IITP-2021-2020-0-01462) supervised by the IITP (Institute for Information & communications Technology Planning & Evaluation).

Institutional Review Board Statement: Not applicable.

Informed Consent Statement: Not applicable.

Data Availability Statement: Not applicable.

Conflicts of Interest: The authors declare no conflict of interest.

References

1. Walcott-Bryant, A.; Kaess, M.; Johannsson, H.; Leonard, J.J. Dynamic Pose Graph SLAM Long-Term Mapping in Low Dynamic Environments. In Proceedings of the 2012 IEEE/RSJ International Conference on Intelligent Robots and Systems (IROS), Vilamoura-Algarve, Portugal, 7–12 October 2012; pp. 1871–1878.
2. Qiping, C.; Yinfei, X.; Shifeng, G.; Jie, B.; Qiang, S. Sensing system of environmental perception technologies for driverless vehicle: A review of state of the art and challenges. *Sens. Actuators A Phys.* **2021**, *102*, 112566.
3. He, K.; Gkioxari, G.; Dollár, P.; Girshick, R. Mask R-CNN. In Proceedings of the 2017 IEEE International Conference on Computer Vision (ICCV), Venice, Italy, 22–29 October 2017; pp. 2980–2988.
4. Davison, A.J. SLAM++: Simultaneous Localisation in addition, Mapping at the Level of Objects. In Proceedings of the IEEE Conference on Computer Vision and Pattern Recognition, Portland, OR, USA, 23–28 June 2013; pp. 1352–1359.
5. Gálvez-López, D.; Salas, M.; Tardós, J.D.; Montiel, J.M.M. Real-time monocular object SLAM. *Robot. Auton. Syst.* **2016**, *75*, 435–449. [CrossRef]
6. Milan, A.; Leal-Taixe, L.; REID, I.; Roth, S.; Schindler, K. MOT16: A benchmark for multi-object tracking. *arXiv* **2016**, arXiv:1603.00831.
7. Tan, W.; Liu, H.; Dong, Z. Robust monocular SLAM in dynamic environments. In Proceedings of the 2013 IEEE International Symposium on Mixed and Augmented Reality (ISMAR), Adelaide, Australia, 1–4 October 2013; pp. 209–218.
8. Wangsiripitak, S.; Murray, D.W. Avoiding moving outliers in visual SLAM by tracking moving objects. In Proceedings of the 2009 IEEE International Conference on Robotics and Automation, Kobe, Japan, 12–17 May 2009; Institute of Electrical and Electronics Engineers (IEEE): Hoboken, TX, USA, 2009; pp. 375–380.
9. Riazuelo, L.; Montano, L.; Montiel, J. Semantic visual SLAM in populated environments. In Proceedings of the 2017 European Conference on Mobile Robots (ECMR), Paris, France, 6–8 September 2017; pp. 1–7.
10. Li, S.; Lee, D. RGB-D SLAM in dynamic environments using static point weighting. *IEEE Robot. Autom. Lett.* **2017**, *2*, 2263–2270. [CrossRef]
11. Alcantarilla, P.F.; Yebes, J.J.; Almazán, J.; Bergasa, L.M. On combining visual SLAM and dense scene flow to increase the robustness of localization and mapping in dynamic environments. In Proceedings of the 2012 IEEE International Conference on Robotics and Automation, Saint Paul, MN, USA, 14–18 May 2012; pp. 1290–1297.
12. Wang, Y.; Huang, S. Motion segmentation based robust RGB-D SLAM. In Proceedings of the 11th World Congress on Intelligent Control and Automation, Shenyang, China, 29 June–4 July 2014; pp. 3122–3127.
13. Kim, D.-H.; Kim, J.-H. Effective background model-based rgb-d dense visual odometry in a dynamic environment. *IEEE Trans. Robot.* **2016**, *32*, 1565–1573. [CrossRef]
14. Sun, Y.; Liu, M.; Meng, M.Q.-H. Improving rgb-d slam in dynamic environments: A motion removal approach. *Robot. Auton. Syst.* **2017**, *89*, 110–112. [CrossRef]
15. Wang, C.C.; Thorpe, C.; Thrun, S. Online Simultaneous Localization and Mapping with Detection and Tracking of Moving Objects: Theory and Results from a Ground Vehicle in Crowded Urban Areas. In Proceedings of the IEEE International Conference on Robotics & Automation, Taipei, Taiwan, 14–19 September 2003; Volume 1, pp. 842–849.
16. Rogers, J.G.; Trevor, A.J.; Nieto, C.-G.; Christensen, H.I. SLAM with expectation maximization for moveable object tracking. In Proceedings of the 2010 IEEE/RSJ International Conference on Intelligent Robots and Systems (IROS), Taipei, Taiwan, 18–22 October 2010; pp. 2077–2082.
17. Bârsan, I.A.; Liu, P.; Pollefeys, M.; Geiger, A. Robust dense mapping for large-scale dynamic environments. In Proceedings of the IEEE International Conference on Robotics and Automation (ICRA), Brisbane, Australia, 21–25 May 2018; pp. 7510–7517.
18. Rosinol, A.; Gupta, A.; Abate, M.; Shi, J.; Carlone, L. 3D dynamic scene graphs: Actionable spatial perception with places, objects, and humans. In Proceedings of the Robotics: Science and Systems 2020, Corvalis, OR, USA, 12–16 July 2020.
19. Li, P.; Qin, T.; Shen, S. Stereo vision-based semantic 3D object and ego-motion tracking for autonomous driving. In Proceedings of the European Conference on Computer Vision (ECCV), Munich, Germany, 8–14 September 2018; pp. 646–661.
20. Yang, S.; Scherer, S. CubeSLAM: Monocular 3-D Object SLAM. *IEEE Trans. Robot.* **2019**, *35*, 925–938. [CrossRef]
21. Huang, J.; Yang, S.; Mu, T.-J.; Hu, S.-M. ClusterVO: Clustering moving instances and estimating visual odometry for self and surroundings. In Proceedings of the IEEE Conference on Computer Vision and Pattern Recognition, Seattle, WA, USA, 13–19 June 2020; pp. 2168–2177.
22. Zhang, J.; Henein, M.; Mahony, R.; Ila, V. VDO-SLAM: A Visual dynamic object-aware SLAM system. *arXiv* **2020**, arXiv:2005.11052.
23. Henein, M.; Kennedy, G.; Mahony, R.; Ila, V. Exploiting rigid body motion for SLAM in dynamic environments. In Proceedings of the IEEE International Conference on Robotics and Automation (ICRA), Brisbane, Australia, 21–25 May 2018; p. 19.
24. Huang, J.; Yang, S.; Zhao, Z.; Lai, Y.-K.; Hu, S.-M. ClusterSLAM: A slam backend for simultaneous rigid body clustering and motion estimation. In Proceedings of the IEEE International Conference on Computer Vision, Seoul, Korea, 27–28 October 2019; pp. 5875–5884.
25. Bescos, B.; Campos, C.; Tardós, J.D.; Neira, J. DynaSLAM II: Tightly-coupled multi-object tracking and SLAM. *IEEE Robot. Autom. Lett.* **2021**, *3*, 5191–5198. [CrossRef]
26. Mur-Artal, R.; Tardós, J.D. ORB-SLAM2: An Open-Source SLMA System for Monocular, Stereo, and RGB-D cameras. *IEEE Trans. Robot.* **2017**, *33*, 1255–1262. [CrossRef]

27. Geiger, A.; Lenz, P.; Urtasun, R. Are we ready for autonomous driving? The KITTI vision benchmark suite. In Proceedings of the 2012 IEEE Conference on Computer Vision and Pattern Recognition, Providence, RI, USA, 16–21 June 2012; pp. 3354–3361.
28. Geiger, A.; Lenz, P.; Stiller, C.; Urtasun, R. Vision meets robotics: The KITTI dataset. *Int. J. Robot. Res.* **2013**, *32*, 1231–1237. [CrossRef]
29. Bescos, B.; Fácil, J.M.; Civera, J.; Neira, J. DynaSLAM: Tracking, mapping, and inpainting in dynamic scenes. *IEEE Robot. Autom. Lett.* **2018**, *3*, 4076–4083. [CrossRef]
30. Ambrus, R.; Folkesson, J.; Jensfelt, P. Unsupervised object segmentation through change detection in a long term autonomy scenario. In Proceedings of the IEEE-RAS International Conference on Humanoid Robots (Humanoids 2016), Cancun, Mexico, 15–17 November 2016; pp. 1181–1187.
31. Wang, C.C.; Thorpe, C.; Thrun, S.; Hebert, M.; Durrant-Whyte, H. Simultaneous localization, mapping and moving object tracking. *Int. J. Robot. Res.* **2007**, *26*, 889–916. [CrossRef]
32. Runz, M.; Buffier, M.; Agapito, L. Maskfusion: Real-Time recognition, tracking and reconstruction of multiple moving objects. In Proceedings of the 2018 IEEE International Symposium on Mixed and Augmented Reality (ISMAR), Munich, Germany, 16–20 October 2018; pp. 10–20.
33. Xu, B.; Li, W.; Tzoumanikas, D.; Bloesch, M.; Davison, A.; Leutenegger, S. Mid-fusion: Octree-based object-level multi-instance dynamic SLAM. In Proceedings of the 2019 International Conference on Robotics and Automation (ICRA), Montreal, QC, Canada, 20–24 May 2019; pp. 5231–5237.
34. Redmon, J.; Farhadi, A. YOLOv3: An Incremental Improvement. *arXiv* **2018**, arXiv:1804.02767.
35. Sualeh, M.; Kim, G.-W. Dynamic Multi-LiDAR Based Multiple Object Detection and Tracking. *Sensors* **2019**, *19*, 1474. [CrossRef] [PubMed]
36. Sualeh, M.; Kim, G.-W. Visual-LiDAR Based 3D Object Detection and Tracking for Embedded Systems. *IEEE Access* **2020**, *8*, 156285–156298. [CrossRef]
37. Zhang, H.; Weerasekera, C.S.; Bian, J.-B.; Reid, I. Visual odometry revisited: What should be learnt? In Proceedings of the IEEE Conference on Computer Vision and Pattern Recognition (CVPR), Seattle, WA, USA, 14–19 June 2020.

sensors

MDPI

Communication

A Redundant Configuration of Four Low-Cost GNSS-RTK Receivers for Reliable Estimation of Vehicular Position and Posture

Jesús Morales *, Jorge L. Martínez and Alfonso J. García-Cerezo

Robotics and Mechatronic Lab, Andalucía Tech, Universidad de Málaga, 29071 Málaga, Spain; jlmartinez@uma.es (J.L.M.); ajgarcia@uma.es (A.J.G.-C.)

* Correspondence: jesus.morales@uma.es; Tel.: +34-951-952-323

Abstract: This paper proposes a low-cost sensor system composed of four GNSS-RTK receivers to obtain accurate position and posture estimations for a vehicle in real-time. The four antennas of the receivers are placed so that every three-antennas combination is optimal to get the most precise 3D coordinates with respect to a global reference system. The redundancy provided by the fourth receiver allows to improve estimations even more and to maintain accuracy when one of the receivers fails. A mini computer with the Robotic Operating System is responsible for merging all the available measurements reliably. Successful experiments have been carried out with a ground rover on irregular terrain. Angular estimates similar to those of a high-performance IMU have been achieved in dynamic tests.

Keywords: vehicle localization; GNSS receivers; RTK corrections; sensor redundancy

Citation: Morales, J.; Martínez, J.L.; García-Cerezo, A.J. A Redundant Configuration of Four Low-Cost GNSS-RTK Receivers for Reliable Estimation of Vehicular Position and Posture. *Sensors* **2021**, *21*, 5853. https://doi.org/10.3390/s21175853

Academic Editor: Aboelmagd Noureldin

Received: 28 July 2021
Accepted: 25 August 2021
Published: 30 August 2021

1. Introduction

Acquiring accurate and reliable three-dimensional (3D) coordinates for a vehicle is of great interest in monitoring its operation for advanced driver assistance systems and for autonomous navigation of mobile robots. Vehicle coordinates include three distances for position and three angles for posture with respect to a global reference system on Earth surface, that usually employs North-East-Down (NED) local axes [1].

A common possibility is the use of an Inertial Measurement Unit (IMU), which contain different sensors such as accelerometers, gyroscopes and magnetometers [2]. For proper operation, these units require calibration once installed on the vehicle and to take into account local magnetic field variations. Knowing the initial position of the vehicle, global 3D coordinates can be obtained with an IMU and odometry, but the estimation of the spatial trajectory tends to deteriorate since the measurements include small deviations that accumulate over time [3].

To avoid the growth of position uncertainty, a Global Navigation Satellite System (GNSS) receiver that make use of various global satellite constellations (North American GPS, Russian GLONASS, European Galileo and Chinese BeiDou) at once can be employed [4]. However, absolute GNSS measurements over the Earth's surface are subject to various types of errors that degrade their accuracy to the order of meters.

GNSS errors can be significantly reduced by incorporating differential corrections provided by a Satellite Based Augmentation Systems (SBAS) or a Continuously Operating Reference Station (CORS) [5]. In this respect, one of the most effective techniques is Real Time Kinematics (RTK) that performs carrier-phase signal synchronization [6] by using the RTCM (Radio Technical Commission for Maritime Services) communication protocol. Thus, GNSS receivers can operate in two different modes: RTK-fixed and RTK-float to indicate when they achieve or not centimeter accuracy, respectively.

Besides, multiple GNSS receivers can be installed onboard to enhance vehicle positioning [5,7]. In addition, a GNSS compass with two antennas can be employed to obtain heading [8,9]. Moreover, by differencing over time the GNSS measurements taken in motion, speed estimations [10] as well as pitch and heading [11] can be deduced.

In field robotics, the combination of GPS and IMU sensors has been a popular option to estimate 3D global coordinates accurately [12]. A different strategy to achieve this objective was to mount the antennas of three high-cost GPS-RTK receivers on the roof of the vehicle [13].

With modern GNSS receivers, the GNSS-IMU sensor combination [14] and the synchronization of three low-cost GNSS-RTK devices [6,15] or three antennas in a single receiver [16] have also been employed in automobiles. Moreover, by tightly coupling three GNSS-RTK receivers and an IMU, accuracy can be improved even more [1,17,18].

This paper proposes a reliable sensor system that provides the position and posture of a vehicle by combining the measurements of four identical GNSS-RTK low-cost receivers. In this way, the main contributions are the following:

1. The best geometrical configuration for three and four antennas to minimize position and posture uncertainty of a vehicle is deduced.
2. A redundant setup with four antennas is analyzed, so when the precision of one receiver degrades, reliable 3D coordinates can be still calculated in real-time.
3. A decentralized node architecture using the Robot Operating System (ROS) that integrates all the available measurements from the receivers is presented.

Regarding the first point, although several antenna configurations have been deployed experimentally on cars, no previous work has performed a theoretical analysis to infer the best layout. This has not prevented two recent papers [6,16] from employing near-optimal configurations for their tests.

With respect to the second contribution, antenna redundancy was previously intended only to improve positioning precision [5,7], but in this paper it also serves to enhance attitude estimations for the vehicle and to tolerate faults on the GNSS receivers.

Regarding the third point, following a low-cost philosophy, it is employed an open-source software of common use in robotics with some already developed nodes in a mini computer instead of programming on specialized boards [6,15,16].

The rest of the paper is organized as follows. Antenna arrangements with three and four GNSS-RTK receivers are discussed in the next Section. Sensor hardware, ROS programming and the optimal calculation of 3D coordinates are described in Section 3. Then, experiments on irregular terrain with the robotic platform Argo XTR are presented in Section 4, including comparisons with measurements from a high-end IMU. Finally, conclusions, acknowledgements, and references complete the paper.

2. Spatial Configurations for the Antennas of the GNSS-RTK Receivers

Let v_i be the actual position of an antenna with respect to a global NED coordinate system:

$$v_i(t) = \begin{pmatrix} x_i(t) \\ y_i(t) \\ z_i(t) \end{pmatrix} = \bar{v}_i(t) + \Delta v_i(t) = \begin{pmatrix} \bar{x}_i(t) \\ \bar{y}_i(t) \\ \bar{z}_i(t) \end{pmatrix} + \begin{pmatrix} \Delta x_i(t) \\ \Delta y_i(t) \\ \Delta z_i(t) \end{pmatrix}, \tag{1}$$

where $\bar{v}_i(t)$ is the measurement produced by the receiver $i = 1,2,3,4$ at instant t and $\Delta v_i(t)$ is its associated error. The x, y, z coordinates correspond to north, east, and down displacements, respectively. The covariance matrix for Δv_i is given by:

$$C_i = E[\Delta v_i(t)\, \Delta v_i^T(t)] =$$

$$E\left[\begin{pmatrix} \Delta x_i(t) \\ \Delta y_i(t) \\ \Delta z_i(t) \end{pmatrix} (\Delta x_i(t), \Delta y_i(t), \Delta z_i(t))\right] = \begin{pmatrix} \sigma_{x_i}^2 & \sigma_{x_i y_i} & \sigma_{x_i z_i} \\ \sigma_{x_i y_i} & \sigma_{y_i}^2 & \sigma_{y_i z_i} \\ \sigma_{x_i z_i} & \sigma_{y_i z_i} & \sigma_{z_i}^2 \end{pmatrix}, \tag{2}$$

where $\sigma_{x_i}^2 = E[\Delta x_i^2(t)]$, $\sigma_{y_i}^2 = E[\Delta y_i^2(t)]$ and $\sigma_{z_i}^2 = E[\Delta z_i^2(t)]$ are the variances, whereas $\sigma_{x_i y_i} = E[\Delta x_i(t)\Delta y_i(t)]$, $\sigma_{x_i z_i} = E[\Delta x_i(t)\Delta z_i(t)]$ and $\sigma_{y_i z_i} = E[\Delta y_i(t)\Delta z_i(t)]$ are the covariances.

For this analysis, it is considered that:

- The mean errors of the receivers along time are null, i.e., $E[\Delta v_i(t)] = 0\,\forall i$.
- The errors of different receivers are independent, i.e., $E[\Delta v_i(t)\Delta v_j^T(t)] = 0$ for $i \neq j$.
- All the receivers share the same covariance matrix, i.e., $C = C_1 = C_2 = C_3 = C_4$.

2.1. Three Receivers Optimal Configuration

To obtain the optimal configuration for the triangle formed by three GNSS-RTK antennas, its associated position and posture uncertainty should be minimized. In this case, it is also assumed that:

- The distances d_a and d_b of the second and third antenna with respect to the first antenna, respectively, are constant values determined without uncertainty.
- The angle θ between the directions given by d_a and d_b is also known certainly on the plane that contains the three antennas.

The centroid of the triangle formed by the antennas:

$$v_0(t) = \frac{v_1(t) + v_2(t) + v_3(t)}{3}, \tag{3}$$

can be estimated from the measurements of the receivers as:

$$\bar{v}_0(t) = \frac{\bar{v}_1(t) + \bar{v}_2(t) + \bar{v}_3(t)}{3} \Rightarrow \Delta v_0(t) = \frac{\Delta v_1(t) + \Delta v_2(t) + \Delta v_3(t)}{3}. \tag{4}$$

The covariance matrix for Δv_0 is calculated as:

$$C_0 = E\left[\Delta v_0(t)\Delta v_0^T(t)\right] = \frac{E[\Delta v_1(t)\Delta v_1^T(t) + \Delta v_2(t)\Delta v_2^T(t) + \Delta v_3(t)\Delta v_3^T(t)]}{9} = \frac{C}{3}, \tag{5}$$

where the position uncertainty of the geometric center of the triangle is reduced by three with respect to each vertex.

Regarding the posture in space of the triangle, the direction cosines of the lines between two antennas are given by the unitary vectors:

$$v_a(t) = \frac{v_2(t) - v_1(t)}{d_a} = \begin{pmatrix} x_a(t) \\ y_a(t) \\ z_a(t) \end{pmatrix} \Rightarrow \tag{6}$$

$$\bar{v}_a(t) = \frac{\bar{v}_2(t) - \bar{v}_1(t)}{d_a}, \quad \Delta v_a(t) = \frac{\Delta v_2(t) - \Delta v_1(t)}{d_a}, \tag{7}$$

$$v_b(t) = \frac{v_3(t) - v_1(t)}{d_b} = \begin{pmatrix} x_b(t) \\ y_b(t) \\ z_b(t) \end{pmatrix} \Rightarrow \tag{8}$$

$$\bar{v}_b(t) = \frac{\bar{v}_3(t) - \bar{v}_1(t)}{d_b}, \quad \Delta v_b(t) = \frac{\Delta v_3(t) - \Delta v_1(t)}{d_b}, \tag{9}$$

where the corresponding covariance matrices are:

$$C_a = E\left[\Delta v_a(t)\Delta v_a^T(t)\right] = \frac{E\left[\Delta v_2(t)\Delta v_2^T(t)\right]}{d_a^2} + \frac{E\left[\Delta v_1(t)\Delta v_1^T(t)\right]}{d_a^2} = \frac{2C}{d_a^2}, \tag{10}$$

$$C_b = E\left[\Delta v_b(t)\Delta v_b^T(t)\right] = \frac{E\left[\Delta v_3(t)\Delta v_3^T(t)\right]}{d_b^2} + \frac{E\left[\Delta v_1(t)\Delta v_1^T(t)\right]}{d_b^2} = \frac{2C}{d_b^2}, \tag{11}$$

$$C_{ab} = E\left[\Delta v_a(t)\Delta v_b^T(t)\right] = C_{ba} = E\left[\Delta v_b(t)\Delta v_a^T(t)\right] = \frac{E\left[\Delta v_1(t)\Delta v_1^T(t)\right]}{d_a\, d_b} = \frac{C}{d_a\, d_b}. \tag{12}$$

Thus, the spatial uncertainty of the direction cosines can be reduced by separating the antennas as much as possible. But d_a and d_b are inherently limited by the available space on the roof of the vehicle. Furthermore, to balance posture uncertainty in both directions, these distances should be selected equal: $d = d_a = d_b$, so that $C_a = C_b = 2C/d^2$ and $C_{ab} = C_{ba} = C/d^2$.

The direction cosine of the normal vector v_c to the plane defined by the three antennas is given by the unitary vector from the cross product of v_a and v_b:

$$v_c(t) = \frac{v_a(t) \times v_b(t)}{\sin(\theta)} = \frac{1}{\sin(\theta)} \begin{pmatrix} y_a(t)z_b(t) - y_b(t)z_a(t) \\ x_b(t)z_a(t) - x_a(t)z_b(t) \\ x_a(t)y_b(t) - x_b(t)y_a(t) \end{pmatrix} \Rightarrow \tag{13}$$

$$\bar{v}_c(t) = \frac{1}{\sin(\theta)} \begin{pmatrix} \bar{y}_a(t)\bar{z}_b(t) - \bar{y}_b(t)\bar{z}_a(t) \\ \bar{x}_b(t)\bar{z}_a(t) - \bar{x}_a(t)\bar{z}_b(t) \\ \bar{x}_a(t)\bar{y}_b(t) - \bar{x}_b(t)\bar{y}_a(t) \end{pmatrix}, \tag{14}$$

as long as the three antennas are not aligned to avoid $\sin(\theta) = 0$, where v_c will be indeterminate.

By using Taylor series expansion [19], v_c can be approximated by:

$$v_c(t) \approx \bar{v}_c(t) + \frac{J(t)}{\sin(\theta)} \begin{pmatrix} \Delta v_a(t) \\ \Delta v_b(t) \end{pmatrix} \Rightarrow \Delta v_c(t) \approx \frac{J(t)}{\sin(\theta)} \begin{pmatrix} \Delta v_a(t) \\ \Delta v_b(t) \end{pmatrix}, \tag{15}$$

where J is the time-dependent Jacobian matrix:

$$J(t) = \begin{pmatrix} 0 & \bar{z}_b(t) & -\bar{y}_b(t) & 0 & -\bar{z}_a(t) & \bar{y}_a(t) \\ -\bar{z}_b(t) & 0 & \bar{x}_b(t) & \bar{z}_a(t) & 0 & -\bar{x}_a(t) \\ \bar{y}_b(t) & -\bar{x}_b(t) & 0 & -\bar{y}_a(t) & \bar{x}_a(t) & 0 \end{pmatrix}. \tag{16}$$

The error Δv_c can be minimized regardless of vehicle posture by choosing $\theta = \pm 90°$. In this case, the covariance matrix for Δv_c can be approximated by:

$$C_c(t) = E[\Delta v_c(t)\Delta v_c^T(t)] \approx J(t) \begin{pmatrix} C_a & C_{ab} \\ C_{ba} & C_b \end{pmatrix} J^T(t) = \frac{J(t)}{d^2} \begin{pmatrix} 2C & C \\ C & 2C \end{pmatrix} J^T(t). \tag{17}$$

Finally, the rotation matrix is obtained from the direction cosines as $R = (\bar{v}_a, \bar{v}_b, \bar{v}_c)$. Roll, pitch and yaw angles can be deduced from R and represent rotations with respect to v_a, v_b and v_c axis, respectively.

To summarize, the best configuration to minimize position and posture uncertainty with three GNSS-RTK antennas is to form a right-angled triangle with two identical sides of the maximum possible length d (see Figure 1).

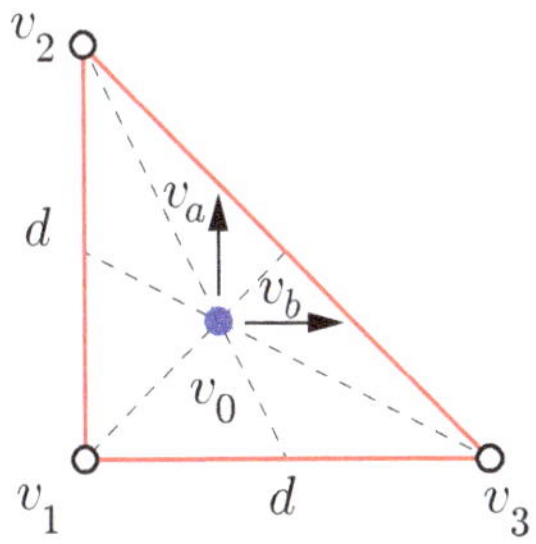

Figure 1. Optimal antenna configuration with three receivers.

2.2. Four Receivers Optimal Layout

The optimum for four receivers would consist of placing the fourth antenna orthogonal to the plane defined by the remaining three at a distance d of the first antenna, where its centroid:

$$v_0(t) = \frac{v_1(t) + v_2(t) + v_3(t) + v_4(t)}{4}, \tag{18}$$

does not coincide with the geometric center of the underlying cube (see Figure 2).

Assuming perfect placement of the fourth receiver with respect to the triangle:

$$v_c(t) = \frac{v_4(t) - v_1(t)}{d}, \tag{19}$$

which can be calculated directly from measurements as:

$$\bar{v}_c(t) = \frac{1}{d}\begin{pmatrix} \bar{x}_4(t) - \bar{x}_1(t) \\ \bar{y}_4(t) - \bar{y}_1(t) \\ \bar{z}_4(t) - \bar{z}_1(t) \end{pmatrix} \Rightarrow \Delta v_c = \frac{\Delta v_4(t) - \Delta v_1(t)}{d}, \tag{20}$$

and can be merged with the estimation (14) to decrease posture uncertainty even more. Furthermore, this reduction can also be applied to $\bar{v}_a$ (7) and $\bar{v}_b$ (9) with their corresponding normal vectors formed by their respective ortogonal planes: $\bar{v}_a(t) = \bar{v}_b(t) \times \bar{v}_c(t)$ and $\bar{v}_b(t) = \bar{v}_c(t) \times \bar{v}_a(t)$, respectively.

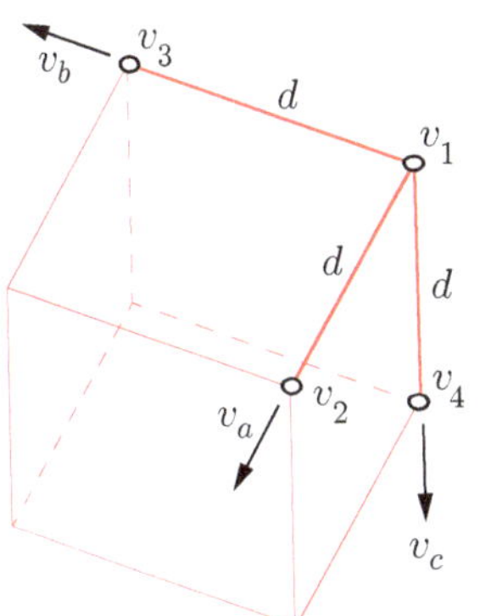

Figure 2. Optimal antenna configuration with four receivers.

2.3. Four Receivers Redundant Configuration

In this paper, an additional fourth receiver is added to the optimal three-receivers configuration to form a square of side d on a planar surface (see Figure 3). This is a redundant arrangement that is easier to mount on the roof of an automobile (see Figure 4) than the optimal one of Figure 2. The local coordinate system has its origin in v_0 with axes v_a, v_b and v_c as defined by (7), (9) and (13), respectively.

In Figure 4, the longitudinal and transverse axes of the car coincide with v_a and v_b, respectively. The vertical axis v_c is not displayed but it would be pointing downwards. This figure also shows the position of a radio antenna, denoted by the letter R, to receive RTK corrections from a CORS.

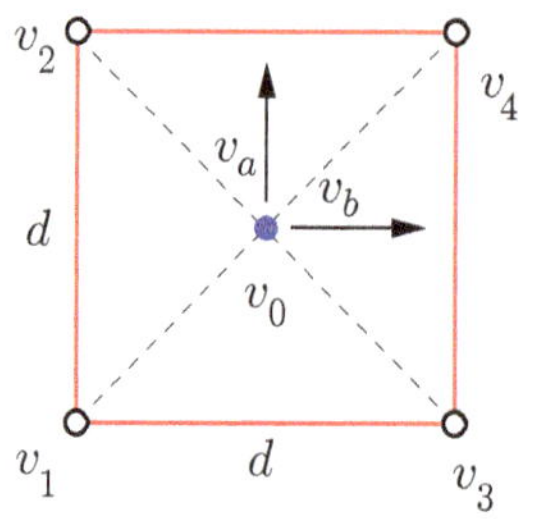

Figure 3. Redundant antenna configuration with four receivers.

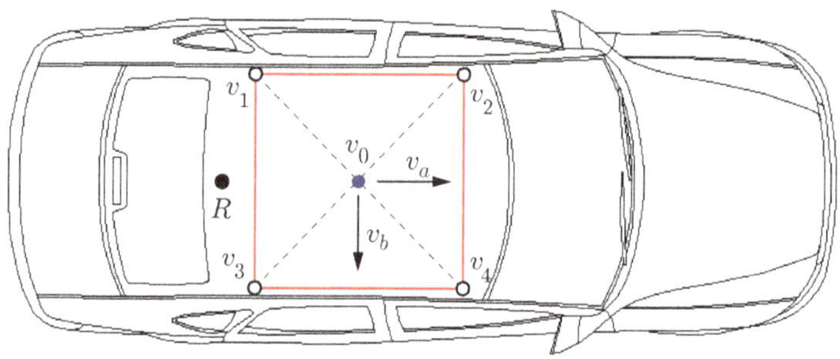

Figure 4. Placement of the antennas on the roof of a car.

The proposed redundancy is useful in two different ways. Firstly, position and posture uncertainty can be reduced further than with three receivers. Secondly, if the precision of one of the receivers deteriorates, the rest of receivers can still provide a reliable position and posture estimation for the vehicle.

Regarding the first advantage, when all the measurements are available, its centroid v_0 (18) can be estimated as:

$$\bar{v}_0(t) = \frac{\bar{v}_1(t) + \bar{v}_2(t) + \bar{v}_3(t) + \bar{v}_4(t)}{4} \Rightarrow \tag{21}$$

$$\Delta v_0(t) = \frac{\Delta v_1(t) + \Delta v_2(t) + \Delta v_3(t) + \Delta v_4(t)}{4} \Rightarrow C_0 = \frac{C}{4}, \tag{22}$$

whose covariance matrix is divided by four, instead of by three (5). Furthermore, the direction cosines v_a and v_b can be estimated by using one side of the square and its opposite:

$$\bar{v}_a(t) = \frac{\bar{v}_2(t) - \bar{v}_1(t) + \bar{v}_4(t) - \bar{v}_3(t)}{2d} \Rightarrow \tag{23}$$

$$\Delta v_a(t) = \frac{\Delta v_2(t) - \Delta v_1(t) + \Delta v_4(t) - \Delta v_3(t)}{2d} \Rightarrow C_a = \frac{C}{d^2}, \tag{24}$$

$$\bar{v}_b(t) = \frac{\bar{v}_3(t) - \bar{v}_1(t) + \bar{v}_4(t) - \bar{v}_2(t)}{2d} \Rightarrow \tag{25}$$

$$\Delta v_b(t) = \frac{\Delta v_3(t) - \Delta v_1(t) + \Delta v_4(t) - \Delta v_2(t)}{2d} \Rightarrow C_b = \frac{C}{d^2}, \tag{26}$$

that represents half uncertainty of (10) and (11). This reduction directly benefits to the covariance matrix C_c (17) of the direction cosine v_c:

$$\Delta v_c(t) \approx J(t)\begin{pmatrix} \Delta v_a(t) \\ \Delta v_b(t) \end{pmatrix} \Rightarrow C_c(t) \approx \frac{J(t)}{d^2}\begin{pmatrix} C & C \\ C & C \end{pmatrix} J^T(t), \tag{27}$$

where $C_{ab} = C_{ba} = C/d^2$.

The second advantage comes from the fact that with a three-antenna configuration, there is no possibility to obtain the complete set of six coordinates for the vehicle when one of the receivers fails. However, the proposed sensor system can keep working with the remaining three receivers. In this case, to obtain the center of the square, instead of the triangle centroid (4), only two measurements from opposite vertices can be employed:

$$\bar{v}_0(t) = \begin{cases} \dfrac{\bar{v}_2(t) + \bar{v}_3(t)}{2}, & \text{when the first or fourth receiver fails,} \quad (28) \\ \dfrac{\bar{v}_1(t) + \bar{v}_4(t)}{2}, & \text{when the second or third receiver fails,} \quad (29) \end{cases}$$

which implies that $C_0 = C/2$, i.e., twice position uncertainty with respect to four available measurements (22).

3. Sensor System

In addition to the four low-cost GNSS receivers and their corresponding antennas, the sensor system includes a mini computer to obtain the 3D position and posture of the vehicle (see Figure 5).

The chosen GNSS-RTK receiver is the SparkFun GPS-RTK2 (https://www.sparkfun.com/products/15136, accessed on 28 July 2021) board, which is based on the compact ZED-F9P module from U-blox. The receiver does not only provide geodetic coordinates (longitude, latitude and height), but also ECEF (Cartesian coordinates with respect to Earth center) and NED coordinates with respect to a nearby CORS to obtain centimeter accuracy with an output rate of 8 Hz.

Each receiver is connected to a multi-band antenna ANN-MB-00 (https://www.sparkfun.com/products/15192, accessed on 28 July 2021) from U-blox. To avoid multi-path problems, the magnetic base of each antenna is mounted on a steel ground plate.

The mini computer is an Intel NUC8i7BEH (https://www.intel.es/content/www/es/es/products/sku/126140/intel-nuc-kit-nuc8i7beh/specifications.html, accessed on 28 July 2021) kit with an Intel Core i7-8559U processor (8M Cache, 4 cores, 2.70 GHz). It has installed the open-source framework ROS [20] on the Linux-based operating system Ubuntu.

The computer communicates with each receiver through different Universal Serial Bus (USB) ports. The Internet connection of the computer is used to get differential correction data via the standard protocol NTRIP (Networked Transport of RTCM via Internet Protocol) through the Andalusian public positioning network [21].

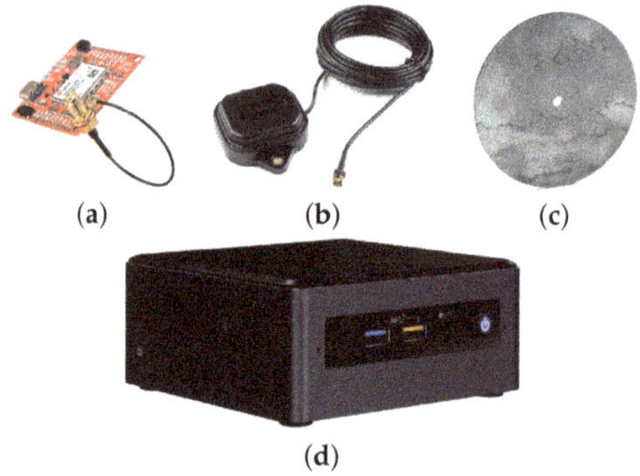

Figure 5. Hardware components of the sensor system: a GNSS-RTK board (**a**), an antenna (**b**), a ground plate (**c**) and the mini computer (**d**).

ROS Programming

The developed ROS software consists of a number of independent nodes, each of which communicates with others using topics under a publish/subscribe messaging model (see Figure 6).

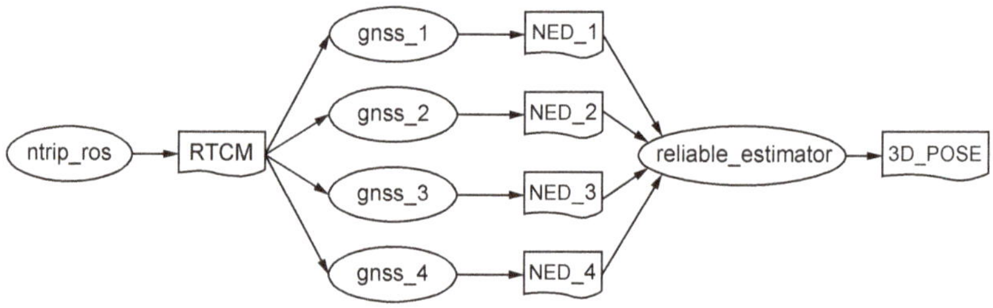

Figure 6. ROS computation graph with nodes (ellipses) and topics.

The `ntrip_ros` (https://github.com/ros-agriculture/ntrip_ros, accessed on 28 July 2021) node connects to a nearby CORS to get RTCM streams through internet and to publish them into the topic `RTCM`. Each receiver i has associated a driver node (https://github.com/KumarRobotics/ublox, accessed on 28 July 2021) named `gnss_i` that is subscribed to this topic to receive differential corrections via callbacks. These nodes publish NED coordinates on its own topic `NED_i` along with a time stamp and three accuracy estimates (each one ≥ 10 mm).

Then, the `reliable_estimator` node receives all the messages from the four `NED_i` topics and computes the six 3D coordinates with three or four synchronized measurements. Finally, it publishes the current pose into the `3D_POSE` topic, making this data available for any navigation node on the ROS system.

All the receivers weight the same to produce vehicular position and attitude in real-time. When the accuracy of one receiver degrades, it is completely discarded from computations.

For calculating the global position of the centroid $\bar{v}_0$ of the square (21), (28) or (29) are employed depending on the number of valid GNSS-RTK measurements. For posture computation, the closed-form method by Horn [22] is applied with a scale factor of 1. Instead of quaternions, an orthonormal rotation matrix R [23] is obtained that minimizes the following cost function:

$$F(t) = \sum_{\forall i} \|(v_i^* - v_0^*) - R(t)(\bar{v}_i(t) - \bar{v}_0(t))\|^2, \tag{30}$$

where v_i^* and v_0^* are the relative positions of the antenna i and the centroid of the square with respect to the local reference system, respectively. This represents a least squares problem that can be solved with three or four valid measurements. Lastly, the roll, pitch and yaw angles with respect to the global NED axes are extracted from the resulting rotation matrix [17].

4. Experiments with the Rover Argo XTR

The rover Argo XTR is a battery-powered unmanned land vehicle that allows extreme mobility with a low center of gravity and amphibious capability (see Figure 7). It features skid-steer traction with eight low-pressure 24-inch tires, a top speed of $16\,\mathrm{km\,h^{-1}}$ and zero turning radius. The robotic rover can be controlled via a follow-me system with a 2D laser scanner or via remote teleoperation with a joystick and a line-of-sight wireless link.

The proposed sensor has been mounted on the rear deck of the vehicle. The four antennas are tied to the side rails forming a square of $d = 1.35$ m on the side (see Figure 7). For comparison purposes, a fifth antenna has been installed at the center of the square together with the high-end AHRS400CC-100 MEMS IMU from Crossbow with an output rate of 60 Hz [24].

The Málaga broadcast station located 4.8 km away is employed to get differential corrections RTCM 3.1 through 4G internet connection and it is considered the global NED reference system in the following experiments.

4.1. Calibration Test

This test was carried out by recording the RTK-fixed measurements of the five GNSS receivers during three hours with the rover stopped on an almost horizontal parking lot. This experiment serves to characterize the covariance matriz C for the positioning errors. To this end, the mean NED coordinates are calculated for each receiver and the difference of each measurement with respect to its mean value is considered as an error. Then, by using (2):

$$C = 10^{-3} \begin{pmatrix} 0.0434 & -0.0025 & -0.0061 \\ -0.0025 & 0.0694 & -0.0083 \\ -0.0061 & -0.0083 & 0.4255 \end{pmatrix} \mathrm{m}^2, \quad (31)$$

where it can be observed almost null covariances and a standard deviation in the z coordinate ($\sigma_z = 21\,\mathrm{mm}$) much greater than in the x ($\sigma_x = 7\,\mathrm{mm}$) and in the y ($\sigma_y = 8\,\mathrm{mm}$) coordinates.

Figure 7. The rover Argo XTR with the sensor system mounted on the rear deck. GNSS antennas are numbered from 1 to 5.

Moreover, the relative location of each antenna v_i^* can also be accurately estimated with the computed mean values (see Table 1). It can be observed small positioning errors on the square with the fifth antenna centered and 166 mm below the rest of antennas. Table 1 also includes the local position of the centroids of the square v_0^* for (30) when using four or three receivers.

Table 1. Relative coordinates of the antennas and of the square centroid.

	a (m)	**b (m)**	**c (m)**
v_1^*	−0.672	−0.675	−0.004
v_2^*	0.673	−0.676	−0.004
v_3^*	−0.672	0.672	−0.004
v_4^*	0.671	0.678	0.011
v_5^*	−0.004	0.005	0.166
v_0^* (21)	0	0	0
v_0^* (28)	0.001	−0.002	−0.004
v_0^* (29)	0	0.002	0.004

4.2. Reliability Test

This test was performed with the vehicle stopped in the countryside as shown in Figure 7. One by one, each GNSS antenna was partially blocked with a metallic cover during one minute approximately to test sensor reliability.

Figure 8 shows the estimation of NED coordinates when using all (21), the first and the fourth (28) or the second and the third (29) receivers. Similarly, Figure 9 displays the estimation of the three angular coordinates with all the combinations of three and four receivers. In both figures, it can be clearly observed significant estimation changes when an antenna was temporarily blocked.

The mean accuracy provided by each receiver is shown in Figure 10 (up), where it can be observed successive antenna covering, in this order: 3, 1, 4 and 2. Apart from checking the RTK-fixed mode, these values can be employed as a fail indicator for each receiver. However, there is a time period between 325 s to 350 s when the indicator for the first receiver does not detect any error but position and posture estimations were inaccurate.

An additional accuracy indicator is the error in calculating the perimeter of the square from measurements with respect to the data of Table 1 (5.389 m). As it is shown in Figure 10 (down), precision degradation can be better detected by using this complementary indicator. Thus, by comparing individually the distances of each vertex with respect to the rest, outlier measurements can be identified adequately when present.

Furthermore, in Figures 8 and 9, it can be observed that the estimations that do not include the failing measurement maintain high accuracy. For example, when the precision of the first receiver degrades in the interval between 290 s to 350 s, good position and posture estimations are provided by the second and third receivers, and by the second, third and fourth receivers, respectively. Therefore, overall accuracy for the sensor system can be maintained by properly detecting a single failure and excluding it from computations.

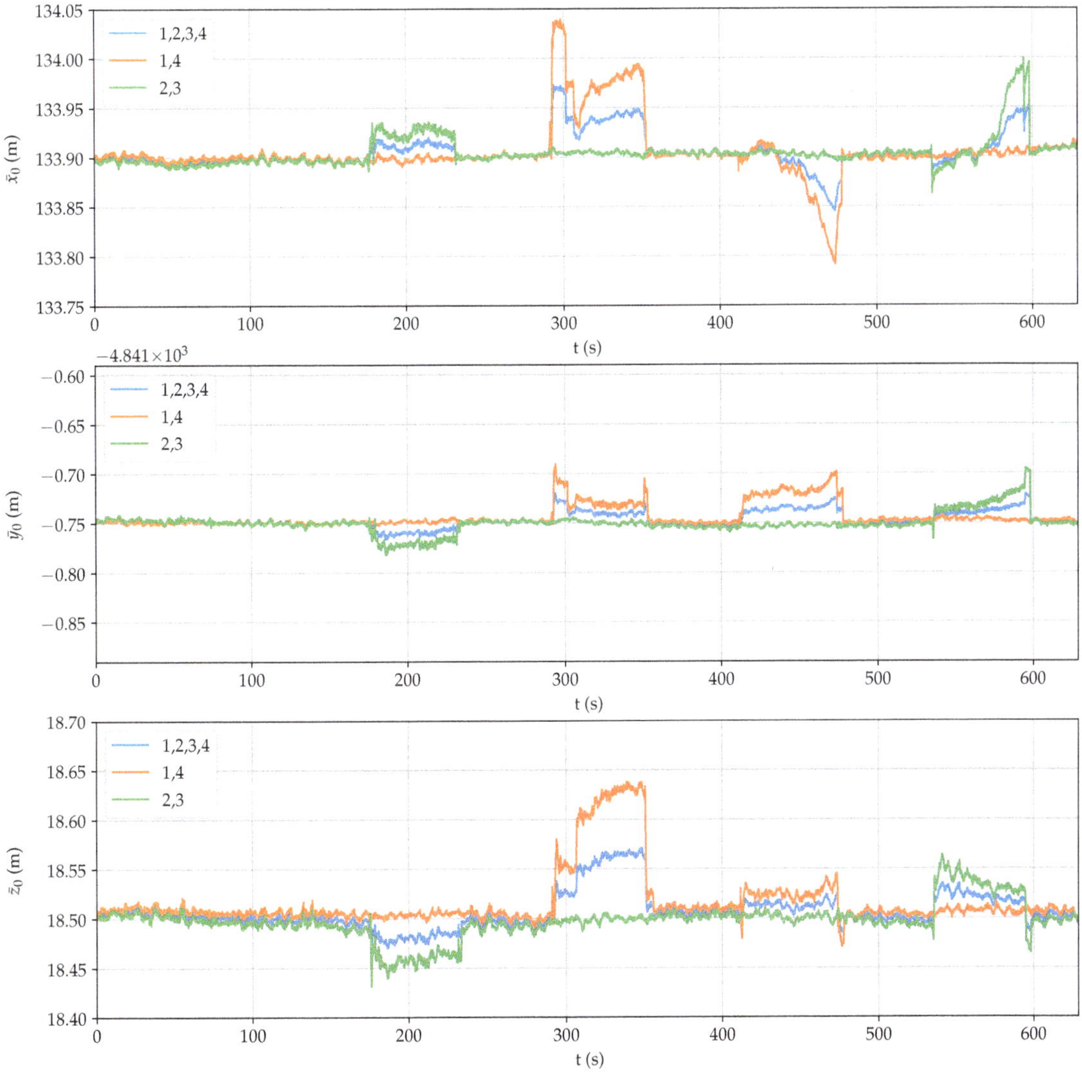

Figure 8. Estimation of position coordinates during the reliability test.

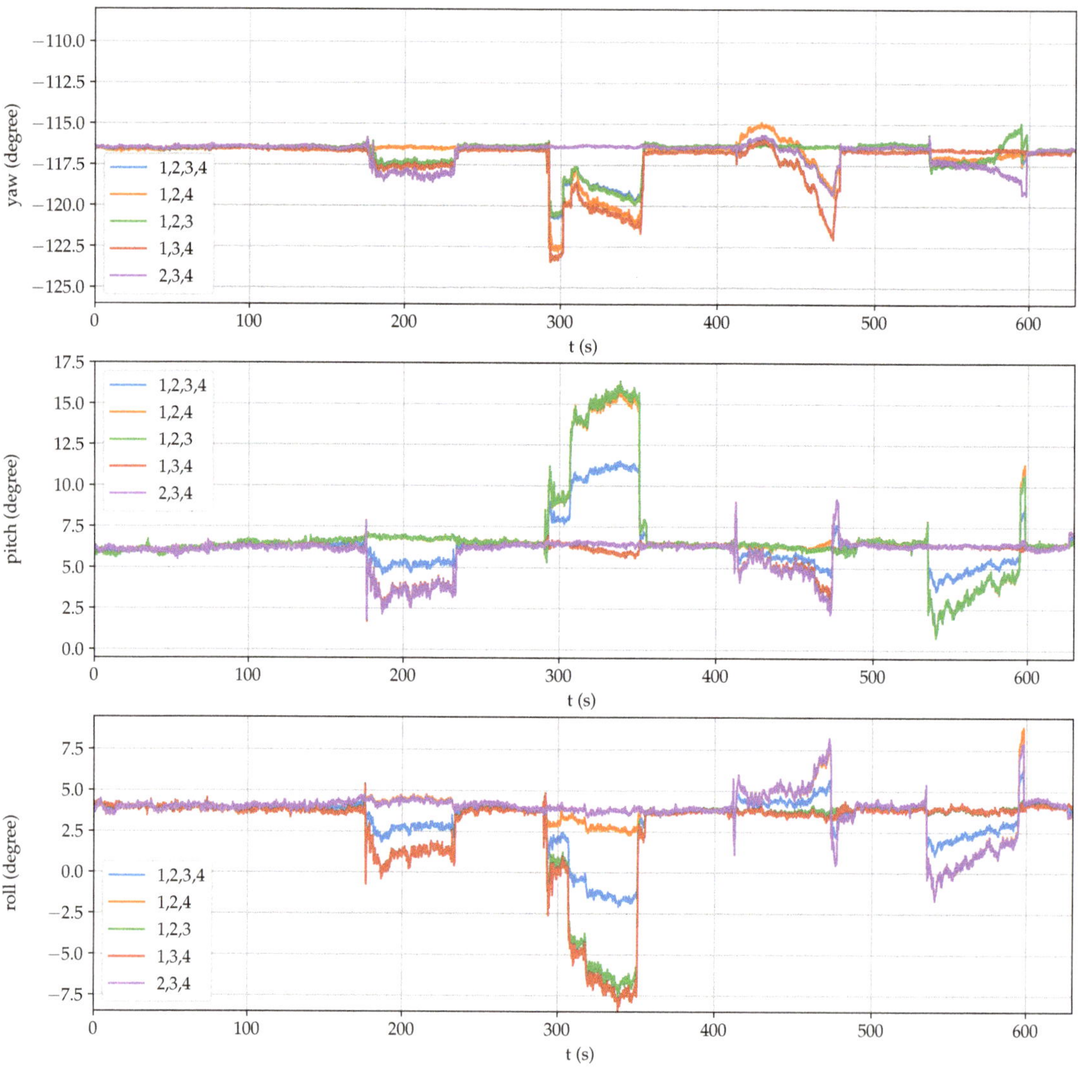

Figure 9. Estimation of posture coordinates during the reliability test.

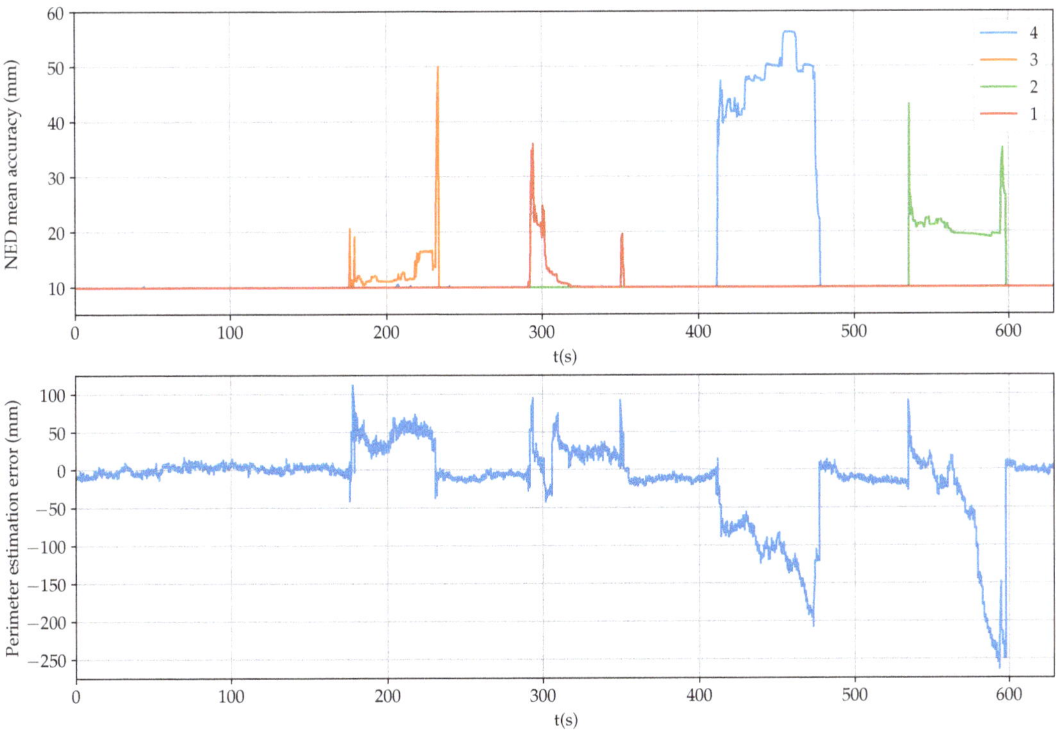

Figure 10. Accuracy estimation during the reliability test.

4.3. *Dynamic Test*

Several experiments were performed by teleoperating the robotic rover on rough countryside. Figure 11 presents an aerial view of one of them using geodetic coordinates for the grid. The beginning and the end of the path, that almost coincide, are marked with a green square and a red circle, respectively. In total, the vehicle travelled 644 m at a mean speed of 4.6 km h^{-1}.

Figure 12 shows the three NED coordinates obtained by the proposed sensor system and by the fifth receiver at the center of the square formed by the antennas. There are no appreciable differences between both estimations, with the exception of the step of 0.166 m in the down coordinate (see Table 1). Altogether, the rover went up and, then, under 12 m.

Figure 13 displays the rover posture obtained by the GNSS setup and by the onboard IMU. It can be observed high peaks in the pitch (above 15°) and the roll (above 25°) angles, as well as several complete turns in the yaw angle during the test. Both estimations are very similar, which confirms the good performance of the proposed sensor system.

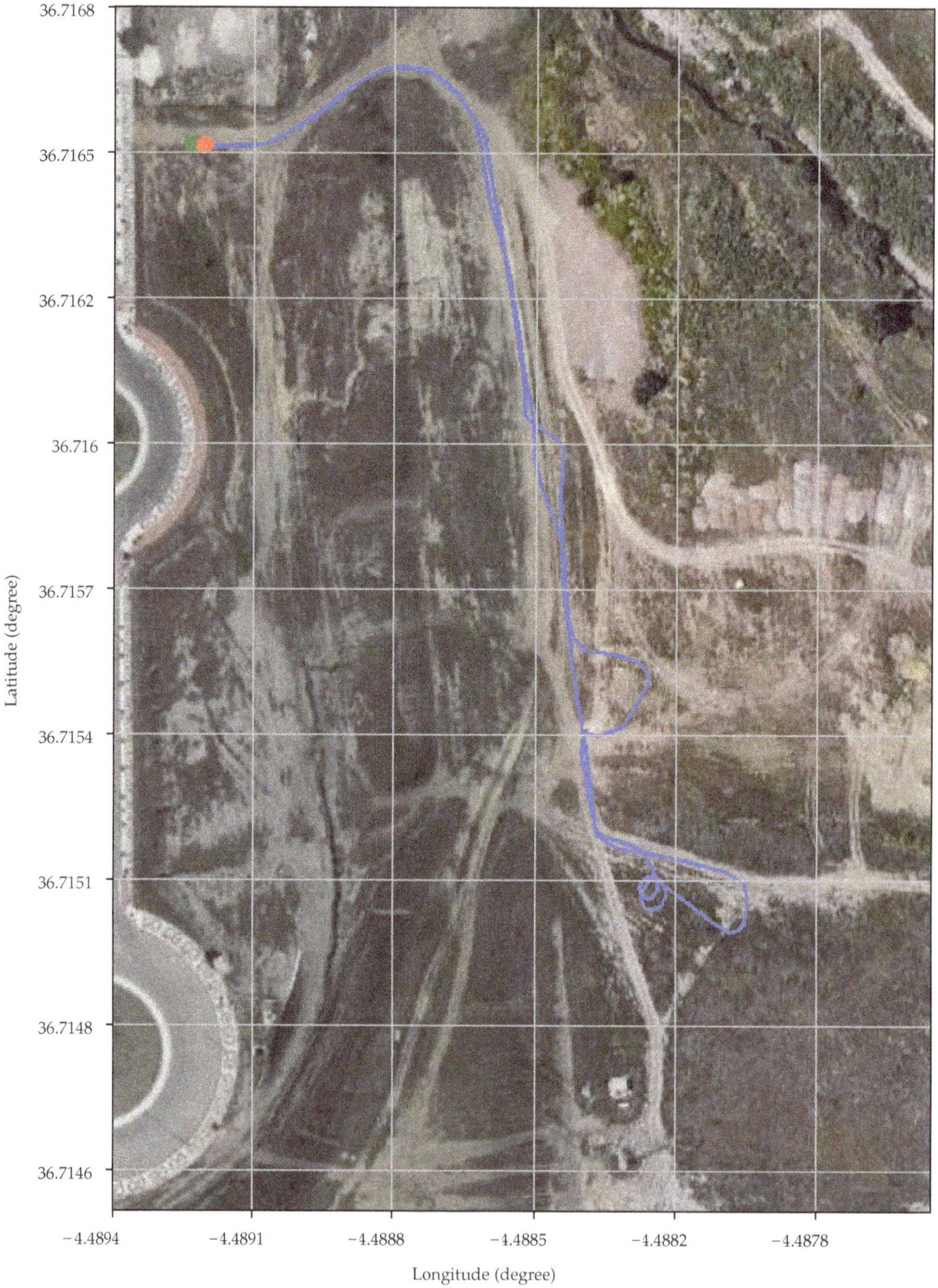

Figure 11. Aerial view of the path followed by the rover on the countryside.

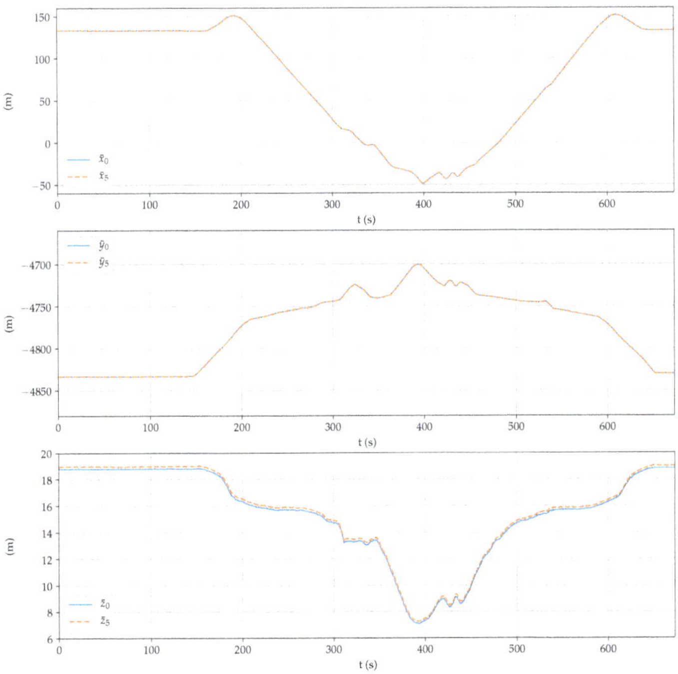

Figure 12. Comparison between the position coordinates obtained by the fifth receiver and the proposed sensor.

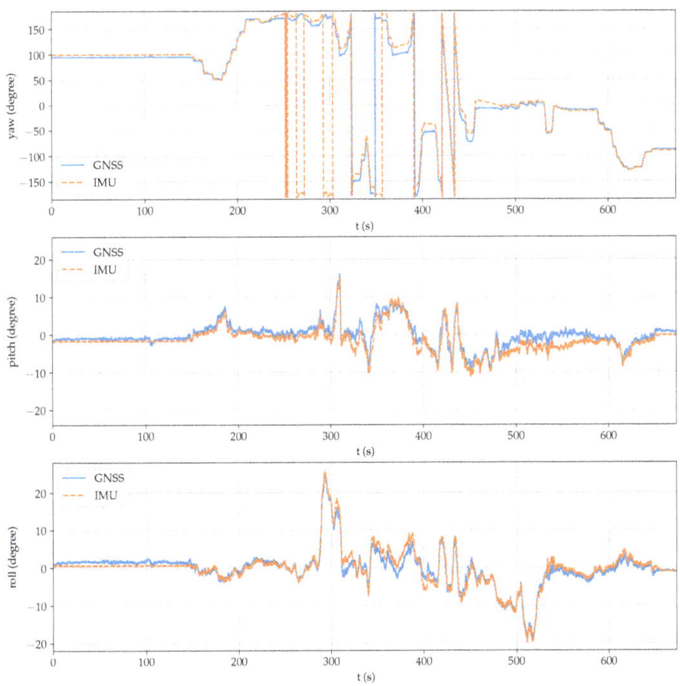

Figure 13. Comparison between the posture coordinates obtained by the IMU and the proposed sensor.

5. Conclusions

A low-cost sensor system composed of four GNSS-RTK receivers connected to a mini computer has been presented in the paper. The placement of three antennas on a vehicle have been analyzed to reduce the uncertainty associated to position and posture estimations with respect to a global reference system. The redundant fourth receiver allows to improve estimations even more and to maintain accuracy when the precision of one of the receivers deteriorates.

Static calibration and reliability tests have been performed with the sensor system mounted on the ground rover Argo XTR. Dynamic experiments on countryside show that this new sensor, in addition to produce reliable positioning, can effectively substitute a high performance IMU to obtain accurate vehicular roll, pitch and yaw angles in real-time.

Future work includes characterizing the achieved pose precision with the robotic rover as well as developing ROS nodes for integrating the proposed sensor system with an IMU for GNSS-denied environments.

Author Contributions: J.M. and J.L.M. conceived the research. J.M. developed the software. J.M. and J.L.M., perform the experiments and analyzed the results. J.L.M. and J.M. wrote the paper and elaborated the figures. J.M. and A.J.G.-C. were in charge of project administration. All authors have read and agreed to the published version of the manuscript.

Funding: This work was partially supported by the Andalusian project UMA18-FEDERJA-090 and the Spanish project RTI2018-093421-B-I00.

Conflicts of Interest: The authors declare no conflict of interest.

Abbreviations

The following acronyms are used in the manuscript:

3D	Three-Dimensional
CORS	Continuously Operating Reference Station
ECEF	Earth-Centered, Earth-Fixed
GPS	Global Positioning System
GNSS	Global Navigation Satellite System
IMU	Inertial Measurement Unit
MEMS	Micro Electro Mechanical Systems
NTRIP	Networked Transport of RTCM via Internet Protocol
ROS	Robot Operating System
RTCM	Radio Technical Commission for Maritime Services
SBAS	Satellite Based Augmentation Systems
USB	Universal Serial Bus

References

1. Henkel, P. Calibration of magnetometers with GNSS receivers and magnetometer-aided GNSS ambiguity fixing. *Sensors* **2017**, *17*, 1324. [CrossRef]
2. Ahmad, N.; Ariffin, R.; Ghazilla, R.; Khairi, N.M.; Kasi, V. Reviews on various Inertial Measurement Unit (IMU) sensor applications. *Int. J. Signal Proc. Syst.* **2013**, *1*, 256–262. [CrossRef]
3. Borenstein, J.; Everett, H.R.; Feng, L.; Wehe, D. Mobile robot positioning: Sensors and techniques. *J. Robot. Syst.* **1997**, *14*, 231–249. [CrossRef]
4. Hein, G.W. Status, perspectives and trends of satellite navigation. *Satell. Navig.* **2020**, *1*, 1–12. [CrossRef]
5. Ciecko, A.; Bakula, M.; Grunwald, G.; Cwiklak, J. Examination of multi-receiver GPS/EGNOS positioning with Kalman filtering and validation based on CORS stations. *Sensors* **2020**, *20*, 2732. [CrossRef] [PubMed]
6. Olivart-Llop, J.M.; Moreno-Salinas, D.; Sánchez, J. Full real-time positioning and attitude system based on GNSS-RTK technology. *Sustainability* **2020**, *12*, 9796. [CrossRef]
7. Fastellini, G.; Radicioni, F.; Schiavoni, A.; Stoppini, A. Comparison of kinematic parameters of a moving vehicle by GNSS measurements and inertial/GPS navigation system. In Proceedings of the 5th International Symposium on Mobile Mapping Technology (MMT), Padua, Italy, 29–31 May 2007; pp. 28–31.

8. Consoli, A.; Ayadi, J.; Bianchi, G.; Pluchino, S.; Piazza, F.; Baddour, R.; Pares, M.E.; Navarro, J.; Colomina, I.; Gameiro, A.; et al. A multi-antenna approach for UAV's attitude determination. In Proceedings of the 2nd International Workshop on IEEE Metrology for Aerospace (MetroAeroSpace), Benevento, Italy, 3–5 June 2015; pp. 401–405. [CrossRef]
9. Liu, S.; Zhang, L.; Li, J. A dual frequency carrier phase error difference checking algorithm for the GNSS compass. *Sensors* **2016**, *16*, 1988. [CrossRef] [PubMed]
10. Ding, W.; Wang, J. Precise velocity estimation with a stand-alone GPS receiver. *J. Nav.* **2011**, *64*, 311–325. [CrossRef]
11. Sun, R.; Cheng, Q.; Wang, J. Precise vehicle dynamic heading and pitch angle estimation using time-differenced measurements from a single GNSS antenna. *GPS Solut.* **2020**, *24*, 1–9. [CrossRef]
12. Sukkarieh, S.; Nebot, E.M.; Durrant-Whyte, H.F. A high integrity IMU/GPS navigation loop for autonomous land vehicle applications. *IEEE Trans. Robot. Autom.* **1999**, *15*, 572–578. [CrossRef]
13. Blanco, J.L.; Moreno, F.A.; González, J. A collection of outdoor robotic datasets with centimeter-accuracy ground truth. *Auton. Robot.* **2009**, *27*, 327–351. [CrossRef]
14. Maddern, W.; Pascoe, G.; Gadd, M.; Barnes, D.; Yeomans, B.; Newman, P. Real-time kinematic ground truth for the Oxford robotcar dataset. *arXiv* **2020**, arXiv:2002.10152.
15. Lourakis, M.; Pateraki, M.; Karolos, I.A.; Pikridas, C.; Patias, P. Pose estimation of a moving camera with low-cost, multi-GNSS devices. In Proceedings of the International Archives of the Photogrammetry, Remote Sensing and Spatial Information Sciences, XXIV ISPRS Congress, Nice, France, 31 August–2 September 2020; pp. 55–62. [CrossRef]
16. Zhg, R. Entanglement difference of GNSS carrier phase for vehicle attitude determination. *Int. J. Transp. Sci. Tech.* **2021**, *10*, 69–82. [CrossRef]
17. Cai, X.; Hsu, H.; Chai, H.; Ding, L.; Wang, Y. Multi-antenna GNSS and INS integrated position and attitude determination without base station for land vehicles. *J. Nav.* **2019**, *72*, 342–358. [CrossRef]
18. Henkel, P.; Sperl, A.; Mittmann, U.; Fritzel, T.; Strauss, R.; Steiner, H. Precise 6D RTK positioning system for UAV-based near-field antenna measurements. In Proceedings of the 14th European Conference on Antennas and Propagation (EuCAP), Copenhagen, Denmark, 15–20 March 2020; pp. 1–5. [CrossRef]
19. Smith, R.C.; Cheeseman, P. On the representation and estimation of spatial uncertainty. *Int. J. Robot. Res.* **1986**, *5*, 56–67. [CrossRef]
20. Quigley, M.; Gerkey, B.; Conley, K.; Faust, J.; Foote, T.; Leibs, J.; Berger, E.; Wheeler, R.; Ng, A. ROS: An open-source robot operating system. In Proceedings of the IEEE International Conference on Robotics and Automation: Workshop on Open Source Software (ICRA), Kobe, Japan, 12–17 May 2009.
21. Berrocoso, M.; Páez, R.; Jijena, B.; Caturla, C. The RAP net: A geodetic positioning network for Andalusia (south Spain). In Proceedings of the EUREF Symposium, Riga, Latvia, 14–16 June 2006; pp. 364–368.
22. Horn, B.K.P. Closed-form solution of absolute orientation using unit quaternions. *J. Opt. Soc. Am.* **1987**, *4*, 629–642. [CrossRef]
23. Bandari, E.; Goldstein, N.; Nesnas, I.; Bajracharya, M. *Realtime Absolute Orientation with Outlier Rejection for Visual Odometry and Mesh Merging*; Technical Report 20040084602; National Aeronautics and Space Administration (NASA): Washington, DC, USA, 2004.
24. Niu, X.; Goodall, C.; Nassar, S.; El-Sheimy, N. An efficient method for evaluating the performance of MEMS IMUs. In Proceeding of the IEEE/ION Position, Location, And Navigation Symposium, San Diego, CA, USA, 25–27 April 2006; pp. 766–771.

sensors

MDPI

Article

Deep Transfer Learning Based Intrusion Detection System for Electric Vehicular Networks

Sk. Tanzir Mehedi [1], Adnan Anwar [2,*], Ziaur Rahman [1] and Kawsar Ahmed [1]

[1] Department of Information and Communication Technology, Mawlana Bhashani Science and Technology University, Tangail 1902, Bangladesh; tanzirmehedi@ieee.org (S.T.M.); zia@iut-dhaka.edu (Z.R.); k.ahmed.bd@ieee.org (K.A.)

[2] Centre for Cyber Security Research and Innovation (CSRI), Deakin University, Geelong 3216, Australia

* Correspondence: adnan.anwar@deakin.edu.au; Tel.: +61-3-522-73679

Abstract: The Controller Area Network (CAN) bus works as an important protocol in the real-time In-Vehicle Network (IVN) systems for its simple, suitable, and robust architecture. The risk of IVN devices has still been insecure and vulnerable due to the complex data-intensive architectures which greatly increase the accessibility to unauthorized networks and the possibility of various types of cyberattacks. Therefore, the detection of cyberattacks in IVN devices has become a growing interest. With the rapid development of IVNs and evolving threat types, the traditional machine learning-based IDS has to update to cope with the security requirements of the current environment. Nowadays, the progression of deep learning, deep transfer learning, and its impactful outcome in several areas has guided as an effective solution for network intrusion detection. This manuscript proposes a deep transfer learning-based IDS model for IVN along with improved performance in comparison to several other existing models. The unique contributions include effective attribute selection which is best suited to identify malicious CAN messages and accurately detect the normal and abnormal activities, designing a deep transfer learning-based LeNet model, and evaluating considering real-world data. To this end, an extensive experimental performance evaluation has been conducted. The architecture along with empirical analyses shows that the proposed IDS greatly improves the detection accuracy over the mainstream machine learning, deep learning, and benchmark deep transfer learning models and has demonstrated better performance for real-time IVN security.

Keywords: electric vehicles; in-vehicle network; controller area network; cybersecurity; intrusion detection; deep learning; transfer learning

Citation: Mehedi, S.T.; Anwar, A.; Rahman, Z.; Ahmed, K. Deep Transfer Learning Based Intrusion Detection System for Electric Vehicular Networks. *Sensors* **2021**, *21*, 4736. https://doi.org/10.3390/s21144736

Academic Editor: Juan A. Cabrera

Received: 18 June 2021
Accepted: 8 July 2021
Published: 11 July 2021

1. Introduction

In recent years, the automotive industry has been undergoing a radical transformation. With the ongoing development of network communication, modern vehicles are rapidly transitioning from fully mechanical to software-controlled technologies [1]. Modern In-vehicle Network (IVN) technologies and services are being integrated with intelligent information systems. As a result, the number of IVN devices is rapidly increasing and becoming more complex. The IVN devices must be seamlessly connected to an external network system in order to receive communication services efficiently. However, this increases the risk of the IVN to potential internal or external threats. The Electronic Control Units (ECUs) are software-controlled technologies that read various sensor data and perform relevant processing, including automatic brake control, pedestrian detection, auto-parking, path-planning, actuators control, and collision avoidance [2]. The sensor and actuator values are transmitted to other ECUs via the IVN protocol, resulting in the formation of a very complex network. There are several IVN protocols in the automotive industry, including Controller Area Network (CAN), Controller Area Network Flexible Data-Rate (CAN FD), Media Oriented Systems Transport (MOST), FlexRay, and Local Interconnect Network

(LIN) [3]. Among all the data communication buses, CAN bus is the most well-known and extensively used protocol in the automotive vehicles industry [4]. Furthermore, the CAN buses are being applied also in other industries, including agriculture, aerospace, medical devices, and commercial machinery [1]. Several other protocols are also available with more security features (e.g., Ethernet). However, in the field of automotive IVN communication, these advanced protocols can not be completely replaced by the CAN bus protocol due to some reasons [4]. First of all, the CAN bus is more design flexible and perfectly appropriate for real-time environments, ensuring secure and fast communication between ECUs with minimal latency time. Secondly, there is a process of prioritization in the CAN bus protocol that prevents lower-priority messages from interfering with higher-priority messages. To cite an example, a message that transmits a more critical function such as an engine control message takes precedence over a door control message. Finally, the CAN bus protocol serves as the backbone of automotive IVN communication in all modern vehicles. To completely replace this protocol with another, the IVN architecture must be completely redesigned. As a result, other protocols will not completely replace the CAN bus's role and application.

However, in-vehicle intrusion detection has become a growing interest field that has been researched across a wide range of disciplines. In the CAN bus protocol, intrusion detection is the method of monitoring normal and abnormal traffic between different ECUs and identifying any abnormal traffic using Traditional Machine Learning (TML) algorithms [5]. With the rapid development of IVNs and evolving threat types, the TML-based IDS has to update to cope with the security requirements of the current environment. Nowadays, regarding the progression of Deep Learning (DL), Deep Transfer Learning (DTL), and its impactful outcome in several areas, these techniques have gained the attention of many researchers in the field of cybersecurity (e.g., IDS, antivirus or malware identification) [6]. In particular, in the field of the automotive industry, recent DL and DTL-based IDS have also gained the attention of many researchers, which is discussed further in the Related Work section.

Automobile manufacturers are working to develop fully autonomous vehicles, which will necessitate the addition of more attack surfaces. Since the CAN bus protocol does not encrypt data, the attackers can use a reverse mechanism to interpret each CAN packet in order to inject malicious messages into in-vehicle networks [7]. This malicious message injection mechanism will cause some abnormal behaviors in the communication traffic, which can be detected by developing an intrusion detection system. The threat of cyber-attacks in the automotive industry and the securing of communication protocols have gotten a lot of attention in recent years. However, due to the complexity of in-vehicle embedded systems and the reality of a real-time experiment with a limited processing unit and memory resources, it is impractical and nearly impossible to apply the standard measures to build a potential IDS for vehicular networks. Therefore, a different mechanism is required to detect normal and abnormal characteristics in a vehicular network. In this manuscript, we propose a deep transfer learning-based intrusion detection model that can efficiently classify the normality and abnormality of a communication traffic and allows the immediate detection of anomalies in the CAN bus protocol. The key contributions of this paper are narrated as follows:

- In this work, a deep transfer learning-based *LeCun Network (LeNet)* model has been proposed for effective intrusion detection in-vehicle network CAN bus protocol. The proposed model enabled to develop effective models that speed up the training process and improve the performance of the deep learning model.
- The experiments have been conducted using an in-vehicle real-time dataset generated from heterogeneous sources that include three types of malicious messages. We have made observations on this practical data to identify the best features in the context of supervised learning for effective intrusion detection.
- In-depth architectural and statistical analyses have been conducted considering traditional machine learning, deep learning, and deep transfer learning algorithms.

Extensive analysis and performance evaluation show that the proposed deep transfer learning-based LeNet model outperforms other approaches.

The rest of this research paper is organized as follows. First of all, Section 2 discusses the background of CAN bus protocol security vulnerabilities and introduces the related work done with in-vehicle networks. Additionally, we present the problem statement and the solution methodology also include the proposed architecture in Section 3. Furthermore, Section 4 discusses a detailed description of the dataset as well as an overview of selected models. Evaluation and experimental results analysis of these methods are discussed in Section 5. Finally, Section 6 includes the summary and feasible future directions for this research.

2. Background and Related Work

While investigating the most recent relevant works in this field, we discovered that several of them share a common motivation in different ways. In terms of security features, various studies show the CAN bus's vulnerabilities and weaknesses [8]. The following subsections illustrate those works before we demonstrate our proposed methodology.

2.1. Background of CAN and Security Vulnerabilities

During the development of the CAN bus protocol, vehicles were considered as isolated objects that did not have a connection with the outside environment [9]. By design, the CAN bus protocol is plagued with various security issues because of the lack of encryption and authentication requirements [10]. Therefore, any malicious or hijacked node can cause disastrous accidents and serious financial loss. For instance, hackers can affect an ECU by injecting malicious messages and various attacks due to the lack of an efficient message authentication method. The CAN node is the combination of a CAN controller and a CAN transceiver that transmits and receives messages but not simultaneously [11]. The architecture of a CAN bus node is shown by Figure 1. The data frame, remote frame, error frame, and overload frame are four different types of frames that have been used in the CAN bus [9,12]. The data frame is used to transmit actual data from a transmitter to receivers (other nodes). A node requests a specific message with a specific identifier using the remote frame. If any of the nodes on the bus detects an error, it will send an error frame. The overload frame adds a delay between the data and remote frames.

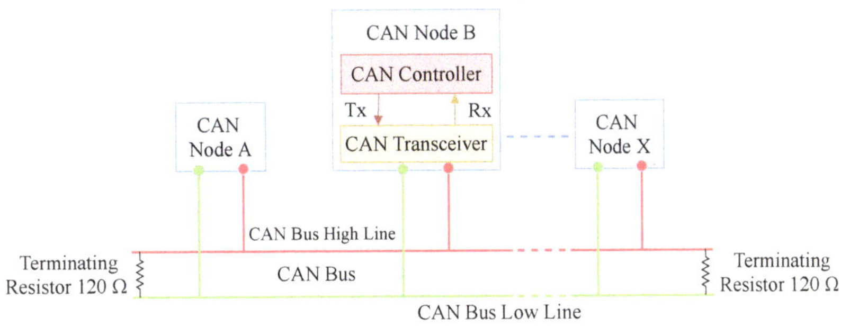

Figure 1. The standard CAN bus node architecture.

The standard CAN message frame format is the composition of header, trailer, and protected data payload field that can be up to 64 bits long. The header field is the combination of 1-bit Start of Frame (SOF), 12-bits arbitration field, and 6-bits control field. Furthermore, the arbitration field divided into an 11-bits identifier and 1-bit Remote Transmission Request (RTR) field. The identifier field represents the message priority. It also consists of Identifier Extension (IDE), Reserved, Data Length Code (DLC), Cyclical Redundancy Check (CRC), Delimiter, Acknowledge (ACK), End of Frame (EOF), and Inter Frame Space

(IFS) fields. Both sides of the message frame end with a bus idle field. Figure 2 shows the standard CAN message frame format [9].

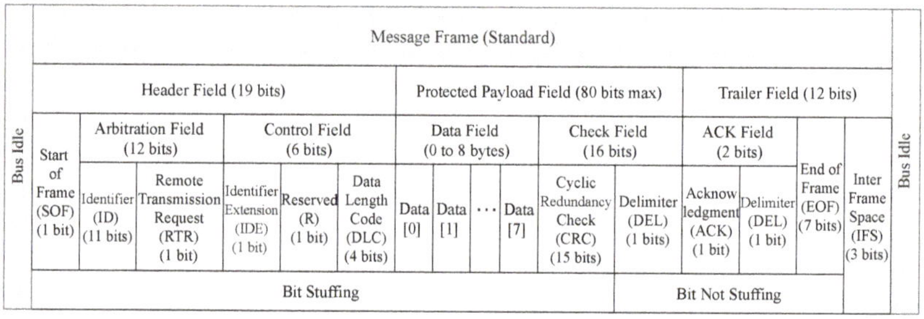

Figure 2. The standard CAN bus message frame format.

Nowadays, with the progression of data-mining techniques, many researchers have addressed this type of attack and are able to detect and ignore any abnormal traffic activities in CAN networks [13–17]. Recently, modern vehicles are not considered only a closed-loop system; instead, they also communicate with the outside world via various intelligent systems. As a result, hackers or attackers can use various internal and external interfaces to inject malicious messages into CAN traffic. Despite the tendency of injecting malicious messages, ECUs can be re-programmed remotely by embracing over-the-air (OTA) updates, which may provide more comfort and advantages to the vehicle owner [15]. However, these mechanisms have also initiated more remote attacks, which can assist attackers or hackers in compromising the ECUs by sending malicious messages.

2.2. TML and DL Based IDSs for IVN

In the existing automotive applications, the TML and DL-based IDSs have obvious advantages in detecting various malicious messages [18–20]. Bozdal et al. [21] and Lokman et al. [22] reviewed the security threats and challenges of the automotive CAN bus system, and discuss some potential security solutions. In 2016, Kang et al. [23] proposed a Deep Neural Network (DNN)-based IDS for IVN security. Using the unsupervised pre-training model of Deep Belief Networks (DBN), the selected parameters of the DNN model were trained with probability-based feature vectors extracted from the IVN packets, followed by the traditional stochastic gradient descent technique. The results of the experiment showed that the model can provide a real-time response to malicious messages, with a detection ratio over 95% on average in the CAN bus. The robustness of the model is high, but detection coverage is not defined. Loukas et al. [24] proposed cloud-based cyber-physical IDS for IVN using the Deep Learning (DL) model to detect Denial-of-Service (DoS), command injection, and malware (Net) attacks. The model had a validation accuracy score of overall 86.9%, which motivates additional research into this field to improve the detection rate, particularly for these attacks. Seo et al. [25] introduced an IDS to identify DoS, Fuzzy, RPM, and Gear attacks in the CAN bus network traffic by applying Generative Adversarial Network (GAN). This DL-based model can detect unknown malicious messages using only normal CAN data for training. The results of various simulations show that each of the four attacks was detected with a high accuracy score of over 95%, demonstrating the robustness of the model.

Lokman et al. [26] developed an IDS for an in-vehicle network using an unsupervised DL-based model, known as Deep Contractive Auto-encoders (DCAEs). The DCAE model outperformed other regularized auto-encoder variants, with a 91.0% detection rate. As the proposed IDS performance is evaluated within a simulated network, further evaluation is necessary to validate the efficiency against a larger array of various cyber-attacks. Zhang et al. [27] in 2019 proposed a DL-based IDSs for in-vehicle security to detect only spoofing and replay attacks. The results were evaluated in a simulated environment, which is hopeful as they can effectively detect only spoofing and replay attacks. The proposed model is capable of adapting to new attacks. The detection accuracy of this model varies between 97.0% and 98.0% when it faces unknown attack types. Zhu et al. [28] proposed a DL-based method to speed up intrusion detection using the LSTM model. They executed the spoofing, replay, and flooding attacks in the CAN network. The authors proposed using a mobile edge-assisted multi-task LSTM model because the computation time with LSTM is so high. The model had an accuracy score of over 80% and a detection latency of 0.61 ms. Avatefipour et al. [29] proposed a new effective IDS based on a modified one-class SVM in the CAN traffic by deploying three attacks (e.g., DoS, fuzzing, and spoofing attacks). The experimental result shows that the proposed model has a high accuracy score of over 90%, demonstrating the robustness of the model. In order to prove the efficiency of the this model, they applied it to other recent popular public datasets in the scope of CAN bus traffic intrusion detection. Xiao et al. [30] proposed a lightweight ML algorithm based on RNN for IDS on the CAN bus network. The experimental evaluation using appropriate hyper-parameters demonstrated that the proposed model had good performance metrics, compared to LSTM and GAN models.

Al-Saud et al. [31] proposed an IDS model based on an improved SVM model for the CAN bus network. The experimental results on the real dataset reveal the good performance metrics and high robustness of the model against only DoS attacks in electric vehicles. Lin et al. [32] proposed a DL-based intrusion detection system for CAN networks to detect DoS, fuzzing, and impersonation attacks particularly. The model is trained with a deep denoising auto-encoder during the training phase, which includes a feature extraction mechanism. Their work had a low detection rate when compared to other ML algorithms. Yang et al. [33] proposed an IDS using a recurrent neural network with long short-term memory (RNN-LSTM). The selected model had a higher validation accuracy score especially for detecting only spoofing attacks in the CAN network traffic, which motivates additional research into this field to detect other cyber-attacks. A Long Short-Term Memory (LSTM) NN-based IDS was proposed by Hossain et al. [34]. The proposed IDS is capable of detecting various attacks on the CAN bus network, such as DoS, fuzzing, and spoofing attacks. Recently, Song et al. [35] proposed an IDS based on a Deep Convolutional Neural Network (DCNN) model called Inception-ResNet to detect various attacks (e.g., DoS, fuzzing, gear, and RPM attacks) to test in a real-time in-vehicle system. The authors also investigated the sequence of messages for intrusion detection. There are two steps to the proposed model. The first is a training step and the last one is a detection step. In the first step, the CNN classifier is trained and, in the last step, real CAN data frames are passed through this proposed model to classify whether they are normal or attack messages. In comparison to previous work, the proposed model had an over 80% detection rate and a low error rate but has high computational cost and memory consumption. However, further analyses are necessary to investigate the performance on new complex types of cyber-attacks in various categories.

The existing IDSs have categorized according to *detection algorithm*, *detection accuracy*, *robustness*, and *detection coverage*. Here, robustness is defined as the ability of the IDS to detect attacks in the CAN bus network. A summary of all the existing IDSs for in-vehicle network is given in Table 1. Several existing IDSs have used data from different small in-vehicle networks, which can not be implemented in a realistic environment. Moreover, existing IDSs concentrated on detecting whether specific cyber-attacks have occurred, but most of them did not classify the type of attack. This limitation of previous approaches is a significant feature for further investigation for in-vehicle security.

Table 1. Overview of recent research on IDSs for in-vehicle networks.

Ref.	Algorithm	Accuracy	Robustness	Detection Coverage
[23]	DL	>95%	High	N/A
[24]	DL	>85%	Medium	DoS, Command Injection, Malware
[25]	GAN	>95%	High	DoS, Fuzzing, RPM, Gear attacks
[26]	DCAE	>90%	Medium	DoS, Fuzzing, Impersonation
[27]	DL	>95%	High	Spoofing, Replay
[28]	LSTM	>80%	Medium	Spoofing, Replay, Flooding
[29]	ML	>90%	High	DoS, Fuzzing, Spoofing
[30]	RNN	>95%	High	DoS, Fuzzing, Impersonation
[31]	ML	>90%	Medium	DoS
[32]	DL	>80%	N/A	DoS, Fuzzing, Impersonation
[33]	RNN-LSTM	>95%	High	Spoofing
[34]	NN-LSTM	>90%	N/A	DoS, Fuzzing, Spoofing
[35]	DCNN	>80%	Medium	DoS, Fuzzing, RPM, Gear attacks
[36]	DTL	>90%	High	Impersonation, ARP, Flooding

N/A means "Not Applicable".

2.3. DTL Based IDSs for IVN

Deep transfer learning (DTL) is a solution that can reuse previous trained-model knowledge and outperform other TML and DL models in terms of intrusion detection [36,37]. Zadrozny et al. [38] proposed a model for intrusion detection that performs better in both labeled and unlabeled data. Another type of transfer learning model called TrAdaBoost was proposed by Dai et al. [39]. This model allows knowledge from the old trained data to be efficiently transferred to the new validation data, resulting in a more efficient classification model. Additionally, Raina et al. [40] also proposed a transfer learning model that builds an informative Bayesian from prior knowledge before validating a new task. Furthermore, Gou et al. [41] proposed a novel transfer learning model for IDS especially to detect the different types of cyber-attacks. The proposed model shows that the detection accuracy of the different types of cyber-attacks has been comprehensively improved than others. Li et al. [36] proposed a transfer learning approach for intrusion detection of different types of attacks on the Internet of Vehicles (IoV). The experimental results show that, when compared to existing TML and DL methods, this model significantly improved detection accuracy by at least 23%. Xu et al. [42] recently proposed an IDS based on DL and transfer learning. To improve the model's efficiency and adaptability, transfer learning is implemented here. The experimental analysis shows that the proposed model outperforms the mainstream TML and DL methods in terms of efficiency and robustness, and it can detect and classify new cyber-attacks more effectively. The deep-computational-intelligence system has recently been applied in transfer-learning to optimize the performance of existing transfer-learning models [37]. As a result, current transfer learning solutions for intrusion detection still need to be updated [36]. A new-generation labeled dataset of an in-vehicle network proposed by Kang et al. [43], which is more suitable for applying transfer learning models because, for time series classification, deep transfer learning approach shows the better performances than other TML or DL models [44–46]. This paper has improved the existing transfer learning model for detecting various complex types of cyber-attacks in CAN bus protocol.

3. Proposed Solution

The proposed deep transfer learning-based *LeCun network (P-LeNet)* approach is presented in this subsection. Following that, we have thoroughly explained the problem statement, solution formulation, the structure of the proposed *P-LeNet* based intrusion detection model, and how we adapted it for deep transfer learning.

3.1. Problem Statement

Automobile manufacturers are working to develop fully autonomous vehicles, which will necessitate the addition of more attack surfaces. Since the CAN bus protocol does not encrypt data, the attackers can use a reverse mechanism to interpret each CAN packet in order to inject various malicious messages into the in-vehicle network. This malicious message injection mechanism will cause abnormal behaviors in the communication traffic, which can be detected by developing an intrusion detection system. Three types of attacks (e.g., *flooding*, *fuzzing*, and *spoofing*) have been considered due to their severely impaired characteristics, the intensity of an attack, and the degree of damage among in-vehicle functions. The three most common attack scenarios against an in-vehicle network are shown in Figure 3. By maintaining an influential situation on the CAN bus, the *flooding* attack allows an ECU node to hold many of the resources allocated to the CAN bus. This attack disrupts normal driving and limits the communication between ECU nodes by sending high frequency and high priority messages (e.g., 0 × 000). Figure 3a shows a scenario of a *flooding* attack on CAN networks. In the *fuzzy* attack, a malicious ECU from IVN transmits random frames with spoofed CAN IDs with arbitrary data values, which caused the vehicle function to be unavailable (e.g., 0 × 4CC, 0 × 7C6). Due to the limited number of valid CAN frames streaming over the bus, this type of attack is easy to implement and does not necessitate reverse engineering. The *fuzzy* attack scenario against an IVN is shown in Figure 3b. *Spoofing* is a type of attack in which a malicious node transmits messages to the receiver with a fake ID (e.g., 0 × 2B0, 0 × 130) that appears identical to that of an original node. As a result, the receiver node considers that the message is from an original node. It is tough to distinguish between malicious and original messages because there is no message authentication mechanism on the CAN bus. Figure 3c shows a scenario for a *spoofing* attack on a CAN network.

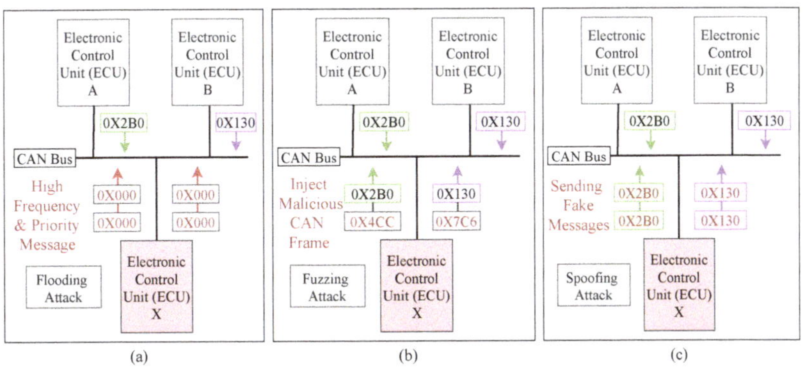

Figure 3. (**a**) Flooding attack scenario against an IVN; (**b**) Fuzzing attack scenario against an IVN; (**c**) Spoofing attack scenario against an IVN.

However, due to the complexity of in-vehicle embedded systems and the realities of a real-time experiment with limited processing and memory resources, applying standard measures to build a potential IDS for vehicular networks is impractical and nearly impossible. As a result, detecting normal and abnormal characteristics in a vehicular network requires a different mechanism. The next subsection discusses the solution formation and the details' architecture of our proposed intrusion detection model that can efficiently classify the normality and abnormality of communication traffic and allows the immediate detection of anomalies in the CAN bus protocol. Figure 4 shows the application of our proposed intrusion detection mechanism for vehicular network traffic.

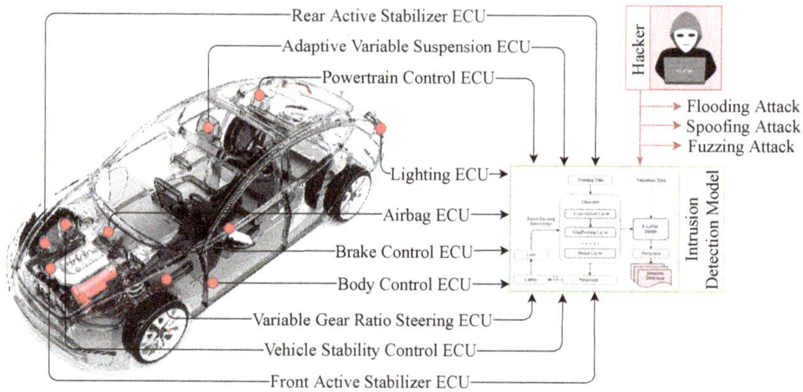

Figure 4. Application of intrusion detection model for IVN traffic. The figure illustrates how the proposed IDS can detect the possible attack vectors within an in-vehicle Network. The car image is adopted from [47].

3.2. Solution Formulation

Table 2 shows the symbols and descriptions, in which we set the initial model with enough labeled data to build an effective intrusion detection model. The source domain data ($D_s : (A_s, B_s)$) is the combination of $(A_{s1}, B_{s1}), (A_{s2}, B_{s2}), (A_{s3}, B_{s3}), \ldots\ldots (A_{sn}, B_{sm})$ and the target domain data ($D_t : (A_t, B_t)$) is the combination of $(A_{t1}, B_{t1}), (A_{t2}, B_{t2}), (A_{t3}, B_{t3}), \ldots\ldots (A_{tn}, B_{tm})$, in which the class of source domain label data (B_s) and target domain label data (B_t) is 0 and 1, where the normal and attack scenario is represented by 0 (zero) and 1 (one), respectively.

Table 2. Symbols and description.

Description	Source (s)	Target (t)
Domain data	$D_s : (A_s, B_s)$	$D_t : (A_t, B_t)$
Domain feature	A_s	A_t
Domain label	B_s	B_t
Number of domain data	n	m

Additionally, both the labels of the source domain (B_s) and the target domain (B_t) data contain only *normal* and *attack* data, although the attackers in the source and target domains may be different. Although the source domain label (B_s) and the target domain label (B_t) share the same feature space, they perform differently in specific features. We used the *Maximum Mean Discrepancy* Equation (1) to calculate the difference between the source and target domains [48]:

$$Distance(A_s, A_t) = \left\| \frac{1}{n} \sum_{i=0}^{n} \phi(A_{s_i}) - \frac{1}{m} \sum_{i=0}^{m} \phi(A_{t_i}) \right\|^2 \tag{1}$$

The detection model trained by the source domain data (D_s) does not have excellent detection accuracy when faced with target domain data (D_t), according to the dependency of TML and DL models, and this has been totally proven by the subsequent experiment. The TML and DL models require a large amount of training data. Thus, it is difficult to train an effective IDS model using a small amount of source domain data (D_s). As a result, we have proposed a deep transfer learning based *P-LeNet* method to transfer the knowledge contained in source domain data (D_s) to the target domain and combine the target domain data (D_t) to build an efficient IDS to improve the detection accuracy for any electric vehicular ecosystems.

3.3. Architecture

The block diagram of the proposed *P-LeNet* model is shown in Figure 5, which contains two parts: the model training part and the intrusion detection part. After pre-processing the raw data, we have applied it to our proposed model for training. Through subsequent empirical experiments, the most important parameters for the selected model have been determined. We used a randomly selected training dataset to train the proposed *P-LeNet* model and a validation dataset to validate the model. The final IDS model has been selected based on its best prediction performance on the validation dataset.

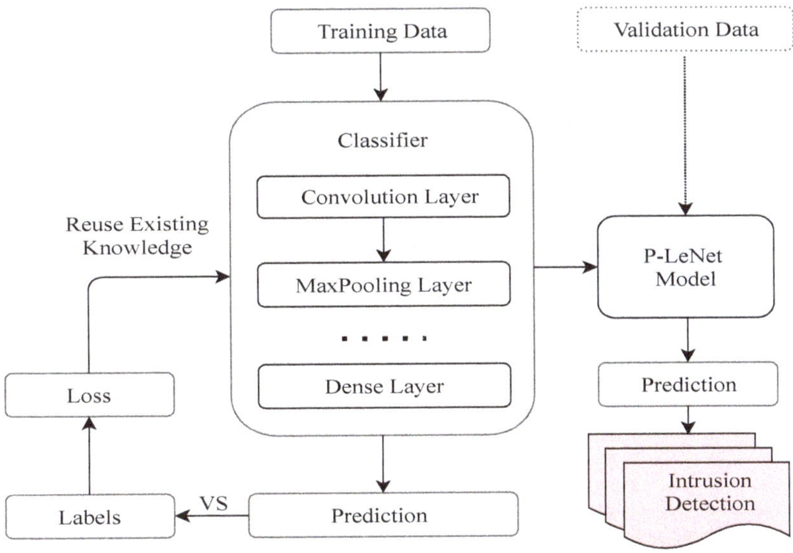

Figure 5. The block diagram of the proposed P-LeNet mode.

The proposed *P-LeNet* architecture is made up of seven layers with a total of 12,052 trainable parameters (weights). The layer is the composition of two *convolutional layers*, two *subsampling layers*, one *flatten layer*, one *fully connected layer*, and one *output layer*. Each layer takes the previous layer outputs as inputs for the current layer and performs some nonlinearity's to transform it into a multivariate series whose dimensions are defined by the number of filters in each layer. The structure of the proposed *LeCun Network (P-LeNet)* model is shown in Figure 6. The first layer is the *Input* layer, which is not considered a network layer because it does not learn anything. The input layer is designed to take dataset and pass it on to the following layer. The dataset has a total of four features including the label feature. The four features are *CAN_ID*, *DLC*, *Data_Field*, and *Label*. The one-dimensional convolutional layer *(Conv1D)* is used in the first, and the third layer respectively to transform the dataset. The first *Conv1D* layer produces as output five feature maps, and has a kernel size of 5, and the second *Conv1D* layer produces as output 20 feature maps, and has a kernel size of 5. The *Rectified Linear-Unit (ReLU)* activation function is used in the both convolution layer. The two *Conv1D* layers contain 30 and 520 trainable parameters, respectively. The first *MaxPooling1D* subsampling layer follows the first *Conv1D* layer, and the second *MaxPooling1D* subsampling layer follows the second *Conv1D* layer shown in Figure 6. The two subsampling layers halves the dimension of the feature maps it receives from the previous layer; this is known commonly as down-sampling. The two subsampling layers also produce 5 and 20 feature maps, respectively, each one corresponding to the feature maps passed as input from the previous layer. The fifth layer of our proposed model is the *Flatten* layer which converts the pooled feature map to a single column that is passed to the next layer. The next is fully-connected *Dense* layer where total trainable parameter is 10,500. This operation reduces drastically the number of trainable parameters in a deep model while enabling the use of a class activation map which allows an interpretation of the learned features [49]. Finally, the output layer whose number of neurons is equal to the number of classes in the dataset. The *softmax* function is used as the activation function in this layer to predict a probability distribution between normal and attack scenarios.

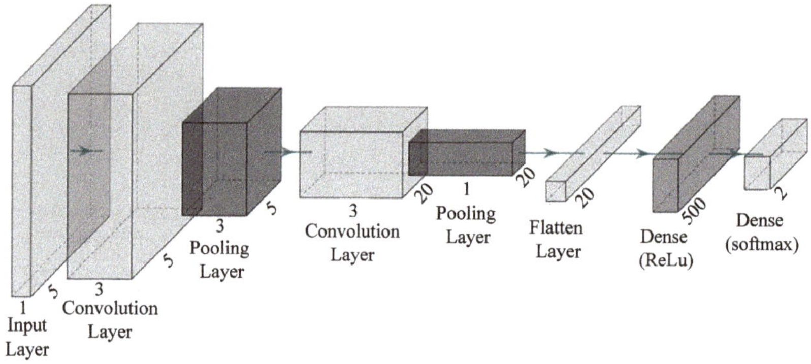

Figure 6. The structure of the proposed P-LeNet model.

Furthermore, the *Compile* function enables the actual building of the model we have implemented with some additional characteristics such as the *loss function*, *optimizer*, *learning rate*, and *metrics*. To train the network, we utilize a loss function called *categorical crossentropy*, which calculates the difference between the network's predicted values and the actual values of the training data. The number of changes made to the weights within the network is facilitated by the loss values accompanied by an optimization algorithm *(Adam)*. During training, we have been used the valuation dataset to validate our proposed model after each epoch. The proposed model has achieved a better validation accuracy. However, we have evaluated the trained model on the test dataset for a more explicit verification of the proposed model's performance on an unknown dataset.

4. Materials and Methods

In this section, we have thoroughly explained the dataset and the transformation process of the dataset to feed the selected models.

4.1. Dataset Description

The dataset has been generated in two different ways. Details of the dataset can be accessed in [43]. The first dataset contained normal driving data without an attack and the second dataset contained abnormal driving data that has been collected during an attack was performed in in-vehicle networks. Each dataset has been combined into one CSV file by a Python script. The class distribution of the combined dataset is shown in Figure 7. The combined dataset has a total of 5 features including the label feature. The five features are *Timestamp*, *CAN_ID*, *DLC*, *Data_Field*, and *Label*. The *Timestamp* feature represents the recorded time in seconds (s). The *CAN_ID* is used to identify the CAN messages in hexadecimal format (e.g., 0 × FA5, 0 × 18F) and assigns its priority. The messages having the lowest *CAN_ID* value represent the highest priority. The *DLC* feature in the control field shows the number of bytes, from 0 to 8, and values change depending on the vehicle categories. The *Data_Filed* feature contains the data to be transferred from one node to another and consists of the data value in a byte that has eight fields in total (e.g., Data[0], Data[5]). Finally, the *Label* feature contains two quantitative values, i.e., 0 and 1, which indicates normal and attack (injected message) scenarios, respectively.

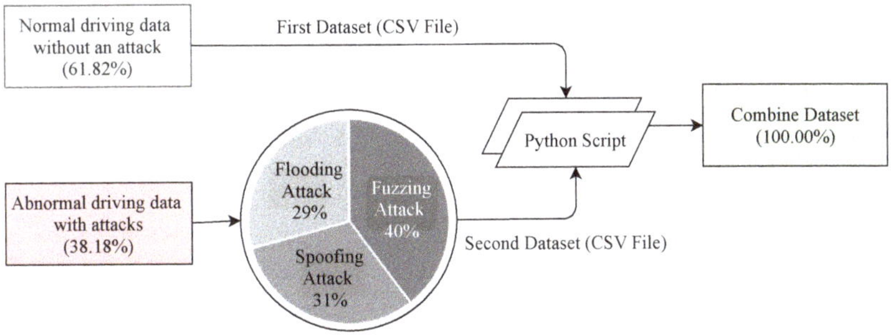

Figure 7. Statistics of the dataset.

4.2. Data Preparation

The dataset must be cleaned and prepared before applying the selected TML, DL, and DTL methods to achieve optimal performance and improve the learning process. Data preparation generally happens by eliminating unnecessary features, checking for changes in independent features, converting non-numeric features, and removing outliers. Three fundamental steps are applied during the data preparation process. The first step is data cleaning, the second is data integration, and the final step is data transformation.

4.2.1. Data Cleaning

This dataset is very sensitive to missing and noisy data because of its large size. There are a total of 1,270,310 instances in the dataset including noisy and inconsistent data. In this subsection, we have discussed the essential steps in prepossessing of data. First of all, we have applied various techniques to remove noise and clean inconsistencies data from the dataset, for example, *Rosner's Test* for outliers checking, and the *Predictive Mean Matching* method for imputing missing values. Then, in order to apply the selected models, we have converted the qualitative values into quantitative values. To cite an example, the *Label* feature in the dataset, which has qualitative values *'Normal'* and *'Attack'*, has been converted into '0' and '1'. These quantitative values have been converted to quantitative values by performing a numerical convolution label-encoding library *numconv*. The *CAN_ID* feature in the dataset, which has hexadecimal values (e.g., 58B, F41), has been converted into decimal values by applying the *hex2dec* function. On the other hand, the *Data_Field* feature in the dataset, which also has hexadecimal values of eight bytes separated by space (e.g., 80 7F 00 73 20 00 0A A1, 14 80 10 80 00 00 0A 73). The space between bytes have been removed by applying *gsub* function and then the hexadecimal values have been converted into decimal values by applying the *Rmpfr* function as most of the data field is over 64 bits (maximum 152 bits). The *Timestamp* feature has been omitted from feature vectors as they may cause overfitting the training data. Furthermore, for some DL and DTL models, the input data shape has been reshaped into three dimensions to feed the models by applying *numpy.reshape* with *swapaxes* and *concatenate* methods.

4.2.2. Data Integration

To improve the accuracy and speed of the training and validation process, the data integration technique helped us by reducing and avoiding redundancies from the resulting dataset. As this dataset originates from two different ways. Thus, it is an essential step to analyze the *redundancy* and *correlation* between the selected features. This analysis has measured how strongly one feature, i.e., *CAN_ID* implies the other, i.e., *Data_Field*. We used cutoff criteria ($p < 0.05$) to find the correlation between different features. The results indicated that the higher the coefficient value, the stronger the relationship between those features [50]. Table 3 shows the correlation between different features. For our analysis, we assessed the correlation between all features by calculating the following *Pearson product-moment coefficient* Equation (2) [51]:

$$r = \frac{\sum_{i=1}^{n}(x_i - \bar{x})(y_i - \bar{y})}{\sqrt{\sum_{i=1}^{n}(x_i - \bar{x})^2}\sqrt{\sum_{i=1}^{n}(y_i - \bar{y})^2}} \quad (2)$$

where n is the number of tuples, x_i and y_i are the respective values in tuple i, and $\bar{x}$ and $\bar{y}$ are the respective mean values of x and y.

Table 3. Pearson product–moment correlation between different features.

	Timestamp	CAN_ID	DLC	Data_Field
Timestamp	0×10^{0}	4.285×10^{-2}	6.525×10^{-6}	1.794×10^{-4}
CAN_ID	4.285×10^{-2}	0×10^{0}	1.663×10^{-1}	3.966×10^{-1}
DLC	6.525×10^{-6}	1.663×10^{-1}	0×10^{0}	2.707×10^{-1}
Data_Field	1.794×10^{-4}	3.966×10^{-1}	2.707×10^{-1}	0×10^{0}

Cutoff criteria: $p < 0.05$ (statistically significant) and the values have been rounded to the four decimal places.

4.2.3. Data Transformation

We have taken this step to achieve more efficient results and to better understand the patterns. Some features are higher than others, leading to wrong performance, though some models may be preferred for larger functional values. We have performed these strategies to re-scale the selected feature values within a range between [0.0, 1.0] without changing the characteristics of original data [52,53]. As shown in the following Equation (3), a technique called *minimum–maximum normalization* has been used to re-scale the selected feature values within the range:

$$N_v = \frac{X - X_{min}}{X_{max} - X_{min}} \quad (3)$$

where N_v is the output normalized values, X is an original value and X_{max}, and X_{min} is the maximum and minimum values of the feature, respectively.

4.3. Training Process

As mentioned in the previous subsection, a Python script combined the two datasets into a single dataset that included both the training and test data. First of all, we have used the *scikit_learn* library's *train_test_split* method to split the combined dataset into the training (80%) and test (20%) datasets. In the raw dataset, the total number of data are 1,270,310. After removing the noisy and inconsistent data, we got a total of 1,257,303 data where the number of training data are 1,005,843 (80%) and the testing data are 251,460 (20%). The training dataset has been used to train the selected models, and the test dataset has been used to further assess the trained classifier. Furthermore, we split again the training data 1,005,843 (80% of the total data) into the new training data 804,674 (80%) for training the selected model and validation data 201,169 (20%) for hyperparameters' optimization. The percentages of 80% for the training dataset and 20% for the test dataset have been chosen as suggested in [54]. To avoid the over-fitting problem, this splitting ratio has been considered as the best ratio between the training and the test dataset [55]. We have used the value of the *random_state* parameter as true, which decided the splitting of dataset into the training and the test dataset randomly [56]. Finally, various performance indicators have used to evaluate the overall performance of the selected models, which have been discussed in the *Results* section. The steps involved to evaluate the performance of all of the selected models are summarized in Figure 8.

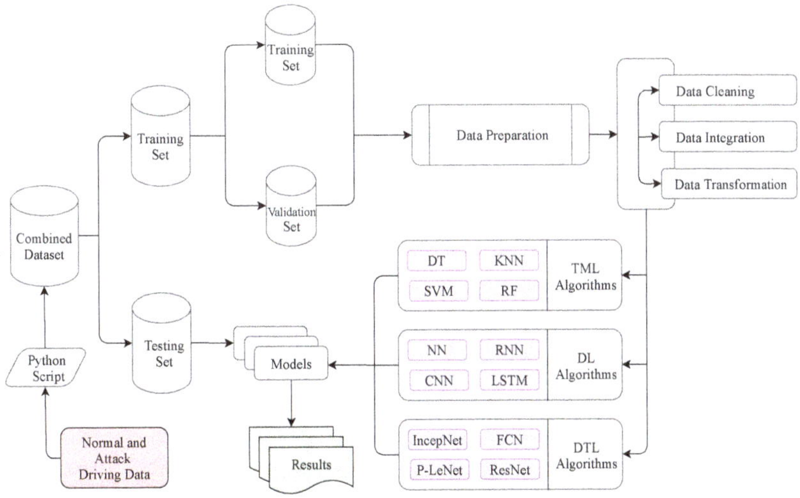

Figure 8. Evaluation process on the datasets with the selected models.

Several supervised TML algorithms have been applied to evaluate their performance for intrusion–detection purposes. The TML algorithms have been chosen based on their extensively used in the security-domain as they have already shown good performance on these scenarios [57]. We have predominantly applied Decision Tree (DT), Random Forest (RF), Support Vector Machine (SVM), and K-Nearest Neighbor (KNN) algorithms for classification analysis. The DT algorithm is the most popular model that is used for the IDS domain, which is showed by the authors in [57,58]. The effectiveness of the RF models in the IDS domain has been shown in a survey conducted by Yang et al. [59]. The SVM algorithm has been considered as it has low computation overheads [60]. Finally, the KNN algorithm has been selected as it achieves good performance in dealing with different sensor data [57].

In recent years, DL algorithms have advanced significantly and some of the variants of DL algorithms have been successfully applied to solve classification tasks related to intrusion detection [61]. Therefore, we have considered simple Neural Network (NN), Recurrent Neural Network (RNN), Convolutional Neural Network (CNN), and Long Short-Term Memory (LSTM) algorithms because of their optimal performance. In the LSTM algorithm, we have used three hidden layers, and the union of hidden layers are 128, 100, and 64, respectively. The *tanh* is the hidden layer activation function and *Adam* is used as an optimizer. In addition, *sigmoid* is used as a network output activation function, and *categorical_crossentropy* is used as a loss function. Furthermore, both NN and CNN algorithms have used the same optimizer and activation function but a different type of hidden layer activation function. Particularly for the RNN model, the *softmax* is used as the network output activation function, and *categorical_crossentropy* is used as the loss function. On the other hand, for the CNN algorithm, the number of hidden layers is four, and *binary_crossentropy* is used as the loss function. The dropout layer has been added after each layer to prevent model overfitting as RNN and LSTM generally have the problem of overfitting [62]. We have evaluated the selected DL models with a wide range of tested hyperparameters. We have obtained the optimal performance when we used these combinations of tested hyperparameters. Table 4 shows the hyper-parameter list of all the selected DL algorithms.

Table 4. The hyper-parameters of the selected DL models.

Parameters	NN	RNN	CNN	LSTM
Number of hidden Layers	2	3	4	3
Units in hidden layers	68, 68	64, 64, 64	32, 64, 256, 128	128, 100, 64
Batch size	64	64	64	16
Hidden layer activation	relu	relu	relu	tanh
Output activation function	sigmoid	softmax	sigmoid	sigmoid
Dropout	N/A	0.1	N/A	0.2
Optimizer	Adam	Adam	Adam	Adam

In addition, we would like to emphasize some DTL models that give better performance than others. We have considered four DTL models and the selected model are Fully Convolutional Networks (FCN), Inception Network (IncepNet), Residual Neural Network (ResNet), and our proposed LeCun Network (LeNet). For all the selected models, the hyper-parameters—batch size, hidden layer activation function, output layer activation function, loss function, and the optimizer are *64*, *ReLu*, *softmax*, *categorical_crossentropy*, and *Adam*, respectively. Furthermore, FCN, IncepNet, and ResNet models have used the same number of hidden layers, but the units in the hidden layer are different. Particularly for the P-LeNet model, the number of the hidden layers is 2, and the units in the hidden layer are 5, 20. We have evaluated the selected DTL models with a wide range of tested hyperparameters. We have obtained the optimal performance when we used these combinations of tested hyperparameters. On the other hand, we have used the Adam optimizer for all the models because it combines the best properties of the *AdaGrad* and *RMSProp* algorithms to provide an optimization algorithm [63]. Furthermore, particularly for this analysis, the *Adam* optimizer has shown the lowest training loss and validation loss among other optimizers. Table 5 shows the hyper-parameters list of all the selected DTL algorithms.

Table 5. The hyper-parameters of the selected DTL models.

Parameters	FCN	IncepNet	ResNet	P-LeNet
Number of hidden Layers	3	3	3	2
Units in hidden layers	128, 256, 128	32, 64, 32	128, 256, 128	5, 20
Batch size	64	64	64	64
Hidden layer activation	relu	linear	relu	relu
Output activation function	softmax	softmax	softmax	softmax
Dropout	N/A	N/A	0.1	N/A
Optimizer	Adam	Adam	Adam	Adam

5. Results

This section discusses the overall performance of the selected models, starting with an analysis of TML metrics and concluding by explaining the effectiveness of the DL and DTL models.

5.1. Experimental Evaluation Indicators

Various measurement indicators (e.g., accuracy, precision, F1-score) are used to illustrate the results where the obtained accuracy shows the overall effectiveness of the proposed model. We evaluated the performance of all the selected models using the following four terms:

- *True-positive (TP)* refers to the number of actual attack instances that are correctly detected as attack.
- *True-negative (TN)* is the number of normal instances that are correctly detected as normal.

- *False-positive (FP)* is the number of normal instances that are incorrectly detected as attack.
- *False-negative (FN)* refers to the number of actual attack instances that are incorrectly detected as normal.

Accuracy is the closeness of the measurements to a specific value, which demonstrates the efficiency of the classifier to determine the total instances. Clearly, a higher accuracy means better classification results. The mathematical expression of accuracy can be defined as follows:

$$Accuracy = \frac{TP + TN}{TP + TN + FP + FN} \tag{4}$$

The fraction of the true positive instances in the positive case determined by the classifier is represented by precision, which demonstrates the closeness of the measurements to each other. Precision can be represented by the following equation:

$$Precision = \frac{TP}{TP + FP} \tag{5}$$

The proportion of relevant instances (positive cases) that are correctly judged to the total positive case is referred to as recall. Recall can be defined as follows:

$$Recall = \frac{TP}{TP + FN} \tag{6}$$

The following F1-score computes the harmonic mean of precision and recall, respectively. F1-score can range from 1.0 to 0.0, with 1.0 indicating perfect precision and recall:

$$F1Score = 2 \times \frac{Precision \times Recall}{Precision + Recall} \tag{7}$$

As one of the significant indicators, the ROC AUC determines areas where the proposed model is classified better within normal and attack scenarios. Measuring ROC AUC requires diagnostic accuracy, which depends on the sensitivity, i.e., true positive rate (TPR), and the specificity, i.e., true negative rate (TNR). As demonstrated by the following equations, TPR, often called recall:

$$Sensitivity(TPR) = \frac{TP}{TP + FN} \tag{8}$$

$$Specificity(TNR) = \frac{TN}{TN + FP} \tag{9}$$

In this analysis, we incorporated a wide range of analyses' scenarios with varying parameters. To this end, we conduct experiments considering different numbers of the hidden layers, units in the hidden layers, numbers of the epoch, a range of hidden layer activation functions, output layer activation functions, loss functions, and different optimizers. These critical parameters significantly affect the calculation of the performance metrics for DL and DTL algorithms. Finally, we show the performance comparisons between TML, DL, and DTL models where the performance of the proposed model has a meaningful impact to indicate the predicted label correctly.

5.2. TML Models Analysis

We start with the traditional machine learning (TML) algorithms because these state-of-the-art algorithms provide the optimal performance and take the least amount of time to run. The performance of the selected TML algorithms has been quantitatively evaluated using the fundamental evaluation indicators. Figure 9 shows the performance metrics of the selected TML algorithms. The DT algorithm shows the optimal performance among all considered TML algorithms with an accuracy score of 0.9532, precision score of 0.9463, recall score of 0.9558, F1-score of 0.9608, and ROC AUC score of 0.8408. As shown in Figure 9, the red quadrangle represents the accuracy score for all of the selected TML algorithms and the best one highlighted in blue color text. On the other hand, the KNN algorithm shows the lowest accuracy score of 0.9248 and a precision score of 0.9141. We used an accuracy plot to determine the proper K value for the KNN algorithm, and the highest accuracy was obtained when K = 12. The RF algorithm achieves the second-best performance with an accuracy score of 0.9448 and a precision score of 0.9448. The SVM algorithm achieves an accuracy score of almost 0.9440 and a precision score of 0.8840 by using C = 1.2, epsilon = 0.1, and *Gaussian RBF Kernel* particularly, which is the third-best performance. However, RF and SVM algorithms have an overall accuracy score of almost 0.9444, but the SVM algorithm has the lowest precision score of 0.8840 when compared to the other selected TML algorithms. The lowest precision score of the SVM algorithm indicates that most of the predicted labels are incorrect. In contrast, the precision score of the RF and DT algorithms is around 0.9455, indicating that the majority of predicted labels are correctly classified. Taking into account all aspects of performance indicators, we conclude that the DT algorithm outperforms the other selected TML algorithms. This optimal performance indicates that most of the predicted labels are correctly classified between normal and attack scenarios.

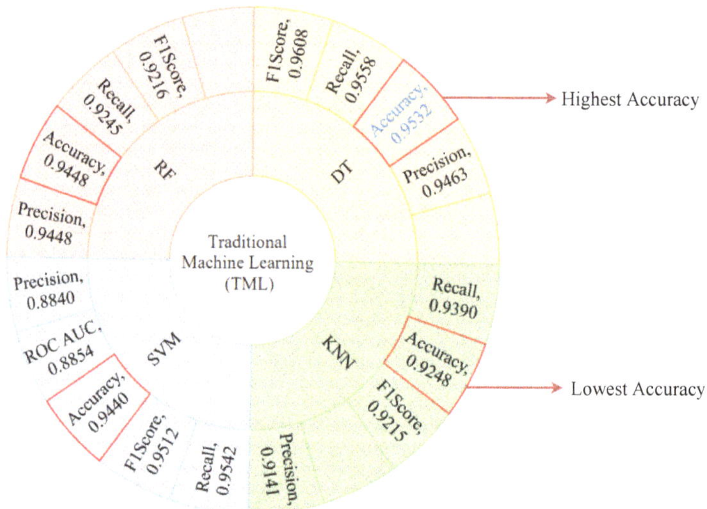

Figure 9. TML algorithms' performance metrics visualization.

5.3. DL Models Analysis

In recent years, DL models have advanced significantly, and several variations of these models have been successfully applied to solve classification tasks related to intrusion detection [61]. Therefore, in this subsection, we consider DL models because of their optimal performance. We have considered LSTM, NN, CNN, and RNN models. The LSTM model shows the highest accuracy score of 0.9762, precision score of 0.9808, recall score of 0.9392, F1-score of 0.8884, and ROC AUC score of 0.9288. Figure 10 shows the performance metrics comparison of the selected DL models. We used three hidden layers for the LSTM model, with units of hidden layers is 128, 100, and 64. The *tanh* is the hidden layer activation function and *Adam* is used as optimizer. In addition, *sigmoid* is used as a network output activation function, and *categorical_crossentropy* is used as a loss function. On the other hand, the NN model has the lowest accuracy score of 0.9563. The performance of the RNN and the CNN models outperform most of the TML models, where these models show an accuracy score of 0.9640 and 0.9590, respectively.

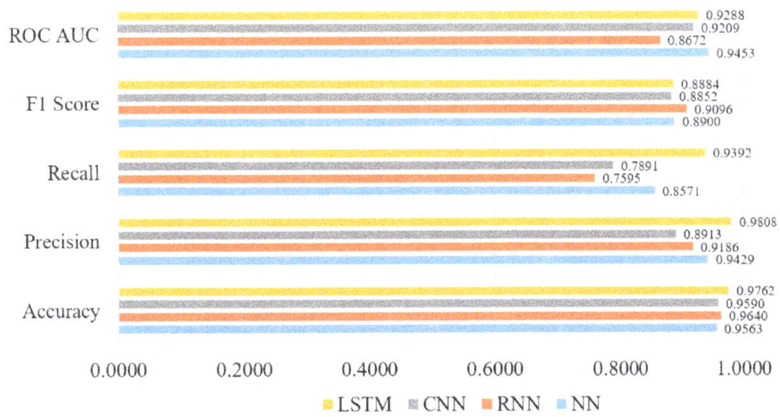

Figure 10. DL models' performance metrics visualization.

Furthermore, both the NN and CNN models used the same optimizer and activation function, but different types of hidden layer activation functions. Particularly for the RNN model, the *softmax* is used as the network output activation function, and *categorical_crossentropy* is used as the loss function. On the other hand, for the CNN model, the number of hidden layers is four, and *binary_crossentropy* is used as the loss function. Finally, taking into account all of the significant parameters of the DL models, the LSTM model demonstrated the best performance, indicating that the majority of the predicted labels are correctly classified.

5.4. DTL Models Analysis

We analyzed the results of TML and DL models to achieve optimal performance. In this subsection, we highlighted four selected DTL models that perform better than others. We considered the number of hidden layers, units in the hidden layers, output layer activation functions, loss functions, and so on, in addition to the fundamental evaluation criteria. First of all, we considered the quantitative performance of DTL models. Table 6 shows the quantitative performance summary of the DTL models, where the proposed *P-LeNet* model shows an optimal performance compared to the other DTL models with an accuracy score of 0.9810, precision score of 0.9814, recall score of 0.9804, F1-score of 0.9783, and ROC AUC score of 0.9542. In this model, we used two hidden-layers where *relu* is the hidden layer activation function. In addition, *softmax* is used as a network output activation function, and *categorical_crossentropy* is used as a loss function along with *adam* optimizer.

Table 6. DTL models' performance comparison metrics.

Algorithm	Accuracy	Precision	Recall	F1-Score	ROC AUC
FCN	0.9786	0.9832	0.9617	0.9488	0.9248
IncepNet	0.9803	0.9152	0.9265	0.9024	0.9129
ResNet	0.9795	0.8958	0.8845	0.9001	0.8703
LeNet	0.9810	0.9814	0.9804	0.9783	0.9542

The values have been rounded to the four decimal places.

Figure 11 shows the accuracy score of every single epoch for both the training and testing phase on the above-mentioned DTL models. We have considered 1000 epochs for our analysis because the flattening characteristics of the curve and the training/testing accuracy are not increasing literally between the epoch number 450 to 1000. The proposed *P-LeNet* model has the highest accuracy score in both the training and testing phases. For a better understanding of our proposed *P-LeNet* model, Figure 12 shows the trend of the accuracy score in both phases. The proposed model's accuracy increases rapidly in epoch number 10 and gradually rises to a point close to 0.9809 at epoch number 400. However, as shown in Figure 12, which remains nearly stable up to the early stopping checkpoint with an accuracy score of 0.9810. In contrast, the FCN model has the lowest accuracy score in the epoch number ranges from 1 to 1000. The FCN model's accuracy begins around 0.9545 for the training phase and 0.9422 for the testing phase in epoch number 46, as shown in Figure 11. However, it rises dramatically around 0.9785 in epoch 554 and 0.9783 in epoch 610 for the training and testing phases, respectively. Furthermore, the accuracy score of 0.9786 remains stable in epochs 555 to 1000 during the training phase. On the other hand, for the testing phase, the accuracy score of 0.9783 remains stable between epoch numbers 611 and 1000.

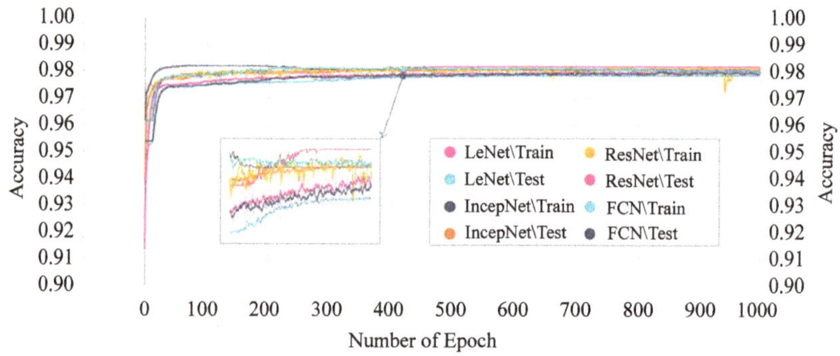

Figure 11. DTL models' training and testing accuracy.

The training and testing accuracy of the IncepNet model remains steady between the epoch number 500 to 1000 as shown in Figure 11. The accuracy of ResNet models starts with a score of 0.9102 for training and 0.9089 for the testing phase. However, as the epoch number increases, this score gradually rises and reaches approximately 0.9745 when the epoch number is 450, and then remains stable between epoch numbers 451 and 1000 for both phases. The remarkable point is that the behavior of the training phase is nearly identical to that of the testing phase. For better understanding, the trend of both phases is shown by zooming in Figure 11.

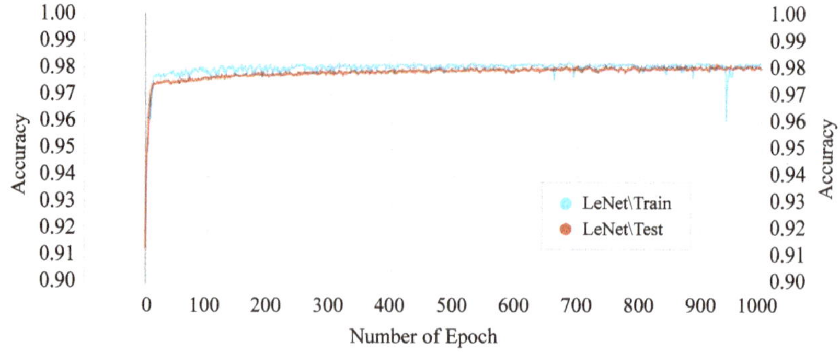

Figure 12. Training and testing accuracy of the proposed P-LeNet model.

Next, we analyzed the losses of each model. The training and testing phases' losses of every single epoch are shown in Figure 13. The FCN model shows the highest loss in both the training and testing phase. The highest loss of this model indicates that the model cannot provide a reliable classification between normal and attack scenarios in the CAN networks.

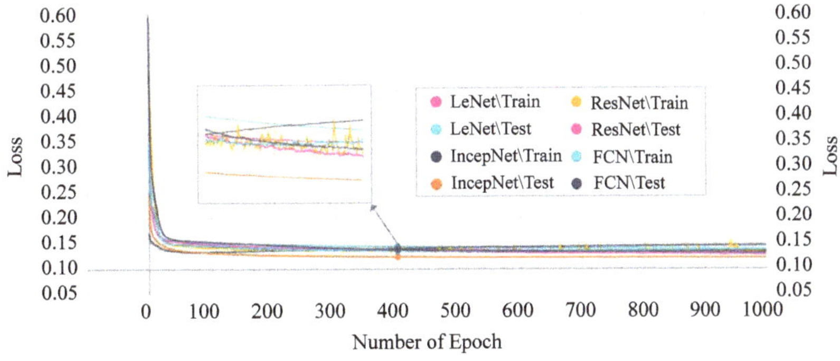

Figure 13. DTL models' training and testing loss.

On the other hand, our proposed *P-LeNet* model shows the lowest loss. In detail, the loss of the proposed model starts around 0.50 for the training phase and 0.35 for the testing phase at the beginning as shown in Figure 14. However, it is decreasing to 0.1689 and 0.1750 for the epoch number 10 for the training and testing phase, respectively. The loss score of 0.1459 remains almost constant between epochs 400 and 1000. Furthermore, the losses of the IncepNet and ResNet models are nearly constant during both the training and testing phases, as shown in Figure 13. The loss of these models is almost 0.5506 at the beginning, which declines gradually to approximately 0.1409 for the epoch number 350 and remains stable for both of the phases. For better understanding, the trend of both phases is shown by zooming in Figure 13.

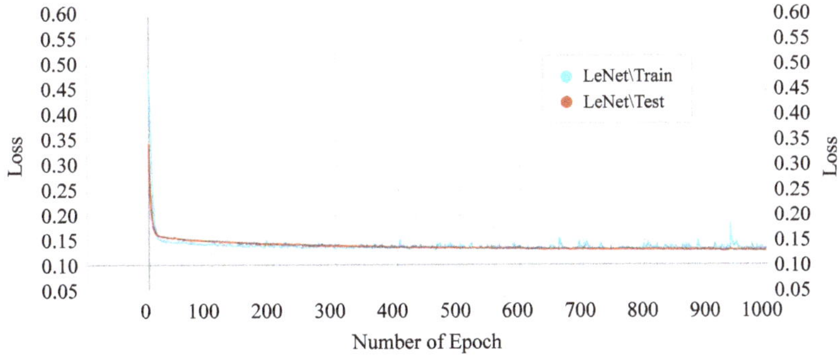

Figure 14. Training and testing loss of the proposed P-LeNet model.

5.5. Performance Comparison

We have considered a variety of strategies for selecting the important features and then applied the chosen algorithms. We have predominantly applied the TML, DL, and DTL approach in the same dataset. The TML algorithms did not show remarkable performance. However, when we have applied the DL algorithms, most of the algorithms perform better than the TML algorithms. Furthermore, when we have applied the DTL algorithms, all the algorithms perform significantly better in most cases. In such cases, accuracy, precision, recall, F1-score, and ROC AUC are better than other experimental scenarios. Moreover, the proposed *P-LeNet* model has adequate stability, low loss, and better classification accuracy than other DTL approaches. Finally, the proposed model can effectively identify and classify the normal and attack scenarios of in-vehicle networks to correctly manage vehicle communications for vehicle security.

6. Conclusions

Automobile manufacturers are working to develop fully autonomous vehicles, which will ensure proper security. In this manuscript, we propose a deep transfer learning-based LeNet model for intrusion detection in electric in-vehicle networks. The proposed detection model has an overall accuracy score of 98.10%. Moreover, the model has precision score 98.14%, recall score 98.04%, F1-score 97.83%, and ROC AUC score by 95.42%, which is a noticeable improvement when compared to the other benchmark ML, DL, and DTL models. These experimental results demonstrated that the proposed *P-LeNet* model efficiently classifies the normality and abnormality and allows the immediate detection of anomalies in the CAN networks. To summarize, it is obvious that the model has proven its potential to efficiently exhibit anomalous data identification to protect the CAN network that can also be extended in other emerging applications within critical infrastructures where automation and secure data processing is the main challenge. In the future, we will try to implement this proposed deep learning model based on decentralized devices or servers. We will also concentrate on improving the performance of the proposed model by optimizing the hyper-parameters.

Author Contributions: Conceptualization, S.T.M. and A.A.; methodology, S.T.M. and A.A.; software, S.T.M., K.A. and Z.R.; validation, S.T.M., A.A., Z.R. and K.A.; formal analysis, S.T.M. and A.A. and Z.R.; investigation, S.T.M., A.A. and K.A.; resources, S.T.M., Z.R. and A.A.; data curation, S.T.M.; writing—original draft preparation, S.T.M.; writing—review and editing, A.A., Z.R. and K.A.; supervision, A.A., Z.R. and K.A.; project administration, A.A.; funding acquisition, A.A. All authors have read and agreed to the published version of the manuscript.

Funding: This research received no external funding.

Institutional Review Board Statement: Not applicable.

Informed Consent Statement: Not applicable.

Data Availability Statement: The data presented in this manuscript are available on request from the corresponding author.

Conflicts of Interest: The authors declare no conflict of interest.

Abbreviations

The following abbreviations are used in this manuscript:

CAN	Controller Area Network
ECU	Electronic Control Unit
IVN	In-Vehicle Network
CAN FD	CAN Flexible Data-Rate
LIN	Local Interconnect Network
MOST	Media Oriented Systems Transport
ML	Machine Learning
IDS	Intrusion Detection System
TML	Traditional Machine Learning
DL	Deep Learning
DTL	Deep Transfer Learning
DT	Decision Tree
RF	Random Forest
SVM	Support Vector Machine
KNN	K-Nearest Neighbor
NN	Neural Network
RNN	Recurrent Neural Network

CNN	Convolutional Neural Network
LSTM	Long Short-Term Memory
FCN	Fully Convolutional Networks
IncepNet	Inception Network
ResNet	Residual Neural Network
LeNet	LeCun Network
ReLu	Rectified Linear-Unit
DLC	Data Length Code
TNR	True Negative Rate
TPR	True Positive Rate
SOF	Start of Frame
RTR	Remote Transmission Request
IDE	Identifier Extension
CRC	Cyclical Redundancy Check
ACK	Acknowledgment
EOF	End of Frame
IFS	Inter Frame Space
DEL	Delimiter
ID	Identifier
OTA	Over the Air
DNN	Deep Neural Network
DBN	Deep Belief network
DoS	Denial of Service
ARP	Address Resolution Protocol
RPM	Radiation Portal Monitors
GAN	Generative Adversarial Network
DCAE	Deep Contractive Auto Encoder
DCNN	Deep Convolutional Neural Network
IoV	Internet of Vehicle
MMN	Minimum Maximum Normalization
MMD	Maximum Mean Discrepancy
HEX2DEC	Hexadecimal to Decimal
R-MPFR	Multiple Precision Floating-Point Reliable
ROC-AUC	Receiver Operating Characteristic-Area Under the Curve

References

1. Han, M.L.; IlKwak, B.; Kim, H.K. Anomaly intrusion detection method for vehicular networks based on survival analysis. *Veh. Commun.* **2018**, *14*, 52–63. [CrossRef]
2. Steger, M.; Boano, C.A.; Niedermayr, T.; Karner, M.; Hillebrand, J.; Roemer, K.; Rom, W.A. Efficient and Secure Automotive Wireless Software Update Framework. *IEEE Trans. Ind. Inform.* **2018**, *14*, 2181–2193. [CrossRef]
3. Zeng, W.; Mohammed, A.S.; Khalid, M.A.S.; Chowdhury, S. In-Vehicle Networks Outlook: Achievements and Challenges. *IEEE Commun. Surv. Tutor.* **2016**, *18*, 1552–1571. [CrossRef]
4. Mathur, R.; Saraswat, R.; Mathur, G. An Analytical Study of Communication Protocols Used in Automotive Industry. *Int. J. Eng. Res. Technol. (IJERT)* **2018**, *2*, 287–292.
5. Kieu, T.; Yang, B.; Jensen, C.S. Outlier Detection for Multidimensional Time Series Using Deep Neural Networks. In Proceedings of the 2018 19th IEEE International Conference on Mobile Data Management (MDM), Aalborg, Denmark, 25–28 June 2018; pp. 125–134. [CrossRef]
6. Sommer, C.; Hoefler, R.; Samwer, M.; Gerlich, D.W. A deep learning and novelty detection framework for rapid phenotyping in high-content screening. *Mol. Biol. Cell* **2017**, *28*, 3428–3436. [CrossRef] [PubMed]
7. Sharma, S.; Krishna, C.R.; Sahay, S.K. Detection of Advanced Malware by Machine Learning Techniques. In *Soft Computing: Theories and Applications*; Springer: Singapore, 2019; pp. 333–342.
8. Avatefipour, O.; Malik, H. State-of-the-Art Survey on In-Vehicle Network Communication "CAN-Bus" Security and Vulnerabilities. *arXiv* **2018**, arxiv:1802.01725.
9. Robert Bosch GmbH CAN Specification. *Postfach* **1991**, *2*, 1–72.
10. Bozdal, M.; Samie, M.; Aslam, S.; Jennions, I. Evaluation of CAN Bus Security Challenges. *Sensors* **2020**, *20*, 2364. [CrossRef]
11. Khatri, N.; Shrestha, R.; Nam, S.Y. Security Issues with In-Vehicle Networks, and Enhanced Countermeasures Based on Blockchain. *Electronics* **2021**, *10*, 893. [CrossRef]
12. Lee, H.; Jeong, S.H.; Kim, H.K. OTIDS: A Novel Intrusion Detection System for In-vehicle Network by Using Remote Frame. In Proceedings of the 2017 15th Annual Conference on Privacy, Security and Trust (PST), Calgary, AB, Canada, 28–30 August 2017.

13. Avatefipour, O.; Hafeez, A.; Tayyab, M.; Malik, H. Linking Received Packet to the Transmitter Through Physical-Fingerprinting of Controller Area Network. In Proceedings of the IEEE Workshop on Information Forensics and Security (WIFS), Rennes, France, 4–7 December 2017; pp. 1–6. [CrossRef]
14. Yuan, Y.; Yuan, H.; Guo, L.; Yang, H.; Sun, S. Resilient Control of Networked Control System Under DoS Attacks: A Unified Game Approach. *IEEE Trans. Ind. Inform.* **2016**, *12*, 1786–1794. [CrossRef]
15. Markovitz, M.; Wool, A. Field classification, modeling and anomaly detection in unknown CAN bus networks. *Veh. Commun.* **2017**, *9*, 43–52. [CrossRef]
16. Theissler, A.; Ulmer, D.; Dear, I. Interactive knowledge discovery in recordings from vehicle tests. In Proceedings of the FISITA World Automotive Congress, Budapest, Hungary, 30 May–4 June 2010; pp. 1–10.
17. Chandola, V.; Banerjee, A.; Kumar, V. Anomaly detection: A survey. *ACM Comput. Surv.* **2009**, *41*, 15. [CrossRef]
18. Alshammari, A.; Zohdy, M.A.; Debnath, D.; Corser, G. Classification Approach for Intrusion Detection in Vehicle Systems. *Wirel. Eng. Technol.* **2018**, *9*, 79–94. [CrossRef]
19. Wang, L.; Jones, R. Big Data Analytics for Network Intrusion Detection: A Survey. *Int. J. Netw. Commun.* **2017**, *7*, 24–31. [CrossRef]
20. Nazakat, I.; Khurshid, K. Intrusion Detection System for In-Vehicular Communication. In Proceedings of the 2019 15th International Conference on Emerging Technologies (ICET), Peshawar, Pakistan, 2–3 December 2019; pp. 1–6. [CrossRef]
21. Bozdal, M.; Samie, M.; Aslam, S.; Jennions, I. A Survey on CAN Bus Protocol: Attacks, Challenges, and Potential Solutions. In Proceedings of the 2018 International Conference on Computing, Electronics Communications Engineering (iCCECE), Southend, UK, 16–17 August 2018; pp. 201–205. [CrossRef]
22. Lokman, S.F.; Othman, A.T.; Abu-Bakar, M.H. Intrusion detection system for automotive Controller Area Network (CAN) bus system: A review. *EURASIP J. Wirel. Commun. Netw.* **2019**, *7*, 24–31. [CrossRef]
23. Kang, M. J.; Kang, J. W. Intrusion Detection System Using Deep Neural Network for In-Vehicle Network Security. *PLoS ONE* **2016**, *11*, e0155781. [CrossRef]
24. Loukas, G.; Vuong, T.; Heartfield, R.; Sakellari, G.; Yoon, Y.; Gan, D. Cloud-Based Cyber-Physical Intrusion Detection for Vehicles Using Deep Learning. *IEEE Access* **2017**, *6*, 3491–3508. [CrossRef]
25. Seo, E.; Song, H.M.; Kim, H.K. GIDS: GAN based Intrusion Detection System for In-Vehicle Network. In Proceedings of the IEEE Access 2018 16th Annual Conference on Privacy, Security and Trust (PST), Belfast, Ireland, 28–30 August 2018; pp. 1–6. [CrossRef]
26. Lokman S.F.; Othman A.T.; Musa S.; Abu-Bakar M.H. Deep Contractive Autoencoder-Based Anomaly Detection for In-Vehicle Controller Area Network (CAN). *Prog. Eng. Technol. Adv. Struct. Mater.* **2019**, *119*._16. [CrossRef]
27. Zhang, J.; Li, F.; Zhang, H.; Li, R.; Li, Y. Intrusion detection system using deep learning for in-vehicle security. *Ad Hoc Netw.* **2019**, *95*. [CrossRef]
28. Zhu, K.; Chen, Z.; Peng, Y.; Zhang, L. Mobile Edge Assisted Literal Multi-Dimensional Anomaly Detection of In-Vehicle Network Using LSTM. *IEEE Trans. Veh. Technol.* **2019**, *68*, 4275–4284. [CrossRef]
29. Avatefipour, O.; Al-Sumaiti, A.S.; El-Sherbeeny, A.M.; Awwad, E.M.; Elmeligy, M.A.; Mohamed, M.A.; Malik, H. An Intelligent Secured Framework for Cyberattack Detection in Electric Vehicles' CAN Bus Using Machine Learning. *IEEE Access* **2019**, *7*, 127580–127592. [CrossRef]
30. Xiao, J.; Wu, H.; Li, X. Internet of Things Meets Vehicles: Sheltering In-Vehicle Network through Lightweight Machine Learning. *Symmetry* **2019**, *11*, 1388. [CrossRef]
31. Al-Saud, M.; Eltamaly, A.M.; Mohamed, M.A.; Kavousi-Fard, A. An Intelligent Data-Driven Model to Secure Intravehicle Communications Based on Machine Learning. *IEEE Trans. Ind. Electron.* **2020**, *67*, 5112–5119. [CrossRef]
32. Lin, Y.; Chen, C.; Xiao, F.; Avatefipour, O.; Alsubhi, K.; Yunianta, A. An Evolutionary Deep Learning Anomaly Detection Framework for In-Vehicle Networks—CAN Bus. *IEEE Trans. Ind. Appl.* **2020**. [CrossRef]
33. Yang, Y.; Duan, Z.; Tehranipoor, M. Identify a Spoofing Attack on an In-Vehicle CAN Bus Based on the Deep Features of an ECU Fingerprint Signal. *Smart Cities* **2020**, *3*, 2. [CrossRef]
34. Hossain, M.D.; Inoue, H.; Ochiai, H.; Fall, D.; Kadobayashi, Y. LSTM-Based Intrusion Detection System for In-Vehicle Can Bus Communications. *IEEE Access* **2020**, *8*, 185489–185502. [CrossRef]
35. Song, H.M.; Woo, J.; Kim, H.K. In-vehicle network intrusion detection using deep convolutional neural network. *Veh. Commun.* **2020**, *21*. [CrossRef]
36. Li, X.; Hu, Z.; Xu, M.; Wang, Y.; Ma, J. Transfer learning based intrusion detection scheme for Internet of vehicles. *Inf. Sci.* **2021**, *547*, 119–135. [CrossRef]
37. Lu, J.; Behbood, V.; Hao, P.; Zuo, H.; Xue, S.; Zhang, G. Transfer Learning using Computational Intelligence: A Survey. *Knowl.-Based Syst.* **2015**, *80*, 14–23. [CrossRef]
38. Zadrozny, B. Learning and Evaluating Classiers under Sample Selection Bias. In Proceedings of the Twenty-First International Conference on Machine Learning, Banff, AB, Canada, 4–8 July 2004. [CrossRef]
39. Dai, W.; Yang, Q.; Xue, G.; Yu. Y. Boosting for transfer learning. In Proceedings of the 24th international conference on Machine Learning (ICML'07), New York, NY, USA, 20–24 June 2007; pp. 193–200. [CrossRef]
40. Raina, R.; Andrew, Y.N.; Koller, D. Constructing Informative Priors using Transfer Learning. In Proceedings of the 23th International Conference on Machine Learning (ICML'06), Pittsburgh, PA, USA, 25–29 June 2006; Association for Computing Machinery: New York, NY, USA, 2006; pp. 713–720. [CrossRef]

41. Gou, S.; Wang, Y.; Jiao, L.; Feng, J.; Yao, Y. Distributed Transfer Network Learning Based Intrusion Detection. In Proceedings of the IEEE International Symposium on Parallel and Distributed Processing with Applications, Chengdu, China, 10–12 August 2009; pp. 511–515. [CrossRef]
42. Xu, Y.; Liu, Z.; Li, Y.; Zheng, Y.; Hou, H.; Gao, M.; Song, Y.; Xin, Y. Intrusion Detection Based on Fusing Deep Neural Networks and Transfer Learning. In Proceedings of the Digital TV and Wireless Multimedia Communication (IFTC 2019), Shanghai, China, 19–20 September 2009; Volume 1181. [CrossRef]
43. Kang, H.; Kwak, B.; Lee, Y.H.; Lee, H.; Lee, H.; Kim, H.K. Car Hacking: Attack and Defense Challenge 2020 Dataset. *IEEE Dataport* **2021**. [CrossRef]
44. Li, F.; Shirahama, K.; Nisar, M.A.; Huang, X.; Grzegorzek, M. Deep Transfer Learning for Time Series Data Based on Sensor Modality Classification. *Sensors* **2020**, *20*, 4271. [CrossRef] [PubMed]
45. Fawaz, H.I.; Forestier, G.; Weber, J.; Idoumghar, L.; Muller, P. A. Transfer learning for time series classification. In Proceedings of the IEEE International Conference on Big Data, Seattle, WA, USA, 10–13 December 2018. [CrossRef]
46. Kimura, N.; Yoshinaga, I.; Sekijima, K.; Azechi, I.; Baba, D. Convolutional Neural Network Coupled with a Transfer-Learning Approach for Time-Series Flood Predictions. *Water* **2020**, *12*, 96. [CrossRef]
47. Engine Structure Image from Pngtree. Available online: https://pngtree.com/so/engine-structure (accessed on 1 June 2021).
48. Borgwardt, K.M.; Gretton, A.; Rasch, M.J.; Kriegel, H.P.; Schölkopf, B.; Smola, A.J. Integrating structured biological data by kernel maximum mean discrepancy. *Bioinformatics* **2006**, *22*, 49–57. [CrossRef]
49. Ioffe, S.; Szegedy, C. Batch Normalization: Accelerating Deep Network Training by Reducing Internal Covariate Shift. In Proceedings of the International Conference on Machine Learning (PMLR), Lille, France, 6–11 July 2015; Volume 37, pp. 448–456.
50. Van Aswegen, A.S.; Engelbrecht, A.S. The relationship between transformational leadership, integrity and an ethical climate in organisations. *SA J. Hum. Resour. Manag.* **2009**, *7*, 9. [CrossRef]
51. Chee, J.D. Pearson's Product-Moment Correlation: Sample Analysis; Available online: https://www.academia.edu/21484289/Pearsons_Product-Moment_Correlation_Sample_Analysis (accessed on 12 November 2020)
52. Aminanto, M.E.; Choi, R.; Tanuwidjaja, H.C.; Yoo, P.D.; Kim, K. Deep Abstraction and Weighted Feature Selection for Wi-Fi Impersonation Detection. *IEEE Trans. Inf. Forensics Secur.* **2018**, *12*, 621–636. [CrossRef]
53. Geller, S. Normalization vs. Standardization Quantitative Analysis. Available online: https://towardsdatascience.com/normalization-vs-standardization-quantitative-analysis-a91e8a79cebf (accessed on 9 November 2020).
54. Geron, A. *Hands-on Machine Learning with Scikit-Learn, Keras, and TensorFlow*; O'Reilly Media: Sebastopol, California, USA 2018.
55. Guyon, I. A scaling law for the Validation-set training-set valid ratio. *AT T Bell Lab.* **1997**, *1*, 1–11.
56. Szajna, A.; Kostrzewski, M.; Ciebiera, K.; Stryjski, R.; Wozniak, W. Application of the Deep CNN-Based Method in Industrial System for Wire Marking Identification. *Energies* **2021**, *14*, 3659. [CrossRef]
57. Wu, X.; Kumar, V.J.; Quinlan, J.R.; Ghosh, J.; Yang, Q.; Motoda, H.; McLachlan, G.J.; Ng, A.; Liu, B.; Yu, P.S.; et al. Top 10 algorithms in data mining. *Knowl. Inf. Syst.* **2008**, *14*, 1–37. [CrossRef]
58. Buczak, A.L.; Guven, E. A Survey of Data Mining and Machine Learning Methods for Cyber Security Intrusion Detection. *IEEE Commun. Surv. Tutor.* **2016**, *18*, 1153–1176. [CrossRef]
59. Yang, Y.; Zheng, K.; Wu, C.; Niu, X.; Yang, Y. Building an Effective Intrusion Detection System Using the Modified Density Peak Clustering Algorithm and Deep Belief Networks. *Appl. Sci.* **2019**, *9*, 238. [CrossRef]
60. Subba, B.; Biswas, S. Intrusion Detection Systems using Linear Discriminant Analysis and Logistic Regression. In Proceedings of the Annual IEEE India Conference (INDICON), New Delhi, India, 17–20 December 2015; pp. 1–6. [CrossRef]
61. Nisioti, A.; Mylonas, A.; Yoo, P.D.; Katos, V. From Intrusion Detection to Attacker Attribution: A Comprehensive Survey of Unsupervised Methods. *IEEE Commun. Surv. Tutor.* **2018**, *20*, 3369–3388. [CrossRef]
62. Goodfellow, I.; Bengio, Y.; Courville, A. *Deep Learning*; MIT Press: Cambridge, MA, USA, 2016.
63. Brownlee, J. Gentle Introduction to the Adam Optimization Algorithm for Deep Learning. Available online: https://machinelearningmastery.com/adam-optimization-algorithm-for-deep-learning (accessed on 9 December 2020).

sensors

MDPI

Article

Service-Centric Heterogeneous Vehicular Network Modeling for Connected Traffic Environments

Ahmad M. Khasawneh [1], Mamoun Abu Helou [2], Aanchal Khatri [3], Geetika Aggarwal [4], Omprakash Kaiwartya [4,*], Maryam Altalhi [5], Waheeb Abu-ulbeh [2] and Rabah AlShboul [6]

1 Department of Cybersecurity, Amman Arab University, Amman 11953, Jordan; a.khasawneh@aau.edu.jo
2 Faculty of Information Technology, Al Istiqlal University, Jericho 4728, Palestine; mabuhelou@pass.ps (M.A.H.); w.abuulbeh@pass.ps (W.A.-u.)
3 Department of Computer Science, Sat Jinda Kalyana PG College, Kalanaur 124113, Haryana, India; aancha84_scs@jnu.ac.in
4 School of Science and Technology, Nottingham Trent University, Nottingham NG11 8NS, UK; geetika.aggarwal@ntu.ac.uk
5 Department of Management Information System, College of Business Administration, Taif University, P.O. BOX 11.99, Taif 21944, Saudi Arabia; marem.m@tu.edu.sa
6 Computer Science Department, Faculty of Information Technology, Al al-Bayt University, Mafraq 25113, Jordan; rabahshboul@aabu.edu.jo
* Correspondence: omprakash.kaiwartya@ntu.ac.uk

Abstract: Heterogeneous vehicular communication on the Internet of connected vehicle (IoV) environment is an emerging research theme toward achieving smart transportation. It is an evolution of the existing vehicular ad hoc network architecture due to the increasingly heterogeneous nature of the various existing networks in road traffic environments that need to be integrated. The existing literature on vehicular communication is lacking in the area of network optimization for heterogeneous network environments. In this context, this paper proposes a heterogeneous network model for IoV and service-oriented network optimization. The network model focuses on three key networking entities: vehicular cloud, heterogeneous communication, and smart use cases as clients. Most traffic-related data–oriented computations are performed at cloud servers for making intelligent decisions. The connection component enables handoff-centric network communication in heterogeneous vehicular environments. The use-case-oriented smart traffic services are implemented as clients for the network model. The model is tested for various service-oriented metrics in heterogeneous vehicular communication environments with the aim of affirming several service benefits. Future challenges and issues in heterogeneous IoV environments are also highlighted.

Keywords: heterogeneous vehicular communication; Internet of connected vehicles; vehicular ad hoc networks; heterogeneous networking; Internet of Things

Citation: Khasawneh, A.M.; Helou, M.A.; Khatri, A.; Aggarwal, G.; Kaiwartya, O.; Altalhi, M.; Abu-ulbeh, W.; AlShboul, R. Service-Centric Heterogeneous Vehicular Network Modeling for Connected Traffic Environments. *Sensors* **2022**, *22*, 1247. https://doi.org/10.3390/s22031247

Academic Editor: Juan A. Cabrera

Received: 29 December 2021
Accepted: 4 February 2022
Published: 7 February 2022

1. Introduction

A universal network architecture is being envisioned considering the significant growth in sensor-enabled digital things in our daily life such as smartphones in our hands, vehicles on roads, entertainment devices in homes, and computing systems in offices [1]. This global network architecture leverages most of the existing networks. It is adopted as the Internet of things (IoT) in academic and industrial research communities. Interoperability is the key feature for achieving seamless integration of heterogeneous networks by utilizing intelligent interfaces [2]. The Internet of connected vehicles (IoV) is a heterogeneous network that has evolved from the existing ad hoc network–oriented vehicular communication. It integrates different vehicular networks in road traffic environments (i.e., vehicle-to-vehicle (V2V), vehicle-to–roadside unit (V2R), vehicle-to–in vehicle sensors (V2S), vehicle-to–mobile infrastructure (V2I), and vehicle-to–personal device (V2P)-enabled vehicular networks) [3,4]. The ad hoc network-oriented conventional vehicular network aimed to enhance traffic safety and efficiency via real-time communication between on-road vehicles utilizing roadside units. Various standards and protocols have been developed to enable ad hoc vehicular networks, including wireless access in vehicular environments (WAVE) and dedicated short-range communication (DSRC) [5].

Ad hoc network-enabled vehicular communication technology lacks commercial interest toward implementation, despite the lower operational cost–driven networks for traffic safety and efficiency services [6]. This is due to the ad hoc vehicular networks' inability to operate compatibly with existing heterogeneous network technologies [7]. Specifically, the issues include pure ad hoc communication architecture, lack of standards for personal devices, intermittent Internet service, and cooperative dependency for network operations. The pure ad hoc network architecture cannot support service-oriented commercial applications [8]. The intelligent decisions based on the enormous amounts of traffic data are far from reality due to the unavailability of cloud support in intermittent Internet service [9]. The growing number of personal devices are dead-ends for vehicles. The compatibility is still a serious challenge considering the heterogeneous personal devices prevalent in traffic environments [10]. Due to cooperative information processing, time-constrained traffic information delivery is far from reality [11]. Moreover, the ever growing connected world era has significantly affected ad hoc vehicular communication in terms of the futuristic, connected-vehicle framework. A vehicle would always remain connected to the Internet via smart-handover-enabled heterogeneous reachable networks.

This paper proposes a heterogeneous network model for enabling the IoV framework. A practice-oriented modeling approach is followed to design and develop the framework. It has significant potential to enable the connected-vehicle paradigm and to spur commercial interest in vehicular communication. Specifically, we aim to answer the following questions:

- What are the key technical components involved in realizing a heterogeneous vehicular network model for the IoV?
- How to realize vehicular cloud-oriented data processing in vehicular environments enabling big traffic data computation for making intelligent traffic decisions?
- How to perform heterogeneous connection management and prioritization in dynamic vehicular traffic environments?
- Is the provisioned heterogeneous vehicular network model for the IoV efficient and scalable considering the growing network heterogeneousness, vehicle speed, and density?

The rest of the paper is articulated as follows: In Section 2, a critical review of the related literature is carried out. Section 3 presents the details of the proposed heterogeneous network model. Section 4 discusses the service-oriented performance evaluation of the network model, followed by the conclusions presented in Section 5.

2. Related Work

Research and development on the connected-vehicle-traffic environment are gaining momentum in the past few years due to the growing government-level support in this the area in most developed countries, particularly in the UK, US, and EU countries [12]. An integrated vehicular network name, i.e., Space–Air–Ground (SAGiven) has been suggested, focusing on heterogeneous network function and network resource identification [13]. It has developed a vehicular communication framework considering mobile-network-connected on-road ground vehicles, unmanned aerial vehicles (UAVs), and satellites-based space vehicles. The framework uses a case-based study rather than a scientific novelty as no new technologies or concepts have developed; instead, existing techniques have been utilized. Another UAV-enabled connected-vehicle framework was investigated, focusing on 6G communication-centric services [14]. A UAV-centric task-offloading technique was developed for edge devices in a vehicular network environment considering the high computing capacity in the 6G communication environment. The edge devices that communicate with a particular UAV were identified as an edge network group. However, the focus of this study was on reducing the energy consumption of edge devices in a vehicular network. The issue of energy consumption is not a potential issue in the vehicular network considering vehicles' battery capacity. A similar UAV-based content distribution vehicular network has been suggested considering 5G-centric IoT services [15]. Initially, an integrated network architecture was developed to optimize the quality of experience (QoE) for vehicle drivers. The integrated network involves a UAV network and a vehicular network for traffic-related content distribution. These UAV-based vehicular network integration frameworks have considered only mobile network integration without focusing on other personal networks' integration with vehicular networks.

The other dimension of research on enabling the Internet of connected vehicle environment is improving the performance of heterogeneous network architecture using innovative techniques [16,17]. A cooperative driving framework has been suggested for the Internet of connected vehicle environment, focusing on velocity prediction of neighboring-vehicle-centric motion planning for path following a driving scenario [18]. The driving data of nearby vehicles were utilized in a neural-network-based framework for generating a safe travel pattern considering the predicted velocity error of all the neighboring drivers. It was validated for lane-changing scenarios in the connected vehicle environment. However, the framework relies on the precision and accuracy of the driving data of neighboring vehicles.

Similarly, another cooperative driving control framework was investigated for the Internet of connected vehicle environment, focusing on collision avoidance at merging roads [19]. The merging road area was divided into three subregions, including delay estimation region, control region, and merging region, for precisely calculating communication delay considering the dynamic mobility data of approaching vehicles and other traffic data in the region. The study assumed roadside infrastructure-based communication, which has practically difficult deployment limitations near all the merging roads. Another cooperative collision avoidance framework has been explored considering trajectory prediction and mobility uncertainty in connected vehicle environments [20]. An edge and cloud server–based vehicle-to-roadside unit reliable communication architecture was considered for improving traffic-data-centric knowledge on mobility uncertainty. The cooperative collision avoidance framework relies on roadside-infrastructure-based communication rather than a connected-vehicle networking environment. The aforementioned studies in the connected-vehicle environment majorly focused on network performance improvement rather than on network prioritization in the heterogeneous vehicular network environment, which is the scope and focus of this paper.

3. Internet of Connected Vehicles

3.1. Heterogeneous Vehicular Networks

The IoV is a global vehicular network leveraging the Internet and various vehicular networks in traffic environments. The proposed heterogeneous vehicular network architecture leverages different kinds of vehicle-oriented networking, including V2V, V2R, vehicle-to–personal devices (V2P), vehicle-to–mobile infrastructure (V2I), and vehicle-to-sensors (V2S) for on-road traffic environments. These vehicular communications are different due to their enabling wireless access technologies. The intervehicle ad hoc type of communication, including V2V and V2R, is supported by WAVE. The long-range V2I web communication is enabled by Wi-Fi and 4G/LTE technologies. The in-vehicle V2P and V2S communications utilize Car-Play and Wi-Fi, respectively. The range of technologies and devices increases the design complexity of the architecture. It is complemented as a market-oriented vehicular communication technology.

The global vehicular network framework has enormous potential not only to guide (with respect to traffic safety and efficiency-related cooperative information sharing among on road vehicles) vehicles but also to supervise (with respect to vehicle-safety-related dedicated information delivery between an intelligent cloud server and vehicles or drivers of vehicles). The abundant traffic applications related to mobile Internet and multimedia services are considered deployable on the heterogeneous vehicular framework using publish–subscribe based architecture. A realistic framework is illustrated in Figure 1, focusing on three advanced traffic information processing scenarios. Firstly, IoV enables the verification of traffic safety information via network coordination. The verified safety information is published over local networks by authorities on global networks. Secondly, the efficiency information available over IoV is near-optimal information, considering the utilization of global traffic scenarios of more significant geographical regions. Thirdly, IoV-based utility information is intelligent due to considering the cloud-oriented market of big-data processing by third-party utility information service providers.

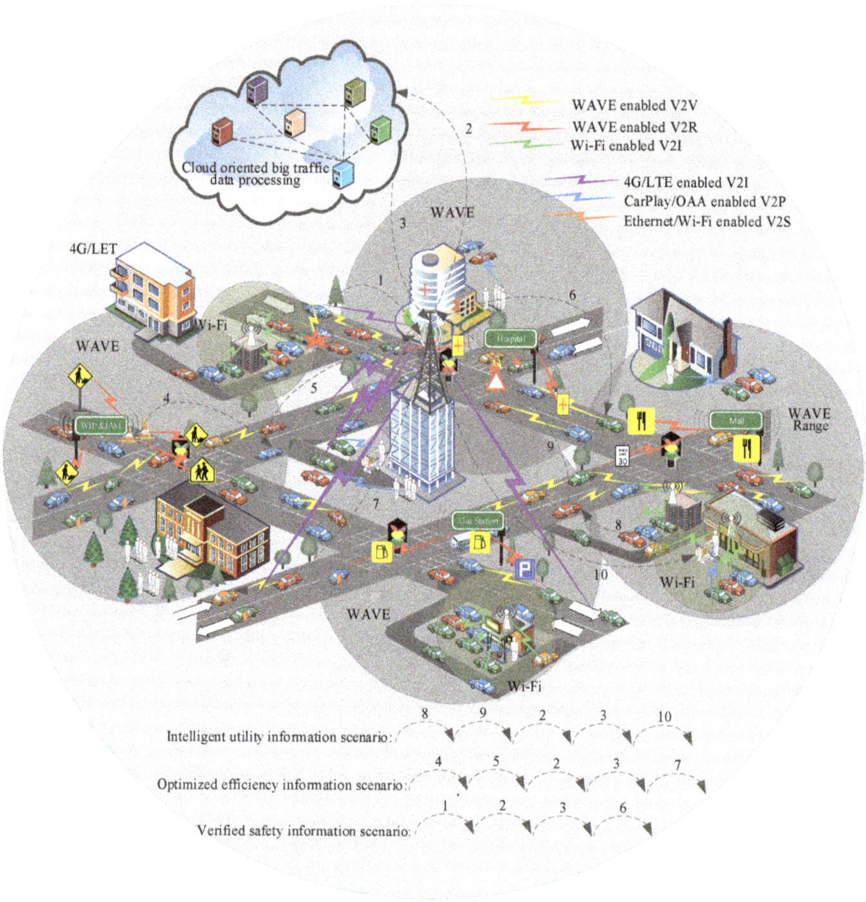

Figure 1. The realization of the IoV scenario with heterogeneous vehicular networks.

3.2. Network Model

The network model of IoV is based on cloud-oriented big traffic data computing and heterogeneous-communication-oriented intelligence. The proposed model considers the concept of graph partitioning in order to achieve quality of service (QoS) flow allocation and prioritization for multitenants. Here, QoS means traffic-service-centric resource allocation in the integrated vehicular network environment. Dijkstra's and Kruskal's algorithms were used to model the procedure of multitenancy QoS path computation. The concept of multitenancy allows sharing of resources and applications. However, its drawback is that some dominating tenants can monopolize the resources, which results in system performance degradation. Therefore, the concept of a software defined network (SDN) was used to overcome this problem. In general, SDN is a network virtualization concept for enabling specific service-centric networks. Here, SDN was used to control the tenants and the network usage in vehicular network integration. The network management layer performed network virtualization, which resulted in the separation of different tenants' flows to increase isolation among tenants. After that, different flows were controlled by SDN dynamically. The SDN handled the virtual network layers and stored subnets of the physical network. SDN based system overview is presented in Figure 2.

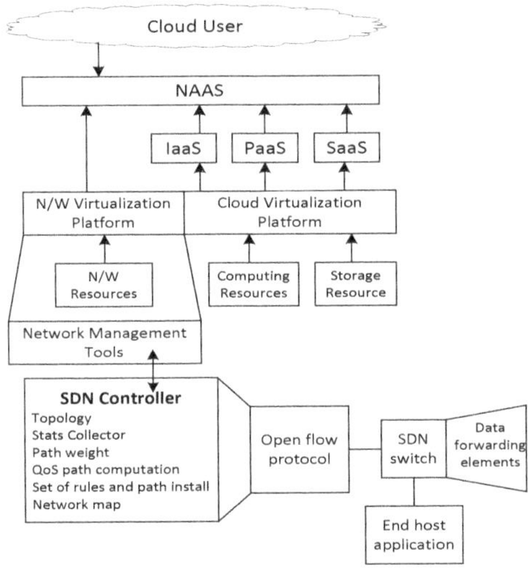

Figure 2. System architecture.

The network controller requested a management tool to assign the best-suited route whenever new flow and resources arrived in the network. The network management tools allocated the best resources to newly arrived flow based on the current status of the network. The concept of graph partitioning was used to achieve QoS flow allocation and prioritization for multitenants. The symbols used in this paper are summarized in Table 1.

Table 1. Symbol description.

Symbol	Description
G	Vehicular network connectivity graph
V	Set of vehicular nodes as vertices of the graph
E	Set of vehicular communication links as edges of the graph
F	Set of vehicular communication flows in the network graph
SP	Shortest communication paths between vehicular nodes
S	Number of segments in a particular path l
P	Number of subpaths in a particular path l
W_l	Weight of a path l used for vehicular path selection
L^{UR}	Link utilization ratio of a vehicular network
λ_l	Link load of shared link in a particular path l
C_l	Link capacity of shared link in aparticular path l

The building blocks of the network model of the IoV include the cloud, the connection, and the clients as key network components (Figure 3). The cloud represents the computing brain, enabling unlimited processing capability in vehicular environments. The cloud-based services are accessible via a reliable vehicle-to-Internet connection. The vehicle-oriented connection is a cooperative combination of various wireless access technologies with vehicular networks. Various traffic-utility-oriented clients utilize the heterogeneous network access technology–enabled Internet connection for making intelligent decisions via cloud-based computing resources. The inner-module-oriented relational structure of the proposed heterogeneous vehicular network model is presented, focusing on key network entities (see Figure 4).

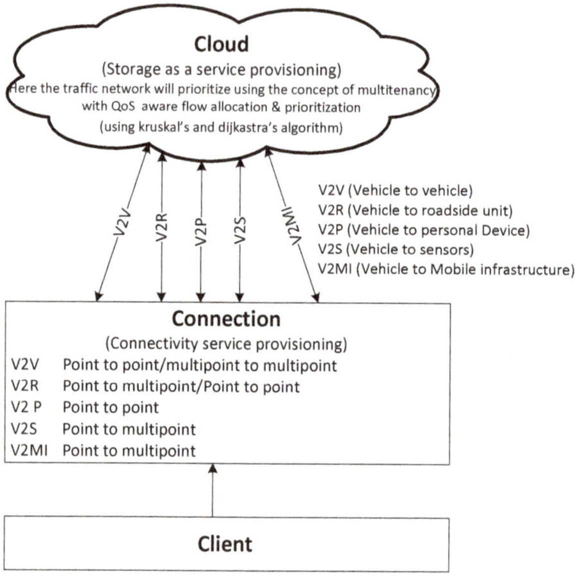

Figure 3. Building blocks of the network model.

(1) Vehicular Cloud

The cloud framework has two key operation levels for realizing cloud-oriented intelligent application servers. The traffic-oriented essential cloud services are considered lower-level, whereas smart IoV servers are developed on top of these essential services. The distinguishable level-wise cloud operation is significant considering the steep growth in the volume of traffic data once the integration of various existing networks with the vehicular network is realized as an IoV. The traffic data uploading, information processing, dissemination, and storage are the basis for the two-level cloud operation on big traffic data. The technical roles of each level of operations are introduced below:

- Traffic-Oriented Cloud Services

The traffic-oriented cloud services are essential to the intelligence processing and analysis of big traffic data (see Figure 3). The implemented services include Computing as a Service (COaaS) on traffic data, Storage as a Service (StaaS) for traffic data, Data as a Service (DaaS) for traffic information re-utilization, Gateway as a Service (GaaS) for heterogeneous network support, and geo-Location as a Service (LoaaS) for vehicle localization. Some other services toward multimedia intelligence are also implementable, including Picture as a Service (PcaaS) for sharing traffic incidence, Platform as a Service (PlaaS) for system-oriented traffic applications, Software as a Service (SoaaS) for traffic analysis, and Network as a Service (NaaS).

- Smart Server

The smart IoV servers consist of two processing engines, namely, internal and external (see Figure 5). These processing engines utilize traffic-oriented cloud services to infer intelligent decisions from traffic data. The responsibilities of an internal processing engine include materializing big traffic data, processing via applying artificial intelligence, and analyzing with a focus on smartness. The external processing engine is majorly responsible for traffic-oriented data collection and dissemination. The coordination between engines to simulate intelligence enables three types of smart IoV servers: verified traffic safety, optimized efficiency, and intelligent utility toward business-oriented servers. The visualization of three IoV services is due to these smart servers' different processing and time-oriented constraints.

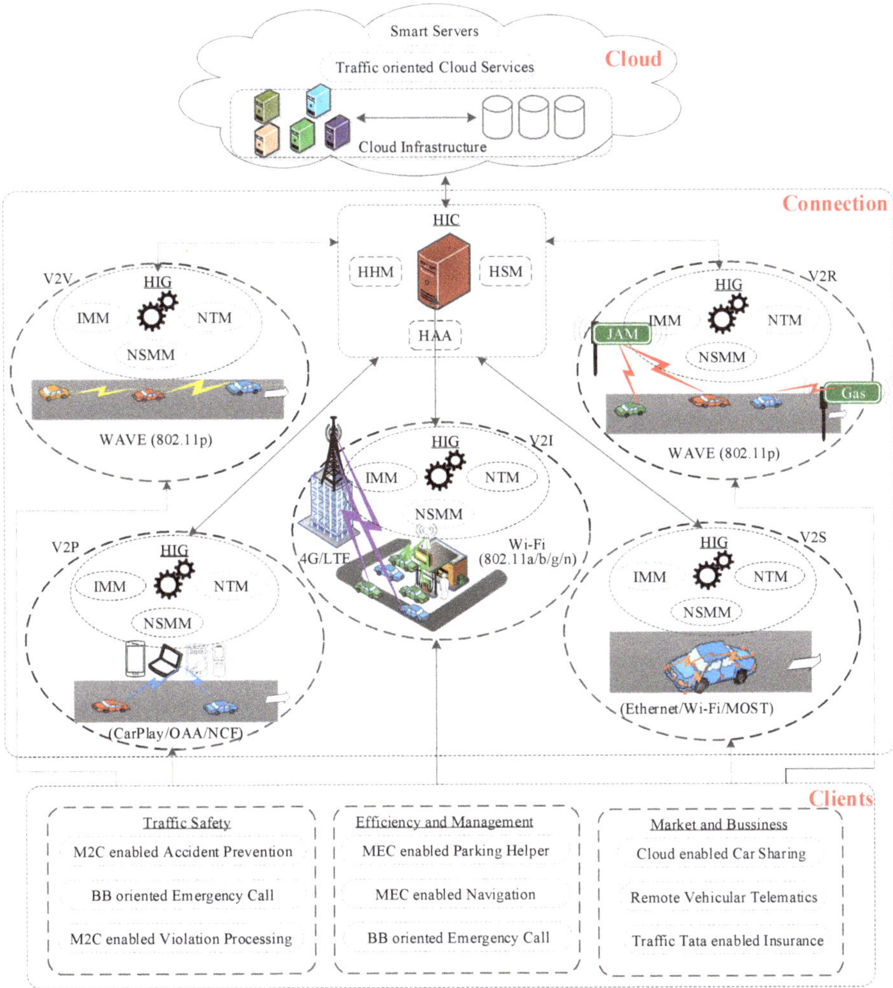

Figure 4. Vehicular cloud-oriented heterogeneous network model for IoV.

(2) Connection for Heterogeneous Vehicular Communication

The vehicular connection between smart IoV server and the vehicular end-user is composed of a third-party heterogeneous internetwork coordinator (HIC) and heterogeneous internetworking gateway (HIG). It is operational cooperation between the cloud server and IoV end-user, including vehicles, personal devices, and roadside infrastructure. The coordination-oriented network management in heterogeneous environments, including 802.11p, Wi-Fi, and 4G/LTE access technology, is the key responsibility of the HIC. The HIG represents the individual network connection. The HIC prioritizes network connections based on wireless access technologies.

- Heterogeneous Internetworking Coordinator (HIC)

The internetwork operator provides a service level agreement to the end-user for the heterogeneous network operators in the IoV. This enables seamless roaming between the heterogeneous networks, enabling internetwork cooperation without compromising the quality of network performance. The HIC eliminates the requirement of a pairwise quality of service agreement between network operators, which is a significantly challenging constraint for any heterogeneous network framework. Three key functional modules were developed in the HIC to carry out the internetworking-oriented connection and service management. The modules include heterogeneous handoff management (HHM), heterogeneous authentication and authorization (HAA), and heterogeneous service management (HSM). These modules interact with two major databases, including heterogeneous network and network operator databases. The functional relationship among key functional modules and databases is presented in Figure 6a.

The HHM module monitors network connections, looking for potential active connections that may require a internetwork handoff shortly. These active connections transform into handoff connections after the operation confirmation from HAA and HSM. The transformation is enabled by a handoff operator responsible for verifying response confirmation from the authentication and service modules. The HAA module maintains end-user credentials across networks for verification. This includes network- and operator-specific access right validation, bypassing the response for initiating handoff operation. It monitors the session-wise network operation and initiates a time-oriented connection closer in case of an idle connection. The third-party-oriented HIC implementation enables smooth end-user authentication and authorization, which is challenging considering heterogeneous network environments.

The HSM module provides an end-user service layer agreement using a service quality rating approach for different operating networks. It uses a list of dedicated services between heterogeneous operating networks. The service quality is rated, with a focus on guarantying service quality to end-users by maintaining a service delivery history for each heterogeneous operating network. The rating is implemented considering user-feedback-oriented service monitoring for the connections initiated in the networks. The heterogeneous network and operator database are accessed as a key information resource in the connection-oriented operations of the three functional modules in HIC.

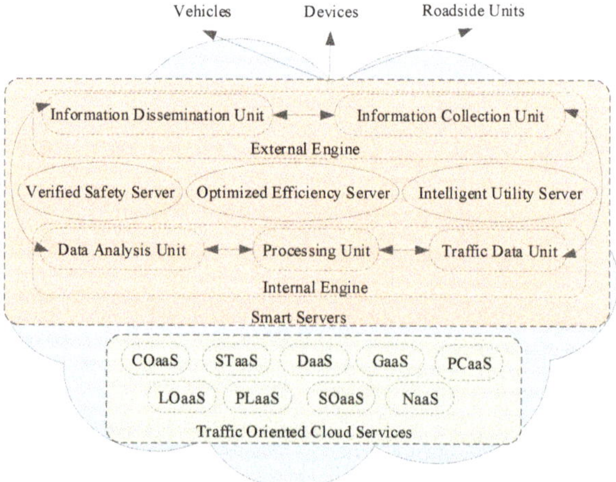

Figure 5. The two-level vehicular cloud engine for IoV.

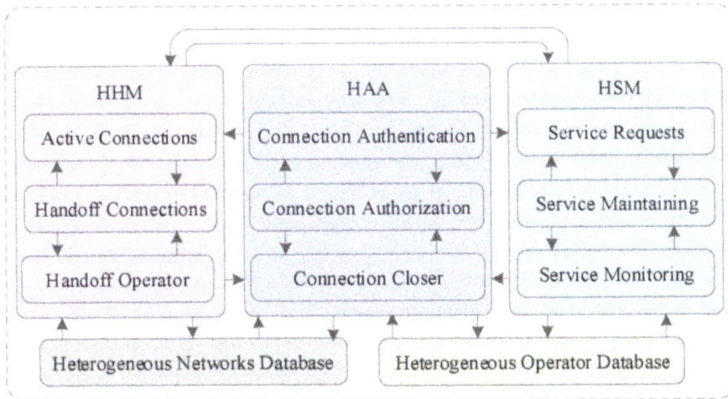

(a)

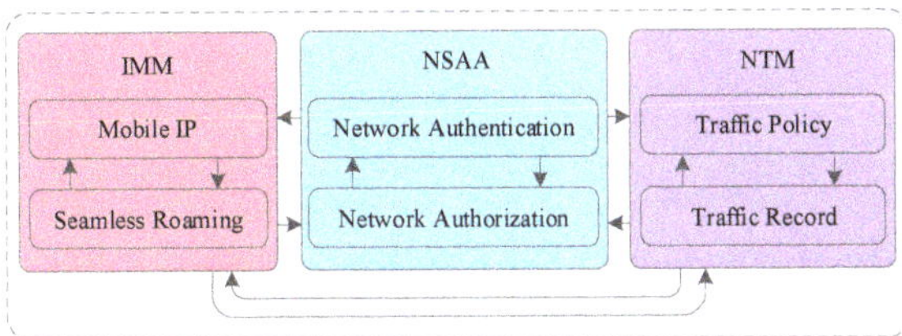

(b)

Figure 6. Key functional modules in heterogeneous connection: (**a**) HIC and (**b**) HIG.

- Heterogeneous Internetworking Gateway

The heterogeneous internetworking gateway of a connection implements network access technologies in IoV to enable effective collaboration with the HIC for initiation and maintenance of heterogeneous vehicular connections. The HIG represents four types of wireless access technologies enabling the five types of vehicular communications. It includes WAVE-enabled V2V and V2I, Wi-Fi- or 4G/LTE-enabled V2I, Car-Play- or android-system-enabled V2P, and media-oriented system transport (MOST)-enabled V2S. The access-technology-oriented HIG consists of three major functional modules, including Internetwork Mobility Management (IMM), Network-Specific Authentication and Authorization (NSAA), and Network Traffic Management (NTM). The operational flow and association among key functional modules in the HIG are presented in Figure 6b.

The IMM module implements mobile IP by utilizing network tunneling between vehicle home agent (VHA) and vehicle foreign agent (VFA). During an on-road journey, the initial operating network enables a home agent, whereas any other network throughout the journey enables a foreign agent for each vehicle. The tunnel-oriented internetwork mobility enabling supports seamless roaming without IP conversion. The NSAA module executes local credential verification for vehicles. It enables HAA to carry out credential verification with the coordination between VFA and VHA. The network traffic management module implements network policies for providing network monitoring services in a particular network. These policies vary with the type of network in heterogeneous network environments. The policy-oriented network monitoring is based on the historical-traffic-usage data and the live traffic data for a specific network connection.

(3) Smart Services as Clients

A client application utilizes access-technology-oriented connection for realizing large cloud-based services in vehicles. Some novel smart client applications are implemented based on cloud service architecture. There are two broader aspects of these client applications in IoV. One is business-focused client applications, majorly oriented toward vehicle insurance, car-sharing, and infotainment. The safety and management-oriented client applications are related to navigation, vehicle diagnostic, and remote telematics in vehicles. Some potential service-oriented clients are materialized below by identifying their parameter-oriented service requirements and corresponding access-technology prioritization (see Figure 7 and Table 2).

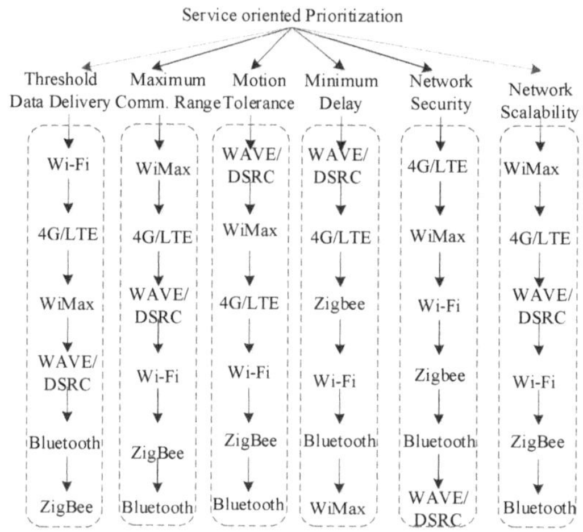

Figure 7. Experimentally validated access-technology prioritization tree.

Table 2. Client-oriented access technology prioritization.

Client	Client-Oriented Priority Order High ⟶ Low
Accident Prevention	WAVE/DSRC → 4G/LTE → ZigBee → Wi-Fi → Bluetooth → WiMax
Emergency Call Guarantee	Bluetooth → ZeeBee → Wi-Fi → WAVE/DSRC → WiMax → 4G/LTE
MEC-Oriented Parking Helper	WiMax → Wi-Fi → 4G/LTE → WAVE/DSRC → Bluetooth → ZigBee
Vehicular Telematics	4G/LTE → WiMax → WAVE/DSRC → Wi-Fi → Bluetooth → ZigBee

- Machine-to-Cloud-Oriented Accident Prevention

The machine-to-cloud (M2C)-oriented traffic safety service in the IoV is implemented considering traffic data inferred knowledge towards accident prevention. It is an advancement in machine-to-machine (M2M)-oriented safety application, majorly relying on locally inferred knowledge from neighboring vehicle's data. The cloud-server-based smart traffic applications utilize global traffic knowledge to improve decision-making for drivers. It focuses on automatic operations on the go, including steering control, speed control, stoppage, and lane change.

- Black-Box-Oriented Emergency Call Guarantee

The emergency call guarantee service is implemented via heterogeneous network cooperation in the IoV. A vehicle uses the nearest and best available network-access technology to call emergency facilities. The call is forwarded between heterogeneous networks to guarantee its quality and completion for the desired facility. The call implementation considers interval-based manual intimation as well as event-based automatic intimation. The call implementation realizes present and past information-sharing regarding emergency incidence, including speed, direction, location, and lane.

- Mobile Edge Computing–Oriented Parking Helper

The parking helper service is implemented, enabling roadside units as mobile edge computing (MEC). The geographical parking space information dissemination and precise localization are the main technical modules involved in realizing MEC-enabled parking helpers. The implementation is based on Wi-Fi-enabled publish–subscribe communication. The roadside units periodically publish parking availability information. The information is accessed by subscribed passing-by vehicles as receiver-initiated information dissemination.

- Remote-Operation-Oriented Vehicular Telematics

The remote vehicular telematics service is implemented considering guaranteed end-to-end communication between vehicle and remote services. The implementation focuses on executing non-driving operations such as password-oriented vehicle authentication and authorization, intimation of vehicle access, and remote vehicle monitoring. These remote operations exploit heterogeneous vehicular communication to transform the existing physical entity-oriented operations into digital entity-oriented forms.

3.3. Network Prioritization in Heterogeneous Vehicular Networks

In the integrated vehicular network environment, the QoS path was computed using a weighted internetwork routing approach. The integrated network path having the greatest weight was chosen as a target for enabling a specific traffic service. The information about the current traffic load was collected from each vehicular communication link using switch port counters traffic history. The concept of multitenancy was used here in order to maximize the QoS requirements in an heterogeneous vehicular network environment. Initially the concept of network virtualization was used to isolate the flow among different vehicular tenants. It was achieved by dividing the vehicular network into layers of local and integrated vehicular networks. Secondly, the weighted internetwork routing algorithm was used to control the allocation of new flows entering into the vehicle network and to prioritize them as per the need of client-specific traffic services.

A software defined network (SDN) controller was used for vehicular communication flow level network prioritization. The SDN controller was modeled using an undirected graph G, where $G = (V, E)$ is undirected graph among vehicular nodes and existing network infrastructure in traffic environment. Here, V is the set of vehicular nodes in the network and E is the set of vehicular communication links between the network nodes. Considering n as the number of vehicular tenants in the vehicular network graph G, the graph can be divided into various subgraphs including $G1$, $G2$, $G3, \ldots Gn$ based on the number of flows and tenants in the traffic-specific network. This graph-centric vehicular network can be represented as expressed in Equation (1).

$$G_n = (V_n, E_n), \text{ where } V_n \in V, \text{ and } E_n \in E \quad (1)$$

where V_n denotes the set of vehicular nodes as vertices and E_n denotes the set of heterogeneous vehicular communication links as edges in the vehicular network subgraph G_n. Each vehicular node in the subgraph was included in the heterogeneous network G so that all possible combination of heterogeneous vehicular communication paths could be explored by network prioritization component from a source vehicular node to any destination client services of another vehicular node. Each subgraph G_n was considered as vehicular nodes with heterogeneous communication links connected to the vehicular graph. Here, two vehicular network subgraphs enabled by different types of local networks were considered edge disjoint. However, it is highlighted that different vehicular communication paths in a subgraph network may have common communication links as common edges. This can be mathematically represented as expressed in Equation (2).

$$G_{n(a,b)} = \begin{cases} E_a \cap E_b = 0, \; if \; G_a \; and \; G_b \; are \; different \; types \; of \; vehicular \; networks \\ e \in l_1 \wedge e \in l_2, \; if \; l_1, \; l_2 \; are \; from \; same \; network, \; l_1, \; l_2 \in G_a \; or \; l_1, \; l_2 \in G_b \end{cases} \quad (2)$$

where E_a and E_b denote set of heterogeneous connectivity links as edges of subgraphs G_a and G_b, respectively, and e denotes a common links of the two communication paths l_1 and l_2 of same network type.

The multitenancy QoS prioritization was implemented by using Dijkstra's and Kruskal's algorithms for managing the shared vehicular communication resources and their utilization. The shortest heterogeneous path between any two vehicular nodes was computed using the algorithms considering two sets of vehicular network space and heterogenous vehicular communication paths. It is highlighted that the two algorithms were used to carry out heterogeneous vehicular network prioritization for enabling smart traffic client services described in the network architecture. These algorithms were used for two-level network prioritizations. Specifically, in first-level prioritization, Dijkstra's algorithm was utilized for localized vehicular network without considering other existing network infrastructure nearby the traffic environment. In second-level prioritization, Kruskal's algorithm was utilized for spanning tree-centric integrated heterogeneous network prioritization where different types of existing networks are considered for enabling vehicular network services. In the two-level prioritization, the shortest heterogeneous vehicular communication path between vehicular nodes was identified considering vehicular flows in the network. The following constraints exist in the two-level prioritization, as expressed in Equations (3) and (4).

$$F \subset V^2 \tag{3}$$

$$SP \subset F \tag{4}$$

where F denotes the set of vehicular node flows between any source and destination, and SP denotes a shortest path between the source and destination vehicular nodes. Here, each heterogeneous communication paths l in the network consists of number of subpaths P_l and segments S_l joining the local vehicular networks. For obtaining the shortest heterogeneous communication path in the shared link resource environment, link utilization ratio L^{UR} of the network was computed as expressed in Equation (5)

$$L^{UR} = \frac{Traffic\ link\ load\ from\ SDN\ switch\ port}{Link\ capacity} = \frac{\lambda_l}{C_l} \tag{5}$$

where λ_l denotes the link load of a shared link in a particular path, and C_l denotes the link capacity of a shared link in a particular path. The shared link utilization ration L^{UR} was further used in computing the weight w_l of a particular path l for making shartest heterogeneous path decsion as expressed in Equation (6).

$$w_l = 1 - \frac{(\sum_m \sum_n L^{UR})/P_l}{\sum_l (\sum_m \sum_n L^{UR})/P_l} \text{ where } w_l \epsilon (0,1) \text{ and } \sum_l w_l = 1 \tag{6}$$

where m and n represents the two vehicular nodes attempting to communicat via heterogeneous vehicular communication links. The path with the highest weight in the heterogenious vehicular network graph was considered the least loaded path and appropriate for establishing prioritized heterogeneous vehicular communication.

4. Performance Evaluation—A Case Study

In this section, the performance of service-oriented clients is evaluated in heterogeneous vehicular networking environments as an IoV implementation. Compared with the traditional ad hoc vehicular system, such as VANETs implementation, where infrastructure support is omitted, including RSU, Wi-Fi, and LTE infrastructure.

4.1. Simulation Setting and Metrics

The heterogeneous networking environment was simulated using network simulator ns-2. A vehicular mobility model generator MOVE and a geographic information system ArcGIS were utilized as supporting applications. Initially, a realistic vehicular mobility network scenario focused on road map editor and vehicle movement editor utility in MOVE. The scenario was executed on a real city road map obtained via a web application open street map (see Figure 8). Two-dimensional location coordinates were precisely embedded on the map using ArcGIS. Notably, the capital city is a real example of dense urban infrastructure where heterogeneous vehicular network environments are a reality. The five types of vehicular communications are implemented, focusing on varying transmission ranges and access technologies. The traditional vehicular communications, including V2V and V2R, were implemented considering IEEE 802.11p-enabled access technology with a [200–300 m] transmission range. The short-range vehicular communications, including V2S and V2P, were implemented considering Wi-Fi-enabled access technology with a lower transmission range, precisely [5–10 m] and [40–80 m] for the respective cases. The more extensive vehicular mobile network was implemented using 4G-enabled access technology considering [500–1000 m]. It is clarified that the tool used to carry out experiments supports 3G and 4G services; therefore, it was mentioned in the paper. However, 5G can be used wherever 4G has been considered. We did not test 5G experimentally in our proposal; therefore, we do not mention it.

Figure 8. Simulation scenario as open street view.

The cooperation among these access technologies was realized by considering heterogeneous network access points in the implementation scenario (see Figure 9). The access points included 6 4G-enabled mobile access points at junctions, 9 Wi-Fi-enabled access points, and 22 RSUs alongside roads. Two major traffic scenarios in urban environments were considered in the implementation: peak and off-peak hours' traffic, where average intervehicle distance and speed are lower and higher, respectively. The values of other general simulation parameters were similar to those considered in vehicular networking implementations [12]. Measuring the performance of the considered IoV clients under these scenarios was attractive for the standardization of heterogeneous vehicular networking and related client implementations.

The performance of M2C-enabled accident prevention was measured via message diversion rate (i.e., the percentage of vehicles receiving an accident intimation distributed via a dedicated cloud server on the point incidence road section) on the road section of interest. The black-box-oriented emergency call guarantee was measured in message drop rate (i.e., the percentage of emergency messages with a failed delivery attempt in point-to-point short message delivery) in point-to-point emergency message delivery. The MEC-enabled parking helper was evaluated in terms of delay in RSU-enabled distributed message delivery. The performance of remote-operation-oriented vehicular telematics was measured via in-stream utilization in video data delivery. It focused on stream density (i.e., the percentage of neighboring vehicles utilized for establishing multiple-stream-oriented communication path between source and destinations) targeting stream-oriented data delivery in the IoV. It is highlighted that wherever we refer to VANET implementation in the paper, it means a localized vehicular network implementation without considering the heterogeneous existing networks' integration. We compared a VANET implementation with an IoV implementation, which means an integrated vehicular network environment considering other existing network infrastructures along with the vehicular network.

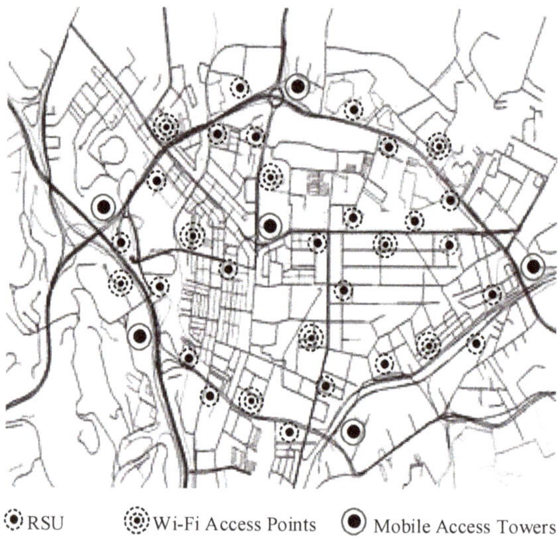

Figure 9. Simulation scenario as simulator view.

4.2. *Analysis of Results*

The comparative investigation between IoV and traditional VANETs implementations presented in Figure 10 focuses on-peak and off-peak traffic-based diversion rate evaluation. It can be observed that the message diversion rate was stable at around 90% in the case of IoV implementation. The stability in message diversion can be attributed to the capability of alternative vehicular network selection in the absence of ad hoc vehicular nodes during off-peak traffic. The diversion rate varies with a more extensive range of 20–55% between peak and off-peak traffic in the case of the traditional implementation. The higher variation in diversion rate can be attributed to the availability of vehicular nodes in ad hoc implementation, which is relatively lower during off-peak traffic and higher during peak traffic. The M2C-oriented accident prevention message diversion rate analysis highlighted the benefits of heterogeneous cooperative vehicular networking.

The comparative investigation in Figure 11 focused on drop reduction with higher vehicle density and drop increment with higher vehicle speed. It can be observed that the drop reduction due to vehicle density was relatively higher in the VANET implementation as compared to that in IoV implementations. It reduced from 35% to 20% in the state-of-the-art VANET implementation and from 12% to 9% in the IoV implementation. This can be attributed to the better opportunistic ad hoc networking probability with higher vehicle density for the traditional implementation and better forwarding network selection in the case of the IoV implementation. It is also clearly visible that the drop increment due to higher vehicle speed was approximately equivalent for both the implementations. This was due to the speed-oriented link failure in communication between vehicles and emergency services, which was quite similar in both the implementation scenarios. The emergency message drop analysis highlights the benefit of IoV implementation as an overall lower message drop rate than the traditional implementation.

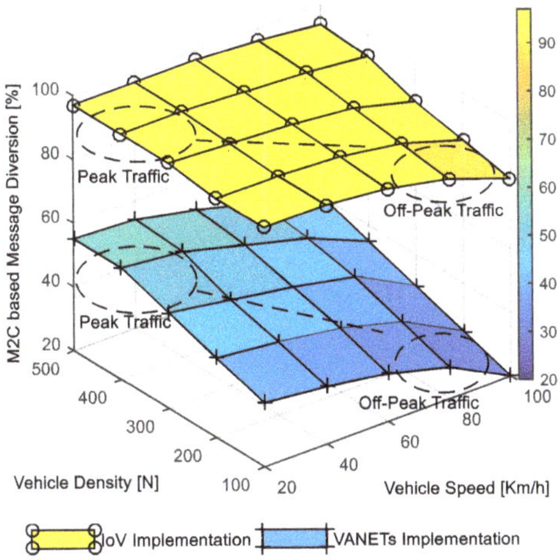

Figure 10. Message diversion in M2C-based accident prevention.

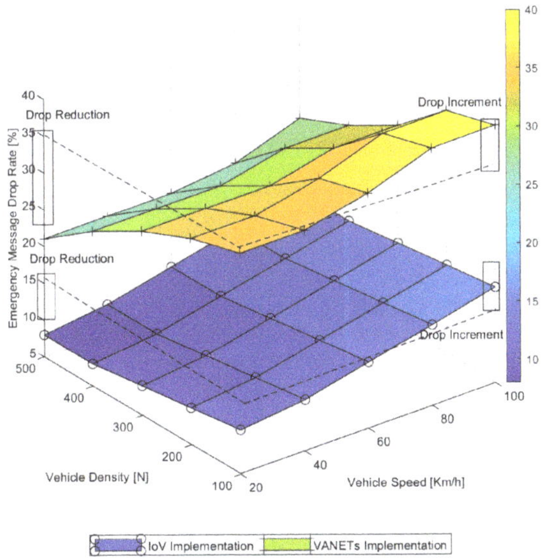

Figure 11. Message drop in back-box-oriented emergency message delivery.

The comparative performance evaluation in Figure 12 focused on outlier delay analysis to enable threshold monitoring for MEC-based clients. It is visible that the impact of speed and density on the distributed operation delay of IoV was comparably negligible. It was in the range of 45–50 ms throughout the density and speed change. This can be attributed to the better operational network availability for distributed operation of clients in heterogeneous IoV environments. The critical impact of higher vehicle speed and density on delay is visible as outlier delay in the case of the VANET implementation. This is an exciting result and provides clear evidence toward poor distributed network management in ad hoc vehicular implementation without considering infrastructure support. The comparative delay analysis for an MEC-based parking helper client attests to the better-distributed network operation capability of the IoV implementation compared to the ad hoc vehicular implementation.

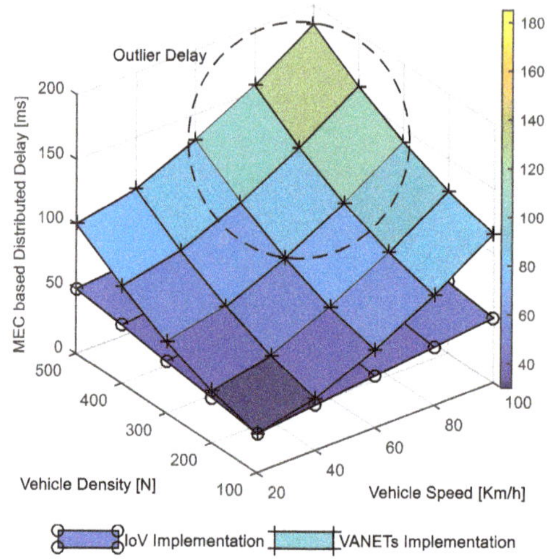

Figure 12. Distributed delay in MEC-based parking helper.

The comparative performance evaluation presented in Figure 13 focused on monitoring utilization bound. It is visible that the stream utilization bound approximately 80% for IoV implementation case was better than the 45% utilization bound observed in the case of VANET implementation. This can be attributed to the durable stream survivability utilizing heterogeneous links in IoV environments, whereas streams are spontaneous in pure ad hoc implementation environments. It is also noteworthy that the utilization increased upward until stream density reached 12–15%. The stream utilization reduced downward with further higher stream density for both the implementation scenarios. The interesting result shows the particular characteristics of multipath streaming where up to 15% of neighbor node usage for streams supports better network resource utilization. Further usage of neighbor nodes for higher stream density degraded utilization. This was due to the duplication of streams with common neighbor nodes resulting in performance degradation in both the implementation scenarios.

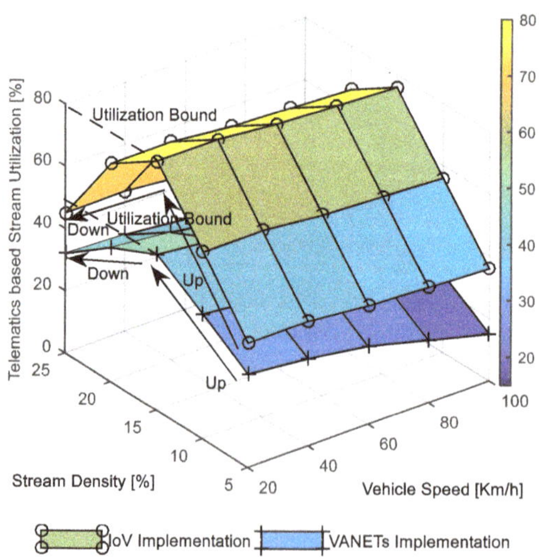

Figure 13. Stream utilization in telematics-based video data delivery.

4.3. Summary of Observations

4.3.1. Network Prioritization in Content-Centric Networking

Content-centric networking has significant potential to address the growing heterogeneousness in vehicular environments [21,22]. It effectively reduces point-to-point network load in heterogeneous environments due to its publish/subscribe-based communication architecture. However, content-oriented network prioritization is a challenging task in heterogeneous network environments.

4.3.2. Virtual Vehicle Hijacking in Vehicular Cyber System

The vehicles are no longer stand-alone entities, particularly with the realization of heterogeneous communication architecture. However, vehicular communication comes with a significant cyber security risk [23]. Specifically, unauthorized wheels access, disabling brakes, locking doors, engine disruption to path forging, location and identity manipulation, and tracking are some examples of virtual vehicle hijacking.

4.3.3. Big Data Analytics in Heterogeneous Traffic Data

The growing heterogeneousness in vehicular traffic data has significantly enlarged traffic data volume towards big data [24]. However, traffic safety and efficiency-oriented intelligent decisions to enable vehicular automation are still based on sensor-based static data. Applying big data analytics in heterogeneous traffic data can bring fundamental changes to the driving experience by inferring sophisticated, intelligent decisions.

4.3.4. Vehicular-Cooperation-Oriented Edge Computing

The realization of cloud computing in heterogeneous vehicular environments is the need of the hour considering the limited computing capability in distributed vehicular networking [25]. However, enabling cooperation-oriented edge computing can significantly enhance the computing scenario considering the overall growth in on-road vehicles and digital things in vehicular environments.

4.3.5. Driver Privacy in Heterogeneous Vehicular Communications

Driver privacy is a potential issue due to the growing advancement toward the connected-vehicle environment [26]. There are different types of privacy concerns in connected vehicle environments, including personal information privacy [27], location privacy [28], driving-data privacy [29], third-party privacy [30], and information sharing consent–related privacy [31]. Location privacy has been gaining more attention from researchers in the past few years of connected-vehicle study due to the suitability of location-based communications technologies and services in vehicular traffic environments [32]. Driving data privacy and information sharing are becoming crucial for today's modern vehicles due to the growing sensor-based technology advancements for connecting vehicles to existing mobile networks and personal gadgets.

5. Conclusions

In this paper, a heterogeneous network model for heterogeneous vehicular communication is presented along with service-oriented implementation. The following conclusions was reached from the design and implementation of the model. The network cooperation enables cloud-oriented computing on big traffic data for realizing intelligent traffic services. The heterogeneous network coordinator and gateway are the key to unambiguous connection management. The service-oriented traffic applications become intelligent with an enlarged traffic data domain and processing capability. The practical simulation verified higher message diversion and stream utilization and lower message drop rate and delays for traffic services in heterogeneous vehicular communication implementation. Mathematical modeling of service-oriented network prioritization and content-centric service implementation in the heterogeneous vehicular environment was also presented to support the heterogeneous vehicular network implemenattion theoretically. In the future, the

UAV-enabled networks will be integrated with the heterogeneous vehicular network for enabling specific service-centric real-time vehicular network infrastructure.

Author Contributions: Conceptualization, A.M.K.; formal analysis, A.M.K.; investigation, A.M.K.; methodology, A.M.K.; resources, M.A.H. and A.K.; supervision, O.K.; validation, M.A.; writing, A.M.K.; review and editing, O.K., G.A., W.A.-u. and R.A. All authors have read and agreed to the published version of the manuscript.

Funding: This study was financially supported via a funding grant by the Deanship of Scientific Research, Taif University Researchers Supporting Project number (TURSP-2020/300), Taif University, Taif, Saudi Arabia.

Institutional Review Board Statement: Not applicable.

Informed Consent Statement: Not applicable.

Data Availability Statement: Research data will be made available on individual requests basis to the corresponding author considering research collaboration possibilities with the researchers or research team and with restrictions that the data will be used only for further research in the related literature progress. As the research data will be used by our team for further research in the particular theme.

Acknowledgments: This study was financially supported via a funding grant by the Deanship of Scientific Research, Taif University Researchers Supporting Project number (TURSP-2020/300), Taif University, Taif, Saudi Arabia.

Conflicts of Interest: The authors declare no conflict of interest.

References

1. Xu, K.; Qu, Y.; Yang, K. A tutorial on the internet of things: From a heterogeneous network integration perspective. *IEEE Netw.* **2016**, *30*, 102–108. [CrossRef]
2. Lloret, J.; Tomas, J.; Canovas, A.; Parra, L. An Integrated IoT Architecture for Smart Metering. *IEEE Commun. Mag.* **2016**, *54*, 50–57. [CrossRef]
3. Kaiwartya, O.; Abdullah, A.H.; Cao, Y.; Altameem, A.; Prasad, M.; Lin, C.T.; Liu, X. Internet of vehicles: Motivation, layered architecture, network model, challenges, and future aspects. *IEEE Access* **2016**, *4*, 5356–5373. [CrossRef]
4. Fangchun, Y.; Shangguang, W.; Jinglin, L.; Zhihan, L.; Qibo, S. An overview of internet of vehicles. *China Commun.* **2014**, *11*, 1–15.
5. Kaiwartya, O.; Kumar, S. Cache agent-based geocasting in VANETs. *Int. J. Inf. Commun. Technol.* **2015**, *7*, 562–584. [CrossRef]
6. Saini, M.; Alelaiwi, A.; El Saddik, A. How Close are We to Realizing a Pragmatic VANET Solution? A Meta-Survey. *ACM Comput. Surv.* **2015**, *48*, 29. [CrossRef]
7. Zheng, K.; Zheng, Q.; Chatzimisios, P.; Xiang, W.; Zhou, Y. Heterogeneous vehicular networking: A survey on architecture, challenges, and solutions. *IEEE Commun. Surv. Tutor.* **2016**, *17*, 2377–2396. [CrossRef]
8. Kaiwartya, O.; Kumar, S. Geocast routing: Recent advances and future challenges in vehicular adhoc networks. In Proceedings of the 2014 International Conference on Signal Processing and Integrated Networks (SPIN), Noida, India, 20–21 February 2014; IEEE: Piscataway, NJ, USA, 2014; pp. 291–296.
9. Aliyu, A.; Abdullah, A.H.; Kaiwartya, O.; Cao, Y.; Usman, M.J.; Kumar, S.; Lobiyal, D.K.; Raw, R.S. Cloud Computing in VANETs: Architecture, Taxonomy, and Challenges. *IETE Tech. Rev.* **2017**, *35*, 523–547. [CrossRef]
10. Applin, S.; Riener, A.; Fischer, M.D. Extending Driver-Vehicle Interface Research Into the Mobile Device Commons: Transitioning to (nondriving) passengers and their vehicles. *IEEE Consum. Electron. Mag.* **2015**, *4*, 101–106. [CrossRef]
11. Chen, J.; Mao, G.; Li, C.; Zafar, A.; Zomaya, A. Throughput of Infrastructure-Based Cooperative Vehicular Networks. *IEEE Trans. Intell. Transp. Syst.* **2017**, *18*, 2964–2979. [CrossRef]
12. Forbes, I. Innovation is Great: Connected and Automated Vehicles. UK, Center for Connected and Autonomous Vehicles. 2020. Available online: https://assets.publishing.service.gov.uk/government/uploads/system/uploads/attachment_data/file/929352/innovation-is-great-connected-and-automated-vehicles-booklet.pdf (accessed on 15 January 2022).
13. Niu, Z.; Shen, X.S.; Zhang, Q.; Tang, Y. Space-air-ground integrated vehicular network for connected and automated vehicles: Challenges and solutions. *Intell. Converg. Netw.* **2020**, *1*, 142–169. [CrossRef]
14. Liu, R.; Liu, A.; Qu, Z.; Xiong, N.N. An UAV-Enabled Intelligent Connected Transportation System with 6G Communications for Internet of Vehicles. *IEEE Trans. Intell. Transp. Syst.* **2021**, 1–15. [CrossRef]
15. Su, Z.; Dai, M.; Xu, Q.; Li, R.; Zhang, H. UAV Enabled Content Distribution for Internet of Connected Vehicles in 5G Heterogeneous Networks. *IEEE Trans. Intell. Transp. Syst.* **2021**, *22*, 5091–5102. [CrossRef]
16. Sharma, P.; Liu, H. A Machine-Learning-Based Data-Centric Misbehavior Detection Model for Internet of Vehicles. *IEEE Internet Things J.* **2020**, *8*, 4991–4999. [CrossRef]

17. Mollah, M.B.; Zhao, J.; Niyato, D.; Guan, Y.L.; Yuen, C.; Sun, S.; Lam, K.-Y.; Koh, L.H. Blockchain for the Internet of Vehicles Towards Intelligent Transportation Systems: A Survey. *IEEE Internet Things J.* **2021**, *8*, 4157–4185. [CrossRef]
18. Chen, Y.; Lu, C.; Chu, W. A Cooperative Driving Strategy Based on Velocity Prediction for Connected Vehicles With Robust Path-Following Control. *IEEE Internet Things J.* **2020**, *7*, 3822–3832. [CrossRef]
19. Fang, Y.; Min, H.; Wu, X.; Wang, W.; Zhao, X.; Mao, G. On-Ramp Merging Strategies of Connected and Automated Vehicles Considering Communication Delay. *IEEE Trans. Intell. Transp. Syst.* **2022**, 1–15. [CrossRef]
20. Tao, L.; Watanabe, Y.; Li, Y.; Yamada, S.; Takada, H. Collision Risk Assessment Service for Connected Vehicles: Leveraging Vehicular State and Motion Uncertainties. *IEEE Internet Things J.* **2021**, *8*, 11548–11560. [CrossRef]
21. Kaiwartya, O.; Cao, Y.; Lloret, J.; Kumar, S.; Aslam, N.; Kharel, R.; Abdullah, A.H.; Shah, R.R. Geometry-Based Localization for GPS Outage in Vehicular Cyber Physical Systems. *IEEE Trans. Veh. Technol.* **2018**, *67*, 3800–3812. [CrossRef]
22. Ahmed, S.H.; Bouk, S.H.; Kim, D.; Rawat, D.B.; Song, H. Named Data Networking for Software Defined Vehicular Networks. *IEEE Commun. Mag.* **2017**, *55*, 60–66. [CrossRef]
23. Eiza, M.H.; Ni, Q. Driving with Sharks: Rethinking Connected Vehicles with Vehicle Cybersecurity. *IEEE Veh. Technol. Mag.* **2017**, *12*, 45–51. [CrossRef]
24. Xu, W.; Zhou, H.; Cheng, N.; Lyu, F.; Shi, W.; Chen, J.; Shen, X. Internet of vehicles in big data era. *IEEE/CAA J. Autom. Sin.* **2017**, *5*, 19–35. [CrossRef]
25. Kaiwartya, O.; Kumar, S. Enhanced caching for geocast routing in vehicular Ad Hoc network. In *Intelligent Computing, Networking, and Informatics*; Springer: New Delhi, India, 2014; pp. 213–220.
26. Zavvos, E.; Gerding, E.H.; Yazdanpanah, V.; Maple, C.; Stein, S.; Schraefel, M. Privacy and Trust in the Internet of Vehicles. *IEEE Trans. Intell. Transp. Syst.* **2021**, 1–16. [CrossRef]
27. Kerrache, C.A.; Lagraa, N.; Hussain, R.; Ahmed, S.H.; Benslimane, A.; Calafate, C.T.; Cano, J.-C.; Vegni, A.M. TACASHI: Trust-Aware Communication Architecture for Social Internet of Vehicles. *IEEE Internet Things J.* **2018**, *6*, 5870–5877. [CrossRef]
28. Sheet, D.K.; Kaiwartya, O.; Abdullah, A.H.; Cao, Y.; Hassan, A.N.; Kumar, S. Location information verification using transferable belief model for geographic routing in vehicular ad hoc networks. *IET Intell. Transp. Syst.* **2016**, *11*, 53–60. [CrossRef]
29. Ni, J.; Zhang, K.; Yu, Y.; Lin, X.; Shen, X.S. Privacy-preserving Smart Parking Navigation Supporting Efficient Driving Guidance Retrieval. *IEEE Trans. Veh. Technol.* **2018**, *67*, 6504–6517. [CrossRef]
30. Azad, M.A.; Bag, S.; Parkinson, S.; Hao, F. TrustVote: Privacy-Preserving Node Ranking in Vehicular Networks. *IEEE Internet Things J.* **2018**, *6*, 5878–5891. [CrossRef]
31. Khatri, A.; Kumar, S.; Kaiwartya, O.; Aslam, N.; Meena, N.; Abdullah, A.H. Towards green computing in wireless sensor networks: Controlled mobility-aided balanced tree approach. *Int. J. Commun. Syst.* **2018**, *31*, e3463. [CrossRef]
32. Zhou, L.; Yu, L.; Du, S.; Zhu, H.; Chen, C. Achieving Differentially Private Location Privacy in Edge-Assistant Connected Vehicles. *IEEE Internet Things J.* **2018**, *6*, 4472–4481. [CrossRef]

sensors

MDPI

Article

Overcoming Challenges of Applying Reinforcement Learning for Intelligent Vehicle Control

Rafael Pina *, Haileleol Tibebu, Joosep Hook, Varuna De Silva and Ahmet Kondoz

Institute of Digital Technologies, Loughborough University London, 3 Lesney Avenue, London E20 3BS, UK; h.tibebu@lboro.ac.uk (H.T.); j.hook@lboro.ac.uk (J.H.); v.d.de-silva@lboro.ac.uk (V.D.S.); a.kondoz@lboro.ac.uk (A.K.)

* Correspondence: r.m.pina@lboro.ac.uk

Abstract: Reinforcement learning (RL) is a booming area in artificial intelligence. The applications of RL are endless nowadays, ranging from fields such as medicine or finance to manufacturing or the gaming industry. Although multiple works argue that RL can be key to a great part of intelligent vehicle control related problems, there are many practical problems that need to be addressed, such as safety related problems that can result from non-optimal training in RL. For instance, for an RL agent to be effective it should first cover all the situations during training that it may face later. This is often difficult when applied to the real-world. In this work we investigate the impact of RL applied to the context of intelligent vehicle control. We analyse the implications of RL in path planning tasks and we discuss two possible approaches to overcome the gap between the theorical developments of RL and its practical applications. Specifically, firstly this paper discusses the role of Curriculum Learning (CL) to structure the learning process of intelligent vehicle control in a gradual way. The results show how CL can play an important role in training agents in such context. Secondly, we discuss a method of transferring RL policies from simulation to reality in order to make the agent experience situations in simulation, so it knows how to react to them in reality. For that, we use Arduino Yún controlled robots as our platforms. The results enhance the effectiveness of the presented approach and show how RL policies can be transferred from simulation to reality even when the platforms are resource limited.

Keywords: vehicle control; reinforcement learning; curriculum learning; sim-to-real world; intelligent mobility

Citation: Pina, R.; Tibebu, H.; Hook, J.; De Silva, V.; Kondoz, A. Overcoming Challenges of Applying Reinforcement Learning for Intelligent Vehicle Control. *Sensors* **2021**, *21*, 7829. https://doi.org/10.3390/s21237829

Academic Editor: Juan A. Cabrera

Received: 1 October 2021
Accepted: 20 November 2021
Published: 25 November 2021

1. Introduction

Reinforcement learning has been well studied in the recent past as it is considered one of the most prominent paradigms in machine learning [1,2]. Inspired by the biological behaviours of humans or animals, RL consists in following a trial-and-error basis to train the agents to learn the best actions for each situation they face [3]. RL quickly proved to be useful in the gaming industry, where famous works have proved how a RL agent was capable of beating the human being in games such as Go or in a set of the famous Atari games [4,5].

Due to its proven success, the areas of application of RL have expanded to diverse fields. For instance, in healthcare studies have found RL to be useful as an assistant for critical decision making [6] or in aiding in the treatment of certain conditions [7,8]. RL has also been used in finance as a tool for portfolio optimization [9] or for analysing different types of stock markets [10]. In the field of robotics control and autonomous systems the applications are also many. As one of the first applications of RL in real robots, in [11] the authors present a robot that learns how to push boxes, trained using RL. More recently, works such as [12,13] propose new methods that use RL in mobile robots and autonomous vehicles to improve on-site learning. Outside the ground domain, [14] show

how RL can be applied also to the air domain by teaching flight control policies to an autonomous helicopter.

Coming back to the ground domain, when it comes to intelligent vehicle control there are multiple aspects that should be considered. Path planning is a key aspect in autonomous navigation where RL can play an important role [15]. Path planning can be described as a process that enables a mobile robot to traverse between the specific start and goal location in an environment with static and dynamic obstacles. The robot is required to find the optimal path with the shortest time, distance, and working cost.

There are three types of path planning algorithms: traditional algorithms, intelligent bionic algorithms, and reinforcement-based path planning algorithms [16]. Traditional path planning algorithms include A* algorithms and Dijkstra algorithms. Genetic algorithms and ant colony algorithms are typical examples of bionic path algorithms. Traditional and bionic algorithms can achieve a good result in a known environment on a global mapping scale. However, real-world road conditions are very complex. In recent years, RL has become more suitable for path planning activity because it requires less prior information about the environment [17,18].

In path planning there are two conventional approaches: offline and online planning. The former approach assumes that the environment is perfectly known, and obstacles are static. The latter approach assumes environments are partially known, and blocks could be dynamic [19]. Traditional offline approaches cannot be directly applied to solve online path planning problems because they assume static obstacles.

Learning-based approaches, such as deep learning and RL, have been studied to tackle online path planning in a dynamic environment [20–22]. With all its shortcomings, RL is best suited for problems such as path planning because it requires low prior knowledge of the environment. In [23], the authors present for the first time how to employ RL for path planning. The focus of this work was to test the usability of RL for collision avoidance in a multi-robot environment. As a combination of deep learning and RL, deep RL has also become a popular research interest for path planning problems [15,24–28]. Deep Q-networks (DQNs) [5] is a deep RL algorithm which combines Q-learning [29] and deep neural networks. In the DQNs approach, it is possible to use the current state and action as an input of the neural network to get the action's Q-value as an output. Once acquiring all possible actions, the one with the maximum Q-value will be chosen. Finally, the corresponding action is selected and executed. In [30] the authors proposed a deep RL method that decreases the learning time by parallelising the algorithm across multiple robots. However, despite Deep Q-networks notable success, DQNs have two major drawbacks: they require high computational power and a long learning time.

While progress has been made about how to use RL in path planning tasks there are still challenges to address. For instance, when the environment is completely unknown and complex, applying RL in path planning can be challenging due to the many possible states that the agent needs to explore in order to become completely aware of the environment around [31,32]. Therefore, in this work we carry out a set of experiments to analyse the impact of the environmental complexity on the learning process, and more specifically on the learning time for path planning tasks. Additionally, we discuss a method for transferring RL policies from simulation to reality. With such setup it is possible to cover multiple situations in simulation that would be unexpected in the real-world and would make the agent confused. By doing it in simulation first, then the agent is able to react to these situations in reality accordingly.

Analysing from a multi-agent perspective, to apply these concepts in situations with multiple agents can be increasingly difficult when the number of agents is increased. Such configuration will have an impact in the complexity and stability of the environment [2]. To overcome the gap between theoretical RL and its applications to real scenarios, in this work we also explore how CL can be applied to multi-agent settings in the context of intelligent vehicle control. More specifically, we use this method in a driving decision making problem that consists of a traffic junction. The contributions of this paper are as follows:

1. We investigate the impact of the environmental complexity in the learning process of RL tasks involving path planning scenarios;
2. We discuss a method for transferring RL policies from the simulation domain to the real-world domain, supported by empirical evidence and a working algorithm for the discussed method;
3. We show how CL can be applied within the context of intelligent vehicle control in tasks involving multiple agents.

The rest of this paper is organized as follows: in section two we introduce the key concepts to the understanding of this paper and we present a review of those concepts. In section three, we describe the methodology and experimental setup used in the experiments that are illustrated and discussed in section four. Finally, we conclude this paper in section five, where we also discuss the future work following the results in this paper.

2. Background

2.1. Q-Learning

Q-learning is one of the most popular RL approaches. Therefore, it has been widely used to solve path planning problems [31,32]. Q-learning attempts to learn the theory of delayed rewards to get better future rewards that are indirectly a consequence of the previous actions. The main objective of this method is to avoid local optimal activities that are not globally aligned. Reference [33] introduced Q-learning as a Temporal Difference approach (TD). In [29], the authors proved the convergence of Q-learning to optima and its relationship with TD and dynamic programming. To overcome one of the limitations of using Q-learning for path planning tasks—the impracticality of storing the Q-table for all states—[34] proposes a real-time Q-learning approach that avoids storing tables in advance. The simulation was conducted in a 10 by 10 grid map with static obstacles. Other algorithms such as Deep Q-Networks (DQNs) [5] have also been proposed to overcome similar limitations.

Reference [31] introduced a Q-learning approach which involves the addition of a distance aspect into the decision of direction. This approach reduced the number of steps taken by the agent. However, the time taken by the model to converge is greater than the traditional Q-learning. Reference [35] proposed a method to reduce the convergence time of Q-learning for a path planning application by introducing the Flower Pollination Algorithm (FPA) to improve the initialisation of Q-learning. This method was only tested in a simple and static environment. Reference [32] investigated the use of a single source transfer and improved Q-learning transfer to acquire better learning. While this method was tested and it reduced learning speed when compared with the conventional methods, it was only tested in a complex static environment.

To demonstrate the use and challenges of RL in path planning applications, in this work we use the Q-learning algorithm in different environmental setups. Q-learning is a model-free representative learning algorithm. An action related to a specific state is defined by a policy. Q-learning uses this policy to select an action for the agent. The policy establishes the reward or penalty for a particular action concerning the state. Q-learning agents learn by updating their Q-function whose values are obtained from the following equation,

$$Q(s,a) = r(s,a) + \gamma max_a(Q(s',a)) \tag{1}$$

where s is the current state, a is the action, r is the reward, s' is the next state resulting from taking an action a at the state s, and γ is the discount factor. The role of the discount factor in this equation is to help defining the agent's preference towards a short-term reward. The value of γ is between [0, 1]. While values closer to 0 make the agent susceptible to a short-term reward, a value closer to 1 makes the agent prone to a long-term reward.

2.2. Multi-Agent Reinforcement Learning and Curriculum Learning

Deep Neural Networks (DNNs) are usually trained using variations of stochastic gradient descent. A subset of training data is randomly sampled from the training set, passed through the DNN, and the loss and the gradients are calculated and propagated through the layers of the DNN. Stochastic gradient descent not only increases computational efficiency for very large datasets, but it is also an unbiased estimator of the true gradient of the DNN. However, we never see human teachers open the textbook, pick random exercises from random topics, and ask their students to solve the exercises in the hopes that after enough time has passed, all the material in the textbook will be covered and the students learned the material. Given enough time, all topics may eventually get covered, but it is more likely that the students quit soon out of sheer frustration. Usually, the teacher follows a curriculum, designed to facilitate the learning of a collection of interrelated topics. For example, it would make sense to teach children addition before multiplication, because you can teach multiplication through addition. This way new knowledge is built on something they already know, minimising the amount of new material needed to learn. CL [36] is a method of structuring the learning process to facilitate learning for DNNs. Informally, CL proposes to first learn and solve simpler versions of the task at hand, gradually building up to more difficult tasks. The curriculum can be thought of as a sequence of tasks, each consecutive task at least as challenging as the previous one. Assuming there is transferable knowledge between any two consecutive tasks, we may be able to re-use existing knowledge to facilitate learning the next task. By training the DNN on intermediate tasks, we assume the learner will be better prepared to learn the target task, compared to a randomly initialised learner. The existing knowledge could help the DNN reach an acceptable level of performance faster than without using a curriculum or be able to learn a complex composite computer vision which cannot be learned well without a curriculum [37].

To create curricula, a way of generating intermediate tasks is required. In Multi-Agent Reinforcement Learning (MARL), multiple agents interact with each other in order to achieve a certain objective. When compared to single-agent RL, MARL differs mainly in the fact that there is more than one single agent involved in the task of the environment. Hence, there is a team reward that corresponds to the performance of all the agents, and they all receive this same team reward. Furthermore, there is also a set of actions corresponding to each agent instead of a single action and each individual agent may receive different state information depending on their positions in the environment. In other words, in MARL different agents may see different parts of the environment at each state. Figure 1 illustrates the main differences between single agent and multi-agent RL. Naturally, in MARL, a simple way of creating new tasks is by changing the number of agents in the system. For example, instead of training robot soccer players in a 11 vs. 11 player environment, they can first be trained in a 5 vs. 5 environment. Changing the number of agents such as this can lower the difficulty of the task (reduce the number of opponents), change the meaning of the task (a 1 vs. 5 soccer game focuses on training the goalkeeper) and lower the resource costs used to operate the (virtual or physical) learning environment. Agent Count Based Curricula have been used with success in MARL systems.

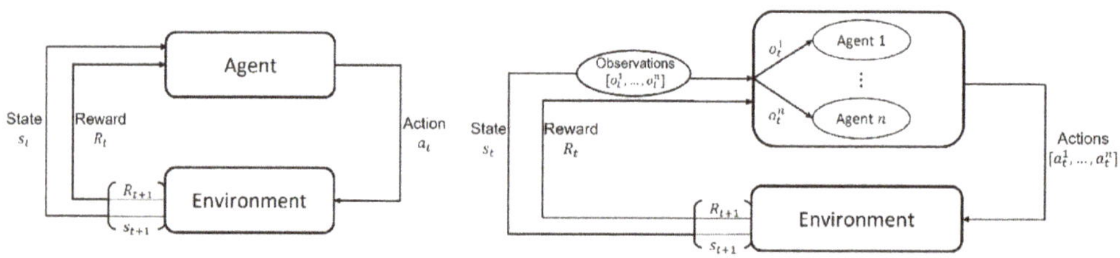

Figure 1. Illustration of the main differences between Reinforcement Learning (RL) (**left**) and Multi-Agent Reinforcement Learning (MARL) (**right**) (adapted from [38]). The figure represents the dynamics of a RL and MARL systems at a given time step t.

Reference [36] defines CL(CL) as a sequence of learning distributions $Q_{(\lambda)}, 0 \leq \lambda \leq 1$, where:

$$\forall \epsilon > 0 : \lambda < \lambda + \epsilon \rightarrow H(Q_\lambda) < H(Q_{\lambda+\epsilon}),$$
$$\forall \epsilon > 0 : \lambda < \lambda + \epsilon \rightarrow W_\lambda(z) < W_{\lambda+\epsilon}(z) \tag{2}$$

where $H(Q_\lambda)$ is entropy H of the distribution Q at step λ, $W_\lambda(z)$ is the weight of training sample z in Q_λ. Intuitively, in each successive distribution, the training samples get re-weighted, with entropy of the distribution $H(Q_\lambda)$ growing monotonically. The initial distribution, Q_0, would favour "easy" samples and place a weight of 0 on "hard" samples. In the distribution sequence, the weights of all samples are gradually increased, culminating in a distribution where all samples have a weight of 1, which is the target training distribution, Q_1. However, in [36] the authors do not specify a measure for the "hardness" of samples, indicating the need for future work in this regard. In addition to formalising CL, the authors found that a learner trained with CL achieved better generalisation compared to the learner trained without a curriculum in their two experiments. Additionally, they hypothesised that CL has a regularising effect, similar to unsupervised pre-training [39], on the final learning task. In [39] the authors theorised that each successive distribution in the curriculum would guide the learner's internal parameters (such as the weights in a neural network) toward regions in parameter space that would serve as a better starting point for the next task in the curriculum.

When it comes to multi-agent systems, recent works have focused on formulating CL specifically in the context of RL. Reference [40] defines a curriculum as a directed acyclic graph, where the edges of the graph introduce a partial ordering over the nodes. Informally, this means that every task has some other task that comes before it in the curriculum. The graph nodes represent subsets of experience (transitions) associated with the intermediate learning tasks. In this general sense, a curriculum defines which intermediate learning tasks should be learned before others to maximise knowledge transfer between tasks. The primary purpose of the curriculum is to help the learner perform better in the final task in the curriculum. Compared to the Supervised Learning formulation of CL in [36], the general idea of ordering experience remains, but the curricula are not limited to a simple sequence of learning tasks. Reference [40] also outlines three key areas of curriculum learning: task generation (how to generate useful intermediate tasks?), sequencing (in what order should we learn the intermediate tasks?), and transfer learning (how should we transfer knowledge between two tasks?).

3. Methodology and Experimental Setup

In this section we describe the environments that are used in the experiments in this work and the platforms used in the proposed sim-to-real approach. The workflow of this project is presented in Figure 2.

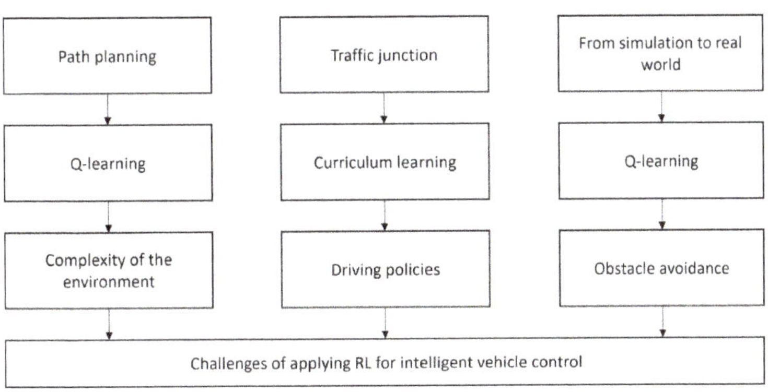

Figure 2. Workflow of the proposed method in this paper.

The environment used in the first set of experiments is represented by set of 2D discrete grid world environments $\mathcal{W} \subseteq \mathbb{R}^2$ with size $H \times W$. There are four possible states for each cell in the environment: free cell, occupied by agent, occupied by a static abstract, or occupied by dynamic obstacles, where a set of static and dynamic obstacles is represented as $\mathcal{C}_s = \{s_1, \ldots, s_{Ns}\}$ and $\mathcal{C}_d(t) = \{d_1(t), \ldots, d_{Nd}(t)\}$, respectively. The free cells are denoted by $\mho = \{f_1, \ldots, f_{Ns}\}$.

The reward function of the model is designed to encourage good moves and discourage wrong actions by the agent. In this task, the agent receives a negative reward when it hits static and dynamic obstacles and then returns to a random starting point. The agent receives a positive reward when it reaches the target point and the episode ends. In this environment, each new episode contains a new unique learning set of static and dynamic obstacles, a new starting point, and a goal point.

Given the (x, y) position of an agent as result of an action a, the reward function R is defined as:

$$R = \begin{cases} -5 & \textit{collision to } C_d \\ -5 & \textit{collision to } C_s \\ 10 & \textit{target reached} \\ -1 & \textit{penalty for each additional move} \end{cases} \tag{3}$$

To compare the transferability of the model to unknown and complex environments, we train the model in different environment settings. The multiple setups are summarized in Table 1. In Section 4 we also present the illustration of the configurations of the experimented environments. In each environment, the number of obstacles is designed in three complexity configurations: easy, moderate, and hard.

Table 1. Configuration of the experimented environments.

Environment	Grid World	Obstacles	Obstacles	Obstacles
Environment 1	100 × 100	easy	moderate	hard
Environment 2	200 × 120	easy	moderate	hard

In the second set of experiments in this paper involving CL, we use the Traffic Junction environment as our learning environment, found in a collection of multi-agent learning environments [41]. In this 2D grid world environment, agents are spawned at different ends of a cross-shaped junction. Each agent is assigned a target they must reach by driving forward towards the four-way intersection, making a turn, and driving forward to reach their destination. The agents can take two actions, move forward, or stop, as they move along the pre-defined route towards their destination. However, the reward function is designed to discourage both collisions with other agents and avoid traffic jams, both of which yield a small punishment to the agents. The size of the grid world is 14 by 14 and the number of cars in the environment is 10. The simpler version of the task used for CL is also 14 by 14, but with only 4 cars. For a visual representation of the environment please see Figure 3.

Finally, the experimental setting used in the proposed sim-to-real approach is composed by both a physical platform and a computer simulation to enable transfer of learning between the physical and simulation domains. With the proposed method we try to address the 'reality gap' [42].

For the physical platform, we assembled multiple robots controlled by the microcontroller Arduino Yún. The model of the robots used is the Pirate 4WD Mobile Platform and they use three HC-SR04 ultrasonic sound sensors to gather information from the environment, placed on the front and both sides of the robots (see Figure 4). These sensors will measure the distances to other obstacles in front of them. We have intentionally limited the robots to use just the sound sensors as their only source of information gathering in order to keep them rudimentary and investigate how learning experience can be transferred from simulation to reality in such limited robots.

Figure 3. Visualisation of the traffic junction environment with 10 cars. From left to right, top to bottom: observe the cars 1 (red) and 2 (blue) as they approach the intersection and pass it without colliding.

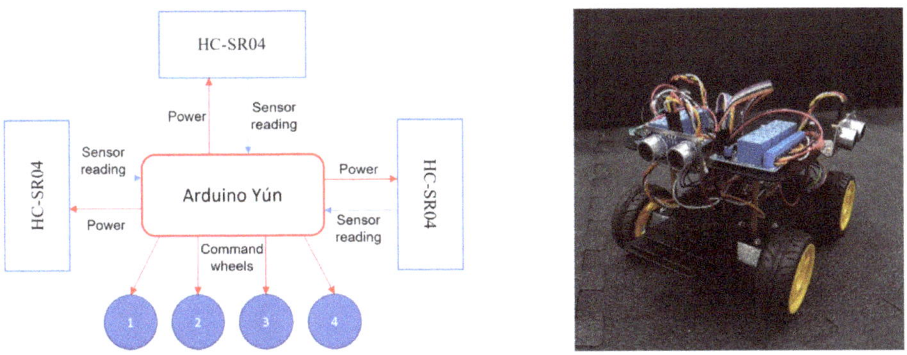

Figure 4. Arduino-controlled Pirate 4WD Mobile Platform (**right**) and diagram illustrating how the main parts of the platform interact with each other (**left**).

It has been discussed before how limited microcontrollers are and how difficult it can be to implement deep neural networks on them due mostly to memory and latency constraints [43,44]. In this sense, we decided to use Q-tables in the presented approach due to their simplicity and ease of transfer and use in the physical platforms controlled by the Arduino Yún. In the simulation implemented we use the Q-learning algorithm [29] to train a Q-table that will be transferred to the physical machines.

The task implemented in the simulated environment is to roam around and avoid obstacles. The environment is composed by two agents that are trained independently and receive a team reward. In the implementation of the simulation, we reproduce subsequent observations to each agent that would correspond to values gathered by the sound sensors in the physical machines. However, it would be impossible to account in simulation all the infinite possible values that the sensors could measure due to the randomness present in the real world or uncertainties related to the sensors [45,46]. Thus, the key to our method is to create a generalisation of the possible observations received by the ultrasonic sensors and map them to a finite number of values. Hence, in the simulated environment we consider the existence of three possible situations: (1) the front of the agent is free and so it can move forward; (2) the front of the agent is blocked and the distance on the right is greater than on the left; (3) the front of the agent is blocked and the distance on the left is greater than on the right. In the simulation, the agents are trained using RL to react to these scenarios and perform the right actions, resulting in two trained Q-tables and one of them will be exported to the microcontroller. Once the trained Q-table is in the robots, they can roam around and, when they receive an observation, that will be mapped to the generalized values defined before and get the right action from the trained Q-table, avoiding obstacles. Algorithm 1 describes the methodology used for transferring the learnt policies.

Algorithm 1. Algorithm used by the Arduino robots to use the transferred policies

1. **Set** static map $m \leftarrow [f_s, r_s, l_s]$
2. **Input** array of measured distances front, right and left, $[f, r, l]$
3. **For** each distance d in $[f, r, l]$ do
4. map d to value in m, $d \sim [f_s, r_s, l_s]$
5. add d to new array of mapped values, $[f_m, r_m, l_m].insert(d)$
6. **End For**
7. **Get** state s corresponding to $[f_m, r_m, l_m]$
8. **Get** Q-value for state s, Q_s
9. **Output** action $\underset{a}{\text{argmax}} Q_s$

4. Results and Discussions

In this section we present the results of the experiments performed in this work. The experiments in this paper are three-fold: first, we use Q-learning in a set of different environments to analyse the impact of the environmental complexity in the learning process of the RL agent in path planning tasks for intelligent vehicle control. Second, we show how CL can be applied to MARL driving decision-making scenarios. Third, we investigate a method for transfer RL policies from simulation to reality supported with empirical evidence.

4.1. Impact of Environmental Complexity in the Learning Process

This experimental setup is designed to examine the impact of the environmental complexity on path planning algorithms using RL or, more specifically in this case, Q-learning. Hence, we experimented with six different environments with different complexity configurations, with a 100 × 100 and 200 × 120 grid world systems. Figures 5 and 6 illustrate the 100 × 100 and 200 × 120 grid world systems, respectively. The model was firstly trained in an accessible environment and then is followed by a moderate and a highly complex setup. As illustrated in Figures 5 and 6, based on the complexity of the environment, the model took a different number of iterations to converge. However, we have not found a significant time difference between easy, moderate, and hard configurations for the same environment. Hence, the results suggest that changing the experimental environment to a more complex one will increase the training time more than just adding obstacles to a fixed environment. As the number of obstacles increases, the learning time will also increase.

The results show that the training time of the model is directly proportional to how big the training environment is. On average, it took 19.12 s to train the 100 × 100 and 26.48 s to train the 200 × 120 environments. Table 2 illustrates the total amount of time taken to train the model for the different configurations in each environment.

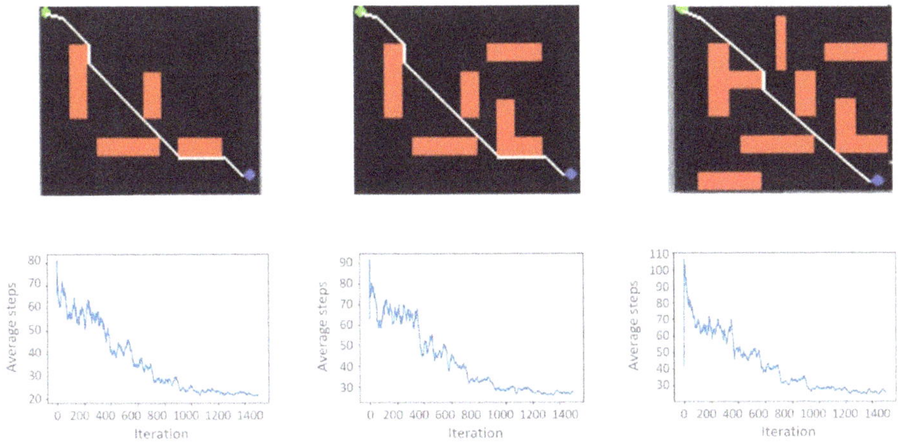

Figure 5. 100 × 100 grid-world environments and corresponding number of steps per learning iteration for each configuration. From left to right, easy, moderate, and hard.

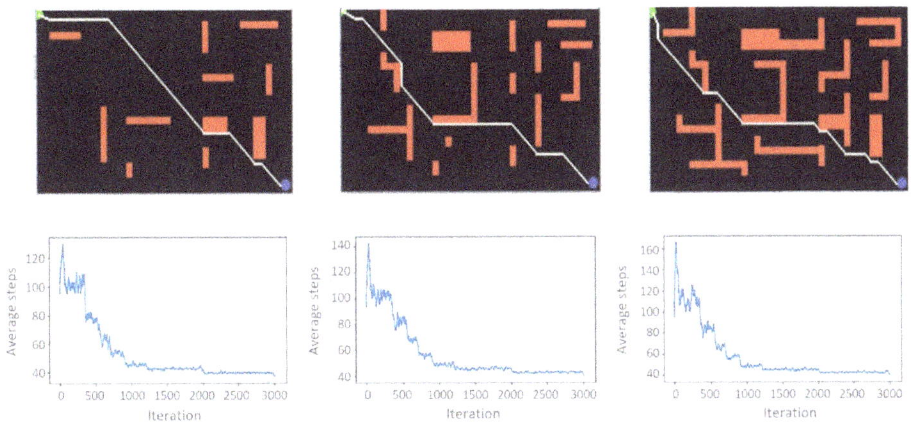

Figure 6. 200 × 120 grid-world environments and corresponding number of steps per learning iteration for each configuration. From left to the right, easy, moderate, and hard.

Table 2. Time taken to train the model for the different configurations in each environment.

Environment	Easy	Moderate	Hard
100 × 100	18.23 s	18.41 s	19.52 s
200 × 120	25.11 s	25.41 s	27.44 s

Environmental contexts in the real world are complex and dynamic. Representing this complexity in simulation is a challenge; however, the real change is the computational cost of training the model. This computational cost is visible in our results. When the environmental complexity increases (Figures 5 and 6), the computational cost also increases. Our paper only considers up to 200 × 120 grid world environments. However, to achieve a more reliable path planning result, the environment needs to be represented in a significantly larger grid world than the one used in this paper.

Figure 7 illustrates the adaptability problem of the Q-learning based path planning model. One of the significant drawbacks of this method is the lack of generalisation; the agents learn the detailed path of the training route. Our test results show that the model tends to always travel in the direction of the training route without considering the target position. For instance, we trained the model from the top left corner start point to the bottom right corner of the destination.

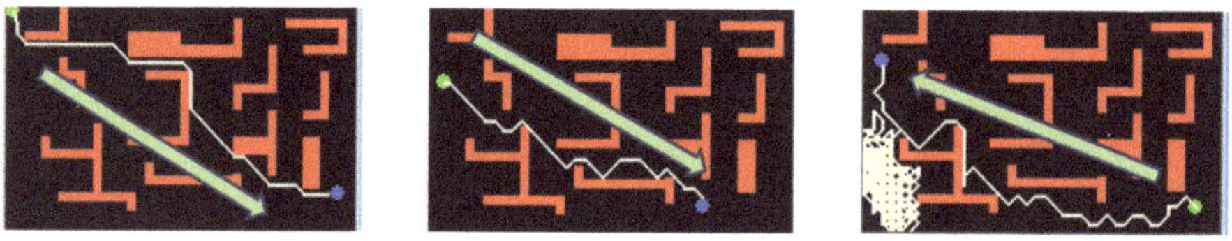

Figure 7. Illustration of the learnt paths for one of the path planning tasks experimented. On the left the training path direction, and on the centre and right the testing learnt path for two different starting and goal places.

Our test results show that the model performs better when the destination and start points are similar to the training environment. When we test the model by inverting the start and end points, the model gives lower performance results, suggesting that researchers in RL-based path planning activities need to find a way to train models in multiple routes within a single environment in a short time. As Figure 7 shows, the agent has more advanced knowledge of the environment in the direction of the training path than in another different path.

4.2. Curriculum Learning for MARL Driving Decision-Making Scenarios

In this set of experiments, we use a no curriculum baseline and two kinds of curricula in order to negotiate driving decision making policies in a traffic junction task in a MARL setup. For the baseline, we train the policy from a random initialisation for 2 million timesteps. The first kind is the forward curriculum, which is a sequence of tasks (in our case, the simple and target task). First, we train in the simple task, evaluating our performance on the target task periodically. If we see no improvement in the target task for 20 evaluations, we stop training in the simple task and switch to training in the target task. Then, we train in the target task for 2 million time steps. The second kind of curriculum is the sampling curriculum. In a sampling curriculum we sample learning tasks from a distribution of tasks. In our case, we sample the simple task with probability 0.2 and the hard task with probability 0.8. The sampling curriculum may perform better in some environments, for example, in [47] the authors found that a forward curriculum didn't accelerate learning in one of their tasks, but a sampling curriculum did. We train in the target task for 2 million time steps in total, so approximately 1.6 million time steps in the target and 0.4 million time steps in the simple task.

We use QMIX [48] as the MARL algorithm. QMIX is capable of training policies in a centralised manner that supports decentralised execution. It works by using a mixing network whose weights are derived from the available state information to estimate the joint action values of a team of cooperative agents. More specifically, each agent produces a set of action values based on their local observations, which the mixing network combines to produce the team's joint action value. The mixing network is designed to make sure the policies behave similarly during both centralised training and decentralised execution phases.

The results in Figure 8 show that after an initial jump in performance, all policies go through a steady decline in performance. This jump and the following decline are a lot more pronounced when training a policy from scratch, with the highest level of performance around −220, with the policy degrading in performance to around −800. After a temporary increase in performance at around the 90th evaluation iteration the policy degrades again, starting a consistent improvement in performance after the 160th evaluation iteration. From now on, the policy keeps consistently improving with slight but short-lived dips every now and then. The forward curriculum starts out at around −440, declining to −530, after which performance temporarily increases, dips again and sharply increases, similar to the no curriculum baseline. Following this jumpy start, we see a slow and jittery increase in performance up until the last iteration. The sampling curriculum follows a similar path to the forward curriculum, but the fluctuations in performance are less pronounced. Overall, the sampling curriculum shows to be the least unstable with regard to the policy's performance. Additionally, the sampling curriculum achieves the best performance level during the last iterations of training, followed by the forward curriculum and no curriculum baseline. These results suggests that the sampling curriculum is more efficient compared to others because it trained in the target task for approximately 80% of the time yet achieving better results. This may be due to training in the simple task often enough causes the DNN parameters to be pulled in two directions in parameter space. 80% of the time Gradient Descent (GD) pulls the DNN parameters towards solving the target task, but 20% of the time GD pulls the parameters towards solving the simple task. Assuming there is knowledge in knowing how to solve the simple task that can be used to help learn how to solve the target task, forcing the policy to learn to solve both problems can provide a regularising effect, constraining the possible DNN parameters to a more specific region of parameter space, aiding optimisation. When applying CL there are some factors that need to be accounted in some cases in order to maintain the desired performances. For instance, when using CL, the difficulty of the tasks and subtasks needs to be considered, which might not always be available. Furthermore, there is still some uncertainty when it comes to creating subtasks on-demand and, in some cases, it might be beneficial to filter the type of knowledge being transferred from task to task [40]. However, in scenarios such as the one experimented CL proves to be successful and capable of taking into account these factors.

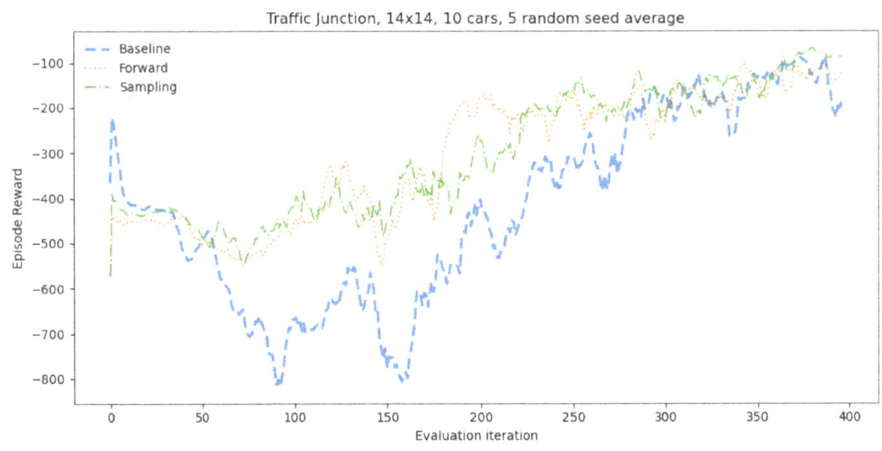

Figure 8. Traffic junction performance using no curriculum, a forward curriculum, and a sampling curriculum.

4.3. Sim-to-Real: Transfer of Reinforcement Learning Policies from Simulation to Reality

4.3.1. Simulation Domain

In this section, we show the results of the implementations of Q-learning [29] and DQN (Mnih, et al., 2015) in the simulated environment described in Section 3 that was used to transfer simulation experience to the physical machines. Although deep neural networks can perform better than simple Q-tables in RL when it comes to map states to the corresponding actions, it has been discussed previously how sometimes it can be difficult to implement neural networks in resource limited microcontrollers [43,44]. Therefore, to support the use of Q-tables in this work we perform a set of experiments in simulation. At the same time, the experimental results show how it is possible to minimize the number of collisions with obstacles in a certain environment by using RL.

In Figure 9 it is illustrated the evolution of the average number of collisions incurred by two agents (the collisions of both agents are added together) per training step over time for simple Q-learning and DQN. As described previously in Section 3, in this task two agents are given successive observations with distances to walls around them. Every time the agents collide with one of the walls each one receives a punishment. In Figure 9 it is possible to see that, although at first both DQN and Q-learning incur in some collisions, after some training time both can minimize the number of collisions, making the performances of the two methods close to each other in further training episodes. Thus, we can conclude that transferring a Q-table instead of a DQN to the physical robots was a reasonable option, since in this setup it would not have impact. Furthermore, in this particular setup, the Q-learning method can minimize the collisions as well as the DQN and in a smaller amount of time.

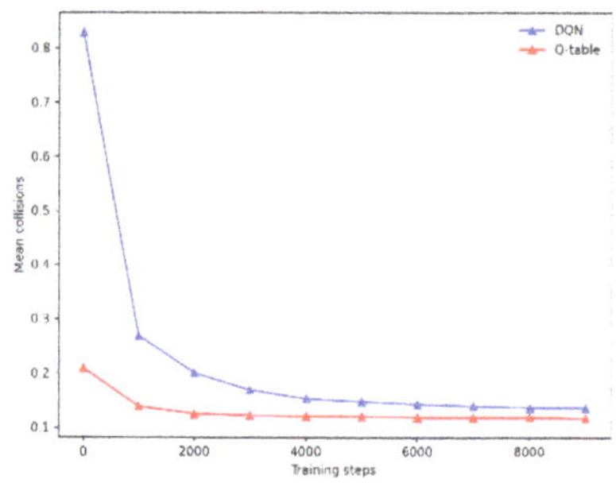

Figure 9. Collisions for the simulation environment described for DQN and Q-table approaches.

4.3.2. Real World Domain

To evaluate the fidelity of the proposed transferability approach from simulation to reality, we have implemented a hardcoded policy in a second Arduino Yún controlled robot to roam around and avoid obstacles. Although transferring complex situations learned in simulation to reality is still a big challenge due to multiple constraints such as the randomness in real world events [45,49], the results demonstrate the success of the approach attempted in the presented scenario. By transferring the learned Q-table from the simulation described to reality it was possible to observe a very similar behaviour between both the hardcoded and the robot following the transferred Q-table. They both were able to roam around and avoid obstacles and consequently, they were able to roam in the same area avoiding colliding to each other.

Figure 10 illustrates an occupancy grid map made using data gathered by the robot with the transferred Q-table. The robot was placed inside a large box where it gathered data using the ultrasonic sound sensors. This was possible without any collisions with the walls of the box, following the policy transferred from simulation and, as the figure shows, the box was mapped with a good level of accuracy.

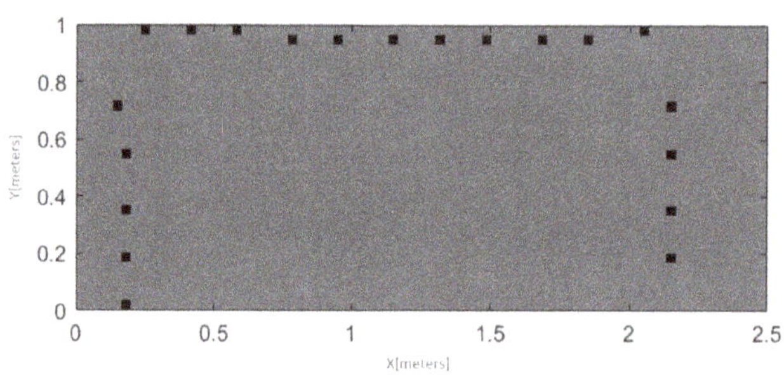

Figure 10. Occupancy grid map built with data collected by one of the Arduino robots corresponding to a rectangular box.

5. Conclusions and Future Work

In this work we have investigated some of the challenges of applying RL within the context of intelligent vehicle control as a real-world application. By referring to common intelligent vehicle control tasks such as path planning using RL, several problems may arise such as the existence of many possible states that the agent should experience, the existence of multiple agents, difficulty of representing states or how should we formulate these safety-critical tasks to be solved by trial-and-error. Considering a use case on path planning, we illustrate how the environmental complexity influences the learning time and the performance of the agents. To mitigate the challenges incurred, we discussed the use of two approaches: a method for transferring RL policies from simulation to reality and a CL based approach to improve learning in driving decision making scenarios. Both these approaches were demonstrated and discussed with reference to use cases pertaining to intelligent vehicle control. Our sim-to-real approach shows to be successful to transfer a collision avoidance policy from a simulated environment to the real world. In a real-world context, our results enhance the importance of simulation to reality experiments by showing that the number of collisions during training time can be minimised in simulation. In our second approach, we demonstrated how CL can be useful when applied to intelligent vehicle control situations. The results showed that using agents with a curriculum to structure the learning process can be beneficial in driving decision making tasks such as a traffic junction task where multiple agents need to negotiate their passages.

In the future, we aim to extend sim-to-real concepts to more complex scenarios, such as dynamic environments and multi decision-making tasks. Furthermore, we intend to investigate how CL can be used together with sim-to-real methods so that the policies trained with CL can be used in reality.

Author Contributions: Investigation, R.P., H.T. and J.H.; Methodology, R.P. and H.T.; Project administration, V.D.S.; Software, R.P., H.T. and J.H.; Supervision, V.D.S. and A.K.; Validation, V.D.S.; Writing—original draft, R.P., H.T. and J.H.; Writing—review & editing, R.P. and V.D.S. All authors have read and agreed to the published version of the manuscript.

Funding: This research was funded by the EPSRC, grant number EP/T000783/1.

Institutional Review Board Statement: Not applicable.

Informed Consent Statement: Not applicable.

Data Availability Statement: Not applicable.

Conflicts of Interest: The authors declare no conflict of interest.

References

1. Chen, G. A New Framework for Multi-Agent Reinforcement Learning—Centralized Training and Exploration with Decentralized Execution via Policy Distillation. In Proceedings of the 19th International Conference on Autonomous Agents and MultiAgent Systems, Auckland, New Zealand, 9–13 May 2020.
2. Hernandez-Leal, P.; Kartal, B.; Taylor, M.E. A survey and critique of multiagent deep reinforcement learning. *Auton. Agents Multi-Agent Syst.* **2019**, *33*, 750–797. [CrossRef]
3. Nguyen, T.T.; Nguyen, N.D.; Nahavandi, S. Deep Reinforcement Learning for Multi-Agent Systems: A Review of Challenges, Solutions and Applications. *IEEE Trans. Cybern.* **2020**, *50*, 3826–3839. [CrossRef]
4. Silver, D.; Huang, A.; Maddison, C.J.; Guez, A.; Sifre, L.; van den Driessche, G.; Schrittwieser, J.; Antonoglou, I.; Panneershelvam, V.; Lanctot, M.; et al. Mastering the game of Go with deep neural networks and tree search. *Nature* **2016**, *529*, 484–489. [CrossRef] [PubMed]
5. Mnih, V.; Kavukcuoglu, K.; Silver, D.; Rusu, A.A.; Veness, J.; Bellemare, M.G.; Graves, A.; Riedmiller, M.; Fidjeland, A.K.; Ostrovski, G.; et al. Human-level control through deep reinforcement learning. *Nature* **2015**, *518*, 529–533. [CrossRef]
6. Liu, S.; See, K.C.; Ngiam, K.Y.; Celi, L.A.; Sun, X.; Feng, M. Reinforcement Learning for Clinical Decision Support in Critical Care: Comprehensive Review. *J. Med. Internet Res.* **2020**, *22*, e18477. [CrossRef]
7. Pineau, J.; Guez, A.; Vincent, R.; Panuccio, G.; Avoli, M. Treating epilepsy via adaptive neurostimulation: A reinforcement learning approach. *Int. J. Neural Syst.* **2009**, *19*, 227–240. [CrossRef]

8. Escandell-Montero, P.; Chermisi, M.; Martínez-Martínez, J.M.; Gómez-Sanchis, J.; Barbieri, C.; Soria-Olivas, E.; Mari, F.; Vila-Francés, J.; Stopper, A.; Gatti, E.; et al. Optimization of anemia treatment in hemodialysis patients via reinforcement learning. *Artif. Intell. Med.* **2014**, *62*, 47–60. [CrossRef]
9. Hu, Y.-J.; Lin, S.-J. Deep Reinforcement Learning for Optimizing Finance Portfolio Management. In Proceedings of the 2019 Amity International Conference on Artificial Intelligence, Dubai, United Arab Emirates, 4–6 February 2019.
10. Liu, X.-Y.; Yang, H.; Chen, Q.; Zhang, R.; Yang, L.; Xiao, B.; Wang, C.D. FinRL: A Deep Reinforcement Learning Library for Automated Stock Trading in Quantitative Finance. In Proceedings of the 34th Conference on Neural Information Processing Systems, Vancouver, BC, Canada, 6–12 December 2020.
11. Mahadevan, S.; Connell, J. Automatic programming of behavior-based robots using reinforcement learning. *Artif. Intell.* **1992**, *55*, 311–365. [CrossRef]
12. Martinez-Marin, T.; Duckett, T. Fast Reinforcement Learning for Vision-guided Mobile Robots. In Proceedings of the 2005 IEEE International Conference on Robotics and Automation, Barcelona, Spain, 18–22 April 2005.
13. Hester, T.; Quinlan, M.; Stone, P. A Real-Time Model-Based Reinforcement Learning Architecture for Robot Control. In Proceedings of the IEEE International Conference on Robotics and Automation, Saint Paul, MN, USA, 14–18 May 2012.
14. Ng, A.Y.; Kim, H.J.; Jordan, M.I.; Sastry, S. Autonomous helicopter flight via reinforcement learning. In Proceedings of the Advances in Neural Information Processing Systems 16, Vancouver, BC, Canada, 8–13 December 2003.
15. Wang, Y.; Fang, Y.; Lou, P.; Yan, J.; Liu, N. Deep Reinforcement Learning based Path Planning for Mobile Robot in Unknown Environment. *J. Phys. Conf. Ser.* **2020**, *1576*, 012009. [CrossRef]
16. Raajan, J.; Srihari, P.V.; Satya, J.P.; Bhikkaji, B.; Pasumarthy, R. Real Time Path Planning of Robot using Deep Reinforcement Learning. *IFAC-PapersOnLine* **2020**, *53*, 15602–15607. [CrossRef]
17. Wulfmeier, M.; Rao, D.; Wang, D.Z.; Ondruska, P.; Posner, I. Large-scale cost function learning for path planning using deep inverse reinforcement learning. *Int. J. Robot. Res.* **2017**, *36*, 1073–1087. [CrossRef]
18. Zheng, Y.; Liu, S. Bibliometric analysis for talent identification by the subject–author–citation three-dimensional evaluation model in the discipline of physical education. *Libr. Hi Tech* **2020**. ahead-of-print. [CrossRef]
19. Wang, B.; Liu, Z.; Li, Q.; Prorok, A. Mobile Robot Path Planning in Dynamic Environments through Globally Guided Reinforcement Learning. *IEEE Robot. Autom. Lett.* **2020**, *5*, 6932–6939. [CrossRef]
20. Sichkar, V.N. Reinforcement Learning Algorithms in Global Path Planning for Mobile Robot. In Proceedings of the 2019 International Conference on Industrial Engineering, Applications and Manufacturing, Sochi, Russia, 25–29 March 2019.
21. Xin, J.; Zhao, H.; Liu, D.; Li, M. Application of deep reinforcement learning in mobile robot path planning. In Proceedings of the 2017 Chinese Automation Congress, Jinan, China, 20–22 October 2017.
22. Liu, X.-H.; Zhang, D.-G.; Yan, H.-R.; Cui, Y.-Y.; Chen, L. A New Algorithm of the Best Path Selection Based on Machine Learning. *IEEE Access* **2019**, *7*, 126913–126928. [CrossRef]
23. Mataric, M.J. Interaction and Intelligent Behavior. Ph.D. Thesis, MIT, Cambridge, MA, USA, 1994.
24. Gao, J.; Ye, W.; Guo, J.; Li, Z. Deep Reinforcement Learning for Indoor Mobile Robot Path Planning. *Sensors* **2020**, *20*, 5493. [CrossRef]
25. Xu, H.; Wang, N.; Zhao, H.; Zheng, Z. Deep reinforcement learning-based path planning of underactuated surface vessels. *Cyber-Phys. Syst.* **2019**, *5*, 1–17. [CrossRef]
26. Yan, C.; Xiang, X.; Wang, C. Towards Real-Time Path Planning through Deep Reinforcement Learning for a UAV in Dynamic Environments. *J. Intell. Robot. Syst.* **2020**, *98*, 297–309. [CrossRef]
27. Luis, S.Y.; Reina, D.G.; Marín, S.L.T. A Multiagent Deep Reinforcement Learning Approach for Path Planning in Autonomous Surface Vehicles: The Ypacaraí Lake Patrolling Case. *IEEE Access* **2021**, *9*, 17084–17099. [CrossRef]
28. Prianto, E.; Park, J.-H.; Bae, J.-H.; Kim, J.-S. Deep Reinforcement Learning-Based Path Planning for Multi-Arm Manipulators with Periodically Moving Obstacles. *Appl. Sci.* **2021**, *11*, 2587. [CrossRef]
29. Watkins, C.; Dayan, P. Technical Note Q-Learning. In *Machine Learning*; Kluwer Academic Publishers: Boston, MA, USA, 1992; Volume 8, pp. 279–292.
30. Gu, S.; Holly, E.; Lillicrap, T.; Levine, S. Deep reinforcement learning for robotic manipulation with asynchronous off-policy updates. In Proceedings of the 2017 IEEE International Conference on Robotics and Automation, Singapore, 29 May–3 June 2017.
31. Ong, P.; Low, E.S.; Low, C.Y. Mobile Robot Path Planning using Q-Learning with Guided Distance and Moving Target Concept. *Int. J. Integr. Eng.* **2020**, *13*, 177–188.
32. Wu, S.; Hu, J.; Zhao, C.; Pan, Q. Path planning for autonomous mobile robot using transfer learning-based Q-learning. In Proceedings of the 2020 3rd International Conference on Unmanned Systems (ICUS), Harbin, China, 27–28 November 2020.
33. Kröse, B.J.A. Learning from delayed rewards. *Robot. Auton. Syst.* **1995**, *15*, 233–235. [CrossRef]
34. Kim, H.; Lee, W. Real-Time Path Planning Through Q-learning's Exploration Strategy Adjustment. In Proceedings of the 2021 International Conference on Electronics, Information, and Communication, Jeju, Korea (South), 31 January–3 February 2021.
35. Low, E.S.; Ong, P.; Cheah, K.C. Solving the optimal path planning of a mobile robot using improved Q-learning. *Robot. Auton. Syst.* **2019**, *115*, 143–161. [CrossRef]
36. Bengio, Y.; Louradour, J.; Collobert, R.; Weston, J. Curriculum Learning. In Proceedings of the 26th Annual International Conference on Machine Learning, Montreal, QC, Canada, 14–18 June 2009.

37. Gülçehre, Ç.; Bengio, Y. Knowledge matters: Importance of prior information for optimization. *J. Mach. Learn. Res.* **2016**, *17*, 226–257.
38. Sutton, R.S.; Barto, A.G. *Reinforcement Learning: An Introduction*; MIT Press: Cambridge, MA, USA, 1998.
39. Bengio, Y.; Lamblin, P.; Popovici, D.; Larochelle, H. Greedy Layer-Wise Training of Deep Networks. In Proceedings of the Advances in Neural Information Processing Systems, Vancouver, BC, Canada, 3–8 December 2007.
40. Narvekar, S.; Peng, B.; Leonetti, M.; Sinapov, J.; Taylor, M.E.; Stone, P. Curriculum Learning for Reinforcement Learning Domains: A Framework and Survey. *J. Mach. Learn. Res.* **2020**, *21*, 1–50.
41. Koul, A. ma-gym: Collection of Multi-Agent Environments Based on OpenAI gym. GitHub Repository. Available online: https://github.com/koulanurag/ma-gym (accessed on 1 September 2021).
42. Lucchi, M.; Zindler, F.; Muhlbacher-Karrer, S.; Pichler, H. robo-gym—An Open Source Toolkit for Distributed Deep Reinforcement Learning on Real and Simulated Robots. In Proceedings of the 2020 IEEE/RSJ International Conference on Intelligent Robots and Systems, Las Vegas, Nevada, 25–29 October 2020.
43. Do, T.-D.; Duong, M.-T.; Dang, Q.-V.; Le, M.-H. Real-Time Self-Driving Car Navigation Using Deep Neural Network. In Proceedings of the 2018 4th International Conference on Green Technology and Sustainable Development (GTSD), Ho Chi Minh City, Vietnam, 23–24 November 2018.
44. Kusupati, A.; Singh, M.; Bhatia, K.; Kumar, A.; Jain, P.; Varma, M. FastGRNN: A Fast, Accurate, Stable and Tiny Kilobyte Sized Gated Recurrent Neural Network. In Proceedings of the 32nd International Conference on Neural Information Processing Systems, Montreal, Canada, 3–8 December 2018.
45. Tobin, J.; Fong, R.; Ray, A.; Schneider, J.; Zaremba, W.; Abbeel, P. Domain randomization for transferring deep neural networks from simulation to the real world. In Proceedings of the 2017 IEEE/RSJ International Conference on Intelligent Robots and Systems (IROS), Vancouver, BC, Canada, 24–28 September 2017.
46. Panda, K.G.; Agrawal, D.; Nshimiyimana, A.; Hossain, A. Effects of environment on accuracy of ultrasonic sensor operates in millimetre range. *Perspect. Sci.* **2016**, *8*, 574–576. [CrossRef]
47. Gupta, J.K.; Egorov, M.; Kochenderfer, M. Cooperative Multi-agent Control Using Deep Reinforcement Learning. In Proceedings of the Autonomous Agents and Multiagent Systems, São Paulo, Brazil, 8–12 May 2017.
48. Rashid, T.; Samvelyan, M.; de Witt, C.S.; Farquhar, G.; Foerster, J.; Whiteson, S. QMIX: Monotonic Value Function Factorisation for Deep Multi-Agent Reinforcement Learning. In Proceedings of the 35th International Conference on Machine Learning, Stockholm, Sweden, 10–15 July 2018.
49. Kalapos, A.; Gór, C.; Moni, R.; Harmati, I. Sim-to-real reinforcement learning applied to end-to-end vehicle control. In Proceedings of the 2020 23rd International Symposium on Measurement and Control in Robotics (ISMCR), Budapest, Hungary, 15–17 October 2020.

sensors

MDPI

Article

On–Off Scheduling for Electric Vehicle Charging in Two-Links Charging Stations Using Binary Optimization Approaches

Rafał Zdunek [1,*], Andrzej Grobelny [1], Jerzy Witkowski [1] and Radosław Igor Gnot [2]

1 Faculty of Electronics, Photonics and Microsystems, Wroclaw University of Science and Technology, Wybrzeze Wyspianskiego 27, 50-370 Wroclaw, Poland; andrzej.grobelny@pwr.edu.pl (A.G.); jerzy.witkowski@pwr.edu.pl (J.W.)
2 Elocity Sp. z o.o., pl. Wolnica 13/10, 31-060 Krakow, Poland; igor@elo.city
* Correspondence: rafal.zdunek@pwr.edu.pl; Tel.: +48-71-320-3215

Abstract: In this study, we deal with the problem of scheduling charging periods of electrical vehicles (EVs) to satisfy the users' demands for energy consumption as well as to optimally utilize the available power. We assume three-phase EV charging stations, each equipped with two charging ports (links) that can serve up to two EVs in the scheduling period but not simultaneously. Considering such a specification, we propose an on–off scheduling scheme wherein control over an energy flow is achieved by flexibly switching the ports in each station on and off in a manner such as to satisfy the energy demand of each EV, flatten the high energy-consuming load on the whole farm, and to minimize the number of switching operations. To satisfy these needs, the on–off scheduling scheme is formulated in terms of a binary linear programming problem, which is then extended to a quadratic version to incorporate the smoothness constraints. Various algorithmic approaches are used for solving a binary quadratic programming problem, including the Frank–Wolfe algorithm and successive linear approximations. The numerical simulations demonstrate that the latter is scalable, efficient, and flexible in a charging procedure, and it shaves the load peak while maintaining smooth charging profiles.

Keywords: electrical vehicles; EV charging scheduling; binary linear programming; binary quadratic programming

Citation: Zdunek, R.; Grobelny, A.; Witkowski, J.; Gnot, R.I. On–Off Scheduling for Electric Vehicle Charging in Two-Links Charging Stations Using Binary Optimization Approaches. *Sensors* **2021**, *21*, 7149. https://doi.org/10.3390/s21217149

Academic Editor: Juan A. Cabrera

Received: 28 September 2021
Accepted: 24 October 2021
Published: 28 October 2021

1. Introduction

The electrification of transportation offers multiple benefits, including the reduction of noise pollution, fumes, and GHG emissions [1,2]. Hence, it has attracted increasing attention in recent decades, both from industry and academia, leading to considerable growth in numerous electric vehicles (EVs) in several countries globally. Consequently, there is a great need to develop an EV charging station infrastructure while maintaining a balanced load of power supply lines [3]. To pursue this goal, EV charging technology must be developed not only for fast charging systems but also for balanced overnight charging strategies with smart management of accessible energy, user energy demands, and cost savings. One of these strategies can be adopted in distributed farms of slow EV charging stations located in residential areas. Such farms are usually established in residential parking lots and are equipped with Level-II AC charging stations, controlled by the centralized coordinated scheduling unit.

In this study, we tackle the problem of smart coordination of the EV charging process in such farms, considering power supply constraints, balanced and smooth load expectations, energy demands, and user transportation habits and needs. The proposed strategies for controlling the EV charging process are mainly addressed for overnight charging, wherein the EVs to be charged and the transportation habits of their owners are usually known to the charging management system. However, vehicles do not have to be parked in the designed spaces. For such EVs, the expected daily energy demands can be easily estimated based on the charging history and parameters of the EV battery. Otherwise, new users or users who wish to modify their statistical daily charging demands should introduce updated data to the system before the initialization of the charging process. The system estimates the expected energy demand for each EV. However, the procedure for pursuing this task is not discussed in this paper.

The charging stations are currently categorized as Levels I, II, and III, depending on the range of power, which considerably affect the charging time. Level-I charging units usually operate on an AC electric power of 120 V and yield a power of approximately 1.4 kW. Level-II chargers offer power ranging from 4.4 kW to 22 kW and operate at 230 V AC. Direct current fast chargers, sometimes referred to as Level-III chargers, can charge with a maximum output of up to 350 kW, and it takes approximately 20–40 min to fill an EV battery to approximately 80% of its capacity.

We assume that the analyzed farm contains three-phase Level-II AC charging stations, each equipped with two charging ports that can be controlled by the central control unit (CCU) and can serve up to two EVs but not simultaneously. The three-phase charging currents are measured at each station, and this information is back-forwarded to the CCU and then to the scheduler located on the cloud platform. The block diagram of the analyzed system architecture of the EV charging stations is shown in Figure 1. The charging stations (referred to as double-socket EVSEs) communicate with the scheduler via the CCU. The user–cloud interface is served by a mobile application that allows users to input its energy and time demand to the system and be informed by the status of its EVs' charging.

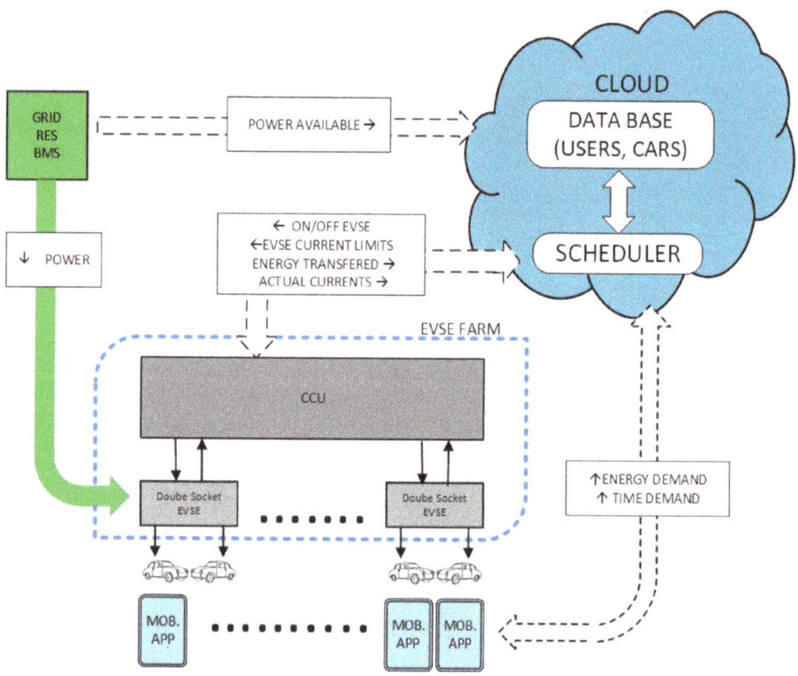

Figure 1. Block diagram of the analyzed farm of EV charging stations.

The scheduler aims to determine an optimal scheduling for a charging process in the farm based on available data (both from the CCU and the database) and to accordingly control the charging stations. In this study, we assume that only a binary control over the charging station is possible, i.e., the charging link between the charging station and the connected EV can be activated or deactivated in a given time slot. This type of scheduling is referred to as on–off scheduling. The period between the plug-in and plug-off time was discretized, thereby allowing us to control the EV charging process using binary variables. The farm is supplied with three-phase power supply lines, and the scheduling should be determined in such a manner that the current consumption of the whole farm from each phase line does not exceed the assumed limits at any time. Moreover, the designed system should ensure a balanced power load over the entire accessible period. Performing this task is difficult because we expect cumulations for energy demands. Many users may plug-in their EVs in a relatively narrow period, possibly leading to a peak of high energy-consuming load if no control over the charging process is applied. The objective of the proposed scheduling scheme is to shave the load peak and ensure that the energy load is approximately constant throughout the entire accessible period. In our approach, we also assume that there is feedback between any working charging station and the scheduler, and the information on the charging currents is back-forwarded to the cloud in the discrete time, for example, every few minutes. The scheduler can update the current binary variables if any significant change in the charging currents is observed, or new EVs are recognized by the system.

Considering the above-mentioned physical limitations and user preferences, we propose three models for scheduling the EV charging process. The models are expressed in terms of numerical optimization problems with binary variables. In the first approach, we assume that the objective function is linear and the weighting coefficients are selected in such a manner as to execute the first-come first-served rule, subject to equality and inequality constraints. The equality constraints express the energy demands, whereas the inequality constraints refer to the power and available time limitations. Next, we extend the objective with a quadratic term that aims to enforce local smoothness to minimize the total number of switching operations. Frequent on/off switching of the battery might lead to a slightly lower battery life expectancy and to an increased number of transient instability effects (voltage ripples, overshooting, current peaks, among others). The quadratic model provides more flexibility in controlling the load profile. In the second approach, the objective function is formulated to maximize the constrained total energy consumption in each time slot, increasing the possibility of fast charging. In the last scenario, the optimization is triggered to prioritize the expected load demand for each EV. The last two models were expressed in terms of binary quadratic programming problems.

A binary linear programming (BLP) problem is not difficult to solve and can be regarded as a particular case of more general integer linear programming. Such problems are usually solved using relaxation or bound methods, including cutting planes, branch-and-bound, branch-and-cut, and many heuristics. In our approach, we need to solve a binary quadratic programming (BQP) problem, which is more challenging than a BLP problem. To address this, we analyze three computational algorithms—Frank–Wolfe (FW) [4], successive linear approximations (SLA) [5], and SLA with gradient descent updates. In the first case, the gradient of the objective function was minimized. It is a linear term for a quadratic objective function that can be easily addressed with the BLP. In the other cases, a BQP problem is reformulated to a BLP subject to quadratic inequality constraints that are subsequently linearized using SLA.

The remainder of this paper is organized as follows. Section 2 reviews related studies on scheduling schemes for EV charging. The proposed models for two-link charging stations and related computational strategies are presented in Section 3. Numerical experiments performed on various EV charging scenarios are presented and discussed in Section 4. The final section provides concluding statements.

2. Related Works

The optimal scheduling of EV charging has been a challenging problem and has been extensively studied in the last decade. There are many approaches and computational strategies for this problem, which are conditioned with a variety of charging station configurations, characteristics, and coordination objectives. A survey of recent trends in EV charging strategies and related optimization techniques can be observed in [6–9].

One of the main strategies, known as a decentralized or distributed strategy for EV charging [10–16], assumes that a major scheduling problem is decomposed into local subproblems individually tackled by charging stations based on the central reference information such as the electricity price or a reference power load. This type of coordination is much more flexible for users and computationally more efficient; however, this system is more prone to a load peak effect, possibly resulting in an overload of power grid components. The suppression of the load peak effect is one of the most important tasks in EV charging management, and it has been extensively discussed in the literature [17–21].

In another approach, the centralized management of a charging process constitutes a fundamental concept. Many studies [22–26] have highlighted the advantages of this type of charging strategy. In this approach, an EV charging control decision is taken globally by an aggregator determining the charging schedule and the rate. Central resource management is much more efficient in the optimization of power consumption and user demands. This strategy requires efficient real-time bidirectional communication between an aggregator, a CCU, and charging stations. It also involves a higher computational cost than a decentralized version; hence, it requires the use of fast scheduling algorithms.

Another issue related to the efficient management of an EV charging process is the selection of objectives and control types. The former may be formulated as an objective function that should be minimized, subject to various constraints. This might be due to the minimization of charging costs [27,28], tardiness [18], power imbalance [29], energy loss [30], grid loss [31], among others. The latter refers to the following options: spatial management (which EVs to charge in a given time instant), time management (when to switch on/off the stations), and available power management (how to determine the charging rate). The latter offers higher flexibility and can be combined with other options. Many studies [22,24,32] refer to a continuous charging rate control; however, this approach, despite its easy algorithmic feasibility owing to continuous variable optimization, is difficult to accomplish in practice. Discrete control over charging currents is easier to perform using digital electronic circuits governed by microcontroller units. Examples of papers discussing scheduling systems with a discrete charging rate can be obtained in [11,33].

A particular case of discrete charging rate control is a binary approach to the charging process. Several studies have demonstrated the efficiency of various on–off scheduling strategies in the context of optimal EV charging. Baek et al. [34] proposed a queuing model with random interruptions of charging EVs to relax the overload problem. Harris et al. [35] used a probabilistic approach for EV chargers modulating on/off, while Nguyen et al. [36] performed a similar task using the BLP combined with the bisection scheme. On–off scheduling schemes have also been studied by Fernandes et al. [37], who developed a dynamic charging scheduler based on greedy and LP algorithms. Another binary scheduling scheme was discussed in [38], where an EV-coordinated discrete charging problem was formulated in terms of a BQP problem and solved using an alternating switching algorithm. Recently, Jawad et al. [39] proposed a real-time EV charge management system that is based on a convex relaxation of on–off scheduling. In this approach, the binary constraint is relaxed to solve the on–off scheduling problem as a convex problem using LP, and then, a modified mapping is used to convert the solution back to binary values. However, this strategy cannot be adapted to the two-link charging stations in our farm because the two-link charging constraint in our model cannot be relaxed in this way.

Binary scheduling may also have some disadvantages. It is thus well known that a large number of charging cycles decreases battery life, as reported in [40–42]; however, Vroey et al. [43] argued that this degradation is marginal. Moreover, there is no evidence that switching between a minimal charging rate and its largest possible value affects the battery life.

In centralized charging systems, fast scheduling is a challenging problem that can be tackled using various computational strategies. Continuous variable optimization methods are usually used, such as LP [22,24,32,44,45] and QP [10,12,17,46]. A discrete charging rate control or any coordination scheme with integer decision variables requires the formulation of a non-convex optimization problem or sometimes even an NP-hard problem [47]. Therefore, integer-variable optimization tools must be used. Examples include BLP [37], BQP [38], MILP [48–50] and various metaheuristics with integer variables, including genetic algorithms [51], ant colony optimization (ASO) [13], particle swarm optimization [52,53], tabu search [54], memetic algorithm [55], artificial bee colony algorithm [56], binary evolutionary programming [57], and other greedy algorithms [33,58]. Other approaches to scheduling can also be obtained in the literature, for example, agent-based scheduling [59,60], deep reinforcement learning [61], and deep learning [62].

Motivated by the advantages of centralized EV charging strategies, we propose a new model and algorithmic approach for scheduling EV charging in such a strategy. The configuration of our charging station is similar to that presented in [56], that is, each charging station has two Type-II charging points, but each EV can be charged from a three-phase power line, and we have only one master or CCU that controls all charging stations in one farm. Motivated by the on–off scheduling in [37], we formulate a similar model with binary decision variables, but our algorithmic approach is completely different. Our algorithmic approach is motivated by the SOCDC algorithm given in [38], but the smoothing is enforced by a regularization term instead of the alternating optimization. We use different numerical approaches to solve QP problems; our model is also different, containing additional physical constraints for two-link charging stations, and our objective function is formulated such as to prioritize user demands.

3. Scheduling Problem

Notations:Boldface uppercase letters (e.g., $\boldsymbol{X}$) denote matrices; lowercase boldface letters represent vectors (e.g., $\boldsymbol{x}$); non-bold letters (e.g., x) are scalars; calligraphic uppercase letters (e.g., $\mathcal{X}$) will be used to denote sets. The j-th column of $\boldsymbol{X}$ is denoted by $\boldsymbol{x}_j$, and $\underline{\boldsymbol{x}}_j$ is the j-th row vector of $\boldsymbol{X}$. The symbol $||\cdot||_F$ denotes the Frobenius norm of a matrix; $||\cdot||$ denotes the 2-nd norm. The sets of real numbers, natural, and binary (0–1) numbers are represented by $\mathbb{R}$, $\mathbb{N}$, and $\mathbb{B}$, respectively. The symbols $\lfloor x \rfloor$ and $\lceil x \rceil$ denote the floor and ceiling functions of x, respectively.

3.1. Problem Formulation

In this section, we formulate a static scheduling problem in the form of a mathematical model with discrete time. The control horizon is expressed in terms of available time, which is partitioned into T time slots of equal length, for example, a few minutes. We assume that the farm contains M charging stations that can serve N EVs. As each station is equipped with two charging points (ports), we have $N = 2M$. Let $\mathcal{T} = \{1, \ldots, T\}$ and $\mathcal{N} = \{1, \ldots, N\}$ contain the indices of time slots and charging points, respectively. The charging points are subsequently indexed, and N charging points can serve up to N EVs. The plug-in and plug-off times of the n-th EV are denoted by $t_n^{(in)}$ and $t_n^{(off)}$, respectively. Obviously, $1 \leq t_n^{(in)} < t_n^{(off)} \leq T$. Without loss of generality, we assume that the first and second EVs are assigned to the first charging station, and the n-th and $(n+1)$-th EVs are assigned to the m-th charging station, where $m = \lfloor \frac{n+1}{2} \rfloor$ for $n = 1, \ldots, N-1$. Let $\boldsymbol{B} = [b_{tn}] \in \mathbb{B}^{T \times N}$ be a binary matrix indicating the available time for each EV, where

$$b_{tn} = \begin{cases} 1 & \text{if } n\text{-th EV is available in } t\text{-th time slot,} \\ 0 & \text{otherwise} \end{cases} \tag{1}$$

The available time of the n-th EV can be computed as $T_n = \sum_{t=1}^{T} b_{tn}$, and it represents the number of time slots between $t_n^{(in)}$ and $t_n^{(off)}$. If $T_n = 0$ for any $n \in \mathcal{N}$, then there is no EV in the n-th charging point.

Let $c_n \leq T_n$ denote the energy demand of the n-th EV, which is the number of time slots required to charge a given EV to its desired state of charge (SoS) level. The charging rate for the n-th EV is represented by the parameter $r_n^{(l)}$, where $l \in \{1,2,3\}$ represents the index of the line in the Type-II three-phase charging cable. In our approach, the changing rate refers to a quasi-stationary current (amperes/time slot) and does not have one constant value, but it is individually set for each EV and for each phase. This setting is motivated by the assumption that the schedule can be adjusted during charging considering the feedback information from charging stations, rates are time-dependent, and EVs can be scheduled in different periods of the available time.

Considering the power supply limit, let $I_t^{(max,l)}$ be the maximal l-th line current supplying the entire farm in the t-th time slot. The scheduler should assure possible fast charging but simultaneously not exceed $I_t^{(max,l)}$ for any time instant and phase. A separate current limit for each line is also required to avoid power imbalance across phases when the lines are unequally loaded.

To control the charging process, that is, which and when charging points to switch on/off, we introduce binary decision variables that form the scheduling matrix $\boldsymbol{U} = [u_{nt}] \in \mathbb{B}^{N \times T}$. The decision variables are defined as follows:

$$u_{nt} = \begin{cases} 1 & \text{if } n\text{-th EV is charging in } t\text{-th time slot,} \\ 0 & \text{otherwise.} \end{cases} \tag{2}$$

The objective of the scheduler is to estimate the matrix $\boldsymbol{U}$ based on given criteria and available data, such as available time matrix $\boldsymbol{B}$, energy demand $\{c_n\}$, charging rates $\{r_n^{(l)}\}$, and maximal line currents $\{I_t^{(max,l)}\}$.

A fundamental requirement from the users is to charge their EVs to satisfy the given SoC level, and considering the variables defined above, this requirement can be determined by the following model:

$$\sum_{t \in [t_n^{(in)},\, t_n^{(off)}]} u_{nt} = c_n, \quad \forall n \in \mathcal{N}, \tag{3}$$

which can be equivalently expressed in the matrix equality constraint:

$$\text{diag}\{\boldsymbol{UB}\} = \boldsymbol{c}, \tag{4}$$

where $\boldsymbol{c} = [c_1, \ldots, c_N]^T \in \mathbb{N}^N$ is the vector of the energy demands. Equation (4) restricts the feasible region to the periods determined by the plug-in and plug-off times. As all entries outside the periods are equal to zero, the energy demands can be reinforced by an additional equality constraint:

$$\boldsymbol{U}\boldsymbol{e}_T = \boldsymbol{c}, \tag{5}$$

where $\boldsymbol{e}_T = [1, \ldots, 1]^T \in \mathbb{R}^T$.

The farm of chargers is supplied with three-phase power with distribution transformers or other suppliers that have limited power. Thus, we assume that the maximal line current $I_t^{(max,l)}$ is not exceeded in any time slot, and this requirement can be modeled as follows:

$$\sum_{n \in \mathcal{N}} r_n^{(l)} u_{nt} + \xi_t^{(l)} \leq I_t^{(max,l)}, \quad \forall t \in \mathcal{T}, \quad \text{and} \quad l \in \{1,2,3\}, \tag{6}$$

where $\xi_t^{(l)}$ is the basic current obtained from the l-th line in the t-th time slot, which is the difference between the cumulative charging phase current and the line current from the transformer. The model (6) can be rewritten in the following matrix form:

$$\boldsymbol{e}_N^T \left(\boldsymbol{D}_R^{(l)} \boldsymbol{U} \right) + \underline{\boldsymbol{\xi}}^{(l)} \leq \underline{\boldsymbol{I}}^{(max,l)}, \quad \text{for} \quad l \in \{1,2,3\}, \tag{7}$$

where $\boldsymbol{e}_N = [1, \ldots, 1]^T \in \mathbb{R}^N$, $\boldsymbol{D}_R^{(l)} = \mathrm{diag}\left\{r_n^{(l)}\right\} \in \mathbb{R}^{N \times N}$, $\underline{\boldsymbol{\xi}}^{(l)} = \left[\xi_1^{(l)}, \ldots, \xi_T^{(l)}\right] \in \mathbb{R}^{1 \times T}$ and $\underline{\boldsymbol{I}}^{(max,l)} = \left[I_1^{(max,l)}, \ldots, I_T^{(max,l)}\right] \in \mathbb{R}^{1 \times T}$.

In our configuration, each charging station has two charging points, which means that two EVs can be connected to one station, but the electronic instrumentation inside the station allows us to charge only one EV at any time instant. Switching between charging points is possible at any time and should be controlled by the CCU. Mathematically, this feature can be considered in a scheduling problem by formulating an inequality constraint that does not allow the setting of more than one decision variable to one for each charging station in any time slot. To tackle this problem, we define the two-link constraint matrix as

$$\boldsymbol{S} = \begin{bmatrix} 1 & 1 & 0 & 0 & \cdots & 0 & 0 \\ 0 & 0 & 1 & 1 & \cdots & 0 & 0 \\ \vdots & \vdots & \vdots & \vdots & \ddots & \vdots & \vdots \\ 0 & 0 & 0 & 0 & \cdots & 1 & 1 \end{bmatrix} \in \mathbb{B}^{N/2 \times N}. \tag{8}$$

Subsequently, the two-link charging constraint can be formulated as follows:

$$\boldsymbol{S}\boldsymbol{U} \leq \boldsymbol{E}_{N/2 \times T}, \tag{9}$$

where $\boldsymbol{E}_{N/2 \times T} = \begin{bmatrix} 1 & \cdots & 1 \\ \vdots & \ddots & \vdots \\ 1 & \cdots & 1 \end{bmatrix} \in \mathbb{B}^{N/2 \times T}$ is a matrix of all ones.

Considering the constraints in (4), (5), (7), and (9), the scheduling problem can be formulated as the following constrained optimization problem:

$$\begin{aligned} \min_{\boldsymbol{U} \in \mathbb{B}^{N \times T}} \quad & \Psi(\boldsymbol{U}), \\ \text{s.t.} \quad & \mathrm{diag}\{\boldsymbol{U}\boldsymbol{B}\} = \boldsymbol{c}, \\ & \boldsymbol{U}\boldsymbol{e}_T = \boldsymbol{c}, \\ & \boldsymbol{e}_N^T\left(\boldsymbol{D}_R^{(l)}\boldsymbol{U}\right) + \underline{\boldsymbol{\xi}}^{(l)} \leq \underline{\boldsymbol{I}}^{(max,l)}, \quad \text{for} \quad l \in \{1,2,3\}, \\ & \boldsymbol{S}\boldsymbol{U} \leq \boldsymbol{E}_{N/2 \times T}, \end{aligned} \tag{10}$$

where $\Psi(\boldsymbol{U})$ is an objective function.

We studied various approaches to define the objective function, which can be stated as follows:

- *Linear:* Motivated by the concept of the objective weighting, given in [38], we formulate the weighted linear function:

$$\Psi(\boldsymbol{U}) = \boldsymbol{e}_N^T\boldsymbol{U}\boldsymbol{w}, \tag{11}$$

where $\boldsymbol{w} = \left[\boldsymbol{e}_{t_w+1}^T, \vartheta + 1, 2\vartheta + 1, \ldots, \vartheta(T - t_w) + 1\right]^T \in \mathbb{R}^T$, where $\vartheta = \frac{w_{max}-1}{T-t_w}$. The parameters t_w and w_{max} were experimentally set to $t_w = \lfloor \frac{T}{10} \rfloor$ and $w_{max} = 10$. Vector $\boldsymbol{w}$ represents a piecewise linear function. The first t_w entries are equal to one, whereas the others linearly increase from one to w_{max}. Such weighting aims to penalize decision variables in later time slots, which should enforce charging as early as possible. Minimization of a linear function can be performed using any BLP solver. This approach is computationally efficient; however, it is not flexible owing to the limited possibility of using multiple penalties.

- *Quadratic I:* In this model, we assume charging of all EVs with a possibly maximum power, which leads to the following objective function:

$$\Psi(\boldsymbol{U}) = \frac{1}{2}\sum_{l=1}^{3}\left\|\underline{\boldsymbol{I}}^{(max,l)} - \left(\boldsymbol{r}^{(l)}\right)^T \boldsymbol{U}\boldsymbol{D}_w\right\|_2^2, \tag{12}$$

where $\boldsymbol{r}^{(l)} = [r_n^{(l)}] \in \mathbb{R}^N$ for $l \in \{1,2,3\}$, and $\boldsymbol{D}_w = \text{diag}\{\boldsymbol{w}\} \in \mathbb{R}^{T\times T}$. Matrix $\boldsymbol{D}_w$ has a task similar to that of the linear function.

- *Quadratic II:* Another possibility is to reinforce the SoC level bilanse with additional weighting of time slots. This task can be achieved using the following objective function:

$$\Psi(\boldsymbol{U}) = \frac{1}{2}\|\boldsymbol{c} - \boldsymbol{U}\boldsymbol{D}_w\boldsymbol{e}_T\|_2^2. \tag{13}$$

- *Penalized quadratic with smoothness constraints:* None of the above-mentioned objective functions assures a smooth solution, indicating that the number of switching on/off charging stations is not controlled within the area of feasibility bounded by the constraints. However, the number of switching operations can be minimized by introducing a trade-off between the model fitting and the local smoothness measure. Taking into account the objective functions (12) and (13), the degradation of model fitting by adding a regularization or penalty term is not a problematic issue because the model constraints are explicitly added to the optimization problem and guarantee feasibility.
The local smoothness of the charging profile for each EV can be measured according to the following function:

$$\Phi(\boldsymbol{U}) = \frac{1}{2}\sum_{n\in\mathcal{N}}\sum_{t=1}^{T-1}(u_{n,t} - u_{n,t+1})^2. \tag{14}$$

Let $\boldsymbol{L}$ be the first-order differential operator defined as:

$$\boldsymbol{L} = \begin{bmatrix} 1 & -1 & 0 & 0 \\ 0 & \ddots & \ddots & 0 \\ \ddots & \ddots & \ddots & -1 \\ 0 & \ddots & 0 & 1. \end{bmatrix} \in \mathbb{R}^{T\times T} \tag{15}$$

The function Φ can be equivalently rewritten using matrix $\boldsymbol{L}$ in the form $\Phi(\boldsymbol{U}) = \frac{1}{2}\|\boldsymbol{U}\boldsymbol{L}^T\|_F^2$. Consequently, the objective function (13) with the additive smoothness penalty term is given by

$$\Psi(\boldsymbol{U}) = \frac{1}{2}\|\boldsymbol{c} - \boldsymbol{U}\boldsymbol{D}_w\boldsymbol{e}_T\|_2^2 + \frac{\alpha}{2}\left\|\boldsymbol{U}\boldsymbol{L}^T\right\|_F^2, \tag{16}$$

where $\alpha \geq 0$ is a penalty term.

For the above objective functions, the scheduling problem in (10) can be regarded as a BLP problem with the objective function (11) or the constrained BQP problem with the other functions. Regardless of the objective function, the problem expressed in the form (10) cannot be directly solved with standard numerical optimization solvers because the solution has the form of a matrix that cannot be successively processed with respect to its rows or columns. This results from the column and row action constraints that must be satisfied simultaneously. However, simple vectorization operations can be applied to transform the existing matrix equations into their equivalent vector forms.

Lemma 1. *Formula (4) can be equivalently expressed in the form:*

$$(\boldsymbol{B} \odot \boldsymbol{I}_N)^T \boldsymbol{u} = \boldsymbol{c}, \tag{17}$$

where $\boldsymbol{u} = \text{vec}(\boldsymbol{U}) \in \mathbb{B}^{NT}$ *is a vectorized version of the matrix* $\boldsymbol{U}$*, and the symbol* $\odot$ *denotes the Khatri–Rao product.*

The proof of Lemma 1 is given in Appendix A.

Definition 1. *Let* $\boldsymbol{A} \in \mathbb{R}^{M\times I}$*,* $\boldsymbol{X} \in \mathbb{R}^{I\times J}$*, and* $\boldsymbol{Y} \in \mathbb{R}^{J\times N}$*. Then:*

$$\text{vec}(\boldsymbol{AXY}) = (\boldsymbol{Y}^T \otimes \boldsymbol{A})\text{vec}(\boldsymbol{X}). \tag{18}$$

Applying Formula (18) to (5), where $\boldsymbol{A} = \boldsymbol{I}_N$, $\boldsymbol{X} = \boldsymbol{U}$, and $\boldsymbol{Y} = \boldsymbol{e}_T$, we obtain:

$$(\boldsymbol{e}_T^T \otimes \boldsymbol{I}_N)\boldsymbol{u} = \boldsymbol{c}. \tag{19}$$

Similarly, the inequalities (7) and (9) can be reformulated using Formula (18) as follows:

$$\left(\boldsymbol{I}_T \otimes (\boldsymbol{r}^{(l)})^T\right)\boldsymbol{u} \leq \tilde{\boldsymbol{I}}^{(max,l)}, \quad \text{for} \quad l \in \{1,2,3\}, \tag{20}$$

$$(\boldsymbol{I}_T \otimes \boldsymbol{S})\boldsymbol{u} = \left(\boldsymbol{I}_T \otimes \boldsymbol{I}_{N/2} \otimes \boldsymbol{e}_2^T\right)\boldsymbol{u} = \left(\boldsymbol{I}_{NT/2} \otimes \boldsymbol{e}_2^T\right)\boldsymbol{u} \leq \hat{\boldsymbol{e}}, \tag{21}$$

where $\tilde{\boldsymbol{I}}^{(max,l)} = \left(\underline{\boldsymbol{I}}^{(max,l)} - \underline{\boldsymbol{\xi}}^{(l)}\right)^T$ and $\hat{\boldsymbol{e}} = \text{vec}(\boldsymbol{E}_{N/2\times T}) \in \mathbb{R}^{NT/2}$ and $\boldsymbol{e}_2^T = [1,1]$. Combining (17) and (19), we obtain the equality constraints:

$$\begin{bmatrix} \tilde{\boldsymbol{A}} \\ \tilde{\boldsymbol{B}} \end{bmatrix} \boldsymbol{u} = \begin{bmatrix} \boldsymbol{c} \\ \boldsymbol{c} \end{bmatrix}, \tag{22}$$

where $\tilde{\boldsymbol{A}} = \boldsymbol{e}_T^T \otimes \boldsymbol{I}_N$, and $\tilde{\boldsymbol{B}} = (\boldsymbol{B} \odot \boldsymbol{I}_N)^T$. The inequality constraints can be presented in the form:

$$\begin{bmatrix} \boldsymbol{R} \\ \boldsymbol{Z}^{(1)} \\ \boldsymbol{Z}^{(2)} \\ \boldsymbol{Z}^{(3)} \end{bmatrix} \boldsymbol{u} \leq \begin{bmatrix} \hat{\boldsymbol{e}} \\ \tilde{\boldsymbol{I}}^{(max,1)} \\ \tilde{\boldsymbol{I}}^{(max,2)} \\ \tilde{\boldsymbol{I}}^{(max,3)} \end{bmatrix}, \tag{23}$$

where $\boldsymbol{R} = \boldsymbol{I}_{NT/2} \otimes \boldsymbol{e}_2^T$ and $\boldsymbol{Z}^{(l)} = \boldsymbol{I}_T \otimes (\boldsymbol{r}^{(l)})^T$ for $l \in \{1,2,3\}$.

The objective functions can also be rewritten using Equation (18). For the linear function, we have:

$$\Psi(\boldsymbol{u}) = \text{vec}(\boldsymbol{e}_N^T \boldsymbol{U} \boldsymbol{w}) = (\boldsymbol{w} \otimes \boldsymbol{e}_N)^T \boldsymbol{u} = \boldsymbol{d}^T \boldsymbol{u}. \tag{24}$$

The other functions can be reformulated to the quadratic form:

$$\Psi(\boldsymbol{u}) = \frac{1}{2}\boldsymbol{u}^T \boldsymbol{Q} \boldsymbol{u} + \boldsymbol{d}^T \boldsymbol{u} + \text{const}, \tag{25}$$

where $\boldsymbol{Q}$ and $\boldsymbol{d}$ are given by:

- *Quadratic I:*

$$\boldsymbol{Q} = \boldsymbol{D}_w^2 \otimes \sum_{l=1}^{3} \boldsymbol{r}^{(l)}(\boldsymbol{r}^{(l)})^T, \quad \boldsymbol{d} = -\begin{bmatrix} \underline{\boldsymbol{I}}^{(max,1)}\left(\boldsymbol{D}_w \otimes (\boldsymbol{r}^{(1)})^T\right) \\ \underline{\boldsymbol{I}}^{(max,2)}\left(\boldsymbol{D}_w \otimes (\boldsymbol{r}^{(2)})^T\right) \\ \underline{\boldsymbol{I}}^{(max,3)}\left(\boldsymbol{D}_w \otimes (\boldsymbol{r}^{(3)})^T\right) \end{bmatrix}^T \boldsymbol{e}_3. \tag{26}$$

The derivations of Q and d in (26) are given in Appendix B.

- *Quadratic II:*

$$Q = D_w e_T e_T^T D_w \otimes I_N, \quad d = -(e_T^T D_w \otimes I_N)^T c. \tag{27}$$

- *Penalized quadratic form with smoothness constraints:*

$$Q = D_w e_T e_T^T D_w \otimes I_N + \alpha(L^T L \otimes I_N), \quad d = -(e_T^T D_w \otimes I_N)^T c. \tag{28}$$

Appendix C contains the derivations of Q and d in (28). By setting $\alpha = 0$, we obtain Q and d in (27).

Considering the constraints in (22) and (23), the set of feasible regions takes the form:

$$\Omega = \left\{ u \in \mathbb{B}^{NT} \middle| \begin{bmatrix} \tilde{A} \\ \tilde{B} \end{bmatrix} u = \begin{bmatrix} c \\ c \end{bmatrix}, \begin{bmatrix} R \\ Z^{(1)} \\ Z^{(2)} \\ Z^{(3)} \end{bmatrix} u \leq \begin{bmatrix} \hat{e} \\ \tilde{I}^{(max,1)} \\ \tilde{I}^{(max,2)} \\ \tilde{I}^{(max,3)} \end{bmatrix} \right\} \tag{29}$$

For objective functions (24) and (25), the scheduling problem (10) can be reformulated to the following single-vector problem:

$$\min_{u \in \Omega} \Psi(u), \tag{30}$$

which can be solved with various binary optimization solvers.

3.2. Algorithmic Approach

We herein do not assume that matrices $\tilde{A}$ and $\tilde{B}$ in (22), and R and $Z^{(l)}$ for $l \in \{1,2,3\}$ in (23) are totally unimodular because $\{\tilde{I}^{(max,l)}\}$ are real-value measured currents; hence, a binary relaxation to the LP problem is not justified. To solve the BLP problem, that is, problem (30) with the objective function in (24), many methods such as the cutting planes, branch-and-bound, branch-and-cut, and heuristic routines can be used. In our approach, we solve this problem using the `intlinprog` function from the Optimization Toolbox in MATLAB 2020b using default settings. This function is addressed to solve a mixed-integer linear programming (MILP) problem. We defined all variables as integers bounded to the range $\{0,1\}$.

Problem (30) with the other objective functions is more challenging, and there is no specific solver in MATLAB 2020b for solving the BQP problem. We studied the various algorithmic approaches described below.

3.2.1. Frank—Wolfe Algorithm

The FW algorithm [4] dates back to the 1950s; however, its popularity is still noticeable in various research areas [63,64]. This algorithm is based on the concept of SLA of the objective function with a first-order Taylor expansion. The original version of the FW algorithm can also be used to solve the BQP problem under the assumption that the objective function is convex and differentiable, a set of feasible regions is convex, and a linearized version of the original problem is easy to solve.

Note that Ω in (29) is a compact convex set in $\mathbb{R}^{NT}$ because it results from the intersection of hyperplanes and closed half-subspaces given by linear equality and inequality constraints. The objective function Ψ is at least weakly convex because it is a quadratic function with a possibly semi-positive defined matrix Q. Hence, problem (30) can be solved using the following linear approximations:

$$\min_{u \in \Omega} \Psi(u_k) + \nabla_u \Psi(u)^T (u - u_k), \quad \text{for} \quad k = 0,1,2,\ldots \tag{31}$$

To solve problem (31), we used the followingAlgorithm 1:

Algorithm 1: FW Algorithm

Input : $Q \in \mathbb{R}^{NT \times NT}$, $\boldsymbol{d} \in \mathbb{R}^{NT}$, τ—tolerance
Output: $\{u_k\}$—approximations
1 **Initialization**: u_0 solution to the BLP problem
2 **for** $k = 1, 2, \ldots$ **do**
3 Compute $\boldsymbol{g}_k = \nabla_{\boldsymbol{u}} \Psi(\boldsymbol{u})|_{\boldsymbol{u}=\boldsymbol{u}_{k-1}} = Q\boldsymbol{u}_{k-1} + \boldsymbol{d}$; `// gradient`
4 Solve $\boldsymbol{s}_k = \arg\min_{\boldsymbol{u} \in \mathbb{B}^{NT}} \boldsymbol{g}_k^T \boldsymbol{u}$, s.t. (22), (23); `// BLP problem`
5 Set $\boldsymbol{u}_k = \boldsymbol{u}_{k-1} + \frac{2}{1+k}(\boldsymbol{s}_k - \boldsymbol{u}_{k-1})$;
6 **if** $\frac{||\boldsymbol{u}_k - \boldsymbol{u}_{k-1}||_2}{||\boldsymbol{u}_{k-1}||_2} \leq \tau$ **then**
7 **break**

The BLP problem in line 4 of Algorithm 1 was solved using the `intlinprog` function from MATLAB. The gradient $\boldsymbol{g}_k$ is Lipschitz continuous with respect to the Frobenius norm, which means that Algorithm 1 has a linear convergence with the rate $\mathcal{O}(1/k)$.

Algorithm 1 can be run with the negative gradient $\boldsymbol{g}_k$, which is equivalent to the following update rule:

$$\boldsymbol{u}_k = \boldsymbol{u}_{k-1} - \eta_k \boldsymbol{s}_k, \tag{32}$$

where η_k is the step length. Formula (32) can be regarded as the standard gradient descent update rule. In the experiments, we set $\eta_k = 1$.

3.2.2. Successive Linear Approximations

SLA [5] is based on a concept similar to that of the FA algorithm. In this approach, the BQP problem given by (30) is reformulated as follows:

$$\min_{\boldsymbol{u} \in \Omega, \zeta \in \mathbb{R}} \zeta + \boldsymbol{d}^T \boldsymbol{u}, \quad \text{s.t.} \quad \frac{1}{2}\boldsymbol{u}^T Q \boldsymbol{u} \leq \zeta, \quad \zeta \geq 0. \tag{33}$$

The objective function in (33) is linear, but the inequality constraints are nonlinear. However, the quadratic constraints can be linearized using the first-order Taylor expansion, similar to the FW algorithm. For the k-th iterative step, the inequality constraint $\frac{1}{2}\boldsymbol{u}^T Q \boldsymbol{u} \leq \zeta$ can be linearized around the point $\boldsymbol{u}_{k-1}$ as follows:

$$\begin{aligned} \Phi(\boldsymbol{u}) &= \frac{1}{2}\boldsymbol{u}^T Q \boldsymbol{u} - \zeta \cong \frac{1}{2}\boldsymbol{u}_{k-1}^T Q \boldsymbol{u}_{k-1} + \nabla_{\boldsymbol{u}} \Phi(\boldsymbol{u})^T|_{\boldsymbol{u}=\boldsymbol{u}_{k-1}}(\boldsymbol{u} - \boldsymbol{u}_{k-1}) - \zeta \\ &= -\frac{1}{2}\boldsymbol{u}_{k-1}^T Q \boldsymbol{u}_{k-1} + \boldsymbol{u}_{k-1}^T Q \boldsymbol{u} - \zeta \leq 0. \end{aligned} \tag{34}$$

The inequality (34) can be expressed in the form of the following matrix inequality:

$$\left[\boldsymbol{u}_{k-1}^T Q, \ -1\right] \left[\begin{array}{c} \boldsymbol{u} \\ \zeta \end{array}\right] \leq \frac{1}{2}\boldsymbol{u}_{k-1}^T Q \boldsymbol{u}_{k-1}. \tag{35}$$

Assuming $\tilde{\boldsymbol{u}} = \left[\begin{array}{c} \boldsymbol{u} \\ \zeta \end{array}\right]$, problem (33) for the k-th iterative step can be rewritten as:

$$\min_{\tilde{\boldsymbol{u}} \in \mathbb{B}^{NT+1}} \tilde{\boldsymbol{d}}^T \tilde{\boldsymbol{u}}, \quad \text{s.t.} \ . \left[\boldsymbol{u}_{k-1}^T Q, \ -1\right] \tilde{\boldsymbol{u}} \leq \frac{1}{2}\boldsymbol{u}_{k-1}^T Q \boldsymbol{u}_{k-1}, \quad \zeta \geq 0, \quad \text{and} \quad \boldsymbol{u} \in \Omega, \tag{36}$$

where $\tilde{\boldsymbol{d}} = \left[\begin{array}{c} \boldsymbol{d} \\ 1 \end{array}\right]$. Problem (36) is a standard BLP problem, and we solve it using the `intlinprog` function.

4. Numerical Simulations

The scheduling algorithms discussed were extensively tested using various topologies of farms and their settings, imitating real EV traffic on EV parking lots. The test cases and a description of the testing environment are presented below.

4.1. Setup

Without loss of generality, we assumed a 12-h time horizon from 6 p.m. to 6 a.m., divided into regular time slots of 7.5 min. Hence, $T = 96$ in the proposed model. For an overnight charging plan, the EVs to be charged usually arrive at a parking lot in the evening. We assumed that the plug-in time for each EV can be modeled with a normal distribution with a mean of 8.30 p.m. and a standard deviation of 75 min. The earliest plug-in time was restricted to 6 p.m. The initial energy demand c_n, that is, the number of time slots required to charge the n-th EV to the desired SoC level is modeled with an integer uniform distribution in the range $[4,\ 30]$ for $n = 1, \ldots, N$. This indicates that the minimum charging time was 1 h, and the longest charging period did not exceed 6 h. The plug-off time was also modeled with the same distribution in the range $[t_{plug-in} + C_n,\ T]$.

The final energy demands were determined using the correction procedure in the preprocessing stage. This procedure aims to correct the initial energy demands in a manner that guarantees the feasibility of problem (30) with a linear objective function (24). If the problem is infeasible, a subset of the EVs with the highest energy demands is selected, and their coefficients $\{c_n\}$ are reduced simultaneously until it is found to be feasible. If this point is obtained, then the energy demand for each EV in this subset is individually upgraded to reach the border of the feasible region. In practice, EVs with the lowest priority of charging will be selected to correct their energy demands.

We analyzed farms containing various numbers of charging stations. For the smallest farm, we assume eight two-link charging stations, which gives us 16 charging points, that is, $N = 16$. We also tested scenarios with N = 32, 64, 128, and 256. Obviously, a number of ports do not have to have a multiplicity of two, but it must be an even number. If there are fewer EVs for scheduling, some ports are empty; consequently, the corresponding rows in the scheduling matrix $\boldsymbol{U}$ will have all-zero entries. For Type-II stations, the charging rate is usually limited to 16 A; however, it is not a constant parameter within a charging period. Moreover, the rates can be different for each phase line, when using three-phase chargers. Several research papers, for example, [65,66], report that the charging current of an EV Li-ion battery has almost a constant value from the plug-in (after a short starting period) to approximately half of the maximum SoC level, and then diminishes exponentially with a negative decay. Following this observation, we assume that half of the randomly selected EVs for each phase in the analyzed system charge with a maximal charging rate of 16 A, and the others have the rates determined by a uniform distribution in the range $[1.6,\ 16)$ A. The phases are treated independently, indicating that there could be an unbalanced load of the three-phase lines.

The proposed scheduling schemes are designed such that the maximal line current $I_t^{((max,l)}$ of the entire farm is not exceeded in each time slot, and this limit can vary with the time horizon. However, considering typical real charging scenarios and for simplicity of simulations, we assume that this limit is neither time nor phase dependent, that is, $I_{max} = I_t^{((max,l)}$ for $\forall t \in \mathcal{T}$ and $l \in \{1,2,3\}$. For each test case, the maximal line current was set according to the data given in Table 1. Owing to the current limit, there is a limited number of EVs that can be charged in one time slot. This number, denoted by L_x, is also given in Table 1 for a constant charging rate of 16 A from each line.

Table 1. Power limit settings: I_{max}—the maximal line current (in Amperes) of the whole farm, L_x—the maximal number of EVs charging in one time slot given a constant charging rate of 16 A.

	$N = 16$	$N = 32$	$N = 64$	$N = 128$	$N = 256$
I_{max}	50	100	200	400	800
L_x	3	6	12	25	50

In this study, we proposed various binary algorithmic approaches to solve the scheduling problem in (30). The results are presented in the following cases.

- BLP: Binary linear programming (BLP) with objective function in (24);
- Q1-FW: Binary quadratic programming (*Quadratic I*), with the objective function expressed by (25) and (26) and solved with the Frank–Wolfe (FW) algorithm;
- SmQ2-FW: Binary quadratic programming (*Quadratic II*), with the objective function expressed by (25) and (28) (including the smoothness), and solved with the FW algorithm;
- SmQ2-NG-FW: Binary quadratic programming (*Quadratic II*), with the objective function expressed by (25) and (28) (including the smoothness), and solved with the negative gradient FW algorithm—rule (32);
- SmSLA: Binary quadratic programming (*Quadratic II*), with the objective function expressed by (25) and (28) (including the smoothness), and solved with the successive linear approximations (SLA);
- FA-FS: First-arrive-first-serve (FA-FS) approach.

Q1-FW is closely related to the scheduling problem given in [38] owing to the formulation of the objective function. However, our two-link constraints are quite specific, and to the best of our knowledge, there is no competitive algorithmic strategy for a simple comparison. FA-FS is a heuristic strategy that turns on charging each EV as quickly as possible. If two EVs are simultaneously plugged into one charging station, their selection is random. This strategy yields the fastest charging; however, it does not have any admissible power constraints.

The algorithms were implemented in MATLAB 2020b and run on a machine supplied with a 4-core Intel Core-i7 CPU, 32-GB RAM, and an SSD drive.

4.2. Results

To statistically validate the algorithms, Monte Carlo (MC) analysis was performed using 30 runs for each algorithm. In each snapshot, the charging rates, energy demands, plug-in, and plug-off times were generated randomly according to the procedures discussed above. The selected single-run results are shown in Figures 2–4, while the MC statistics are presented in Tables 2–4, and in Figures 5 and 6.

The algorithms were validated using various criteria and datasets. Figure 2 presents a graphical visualization of the charging matrices $\boldsymbol{U}$ obtained in one selected MC run with the tested algorithms for the smallest farm containing 16 ports ($N = 16$) and random charging rates. The vertical axis corresponds to the ports, and the horizontal axis represents the time slots. The yellow fields correspond to the switch-on state. The light blue fields show the time slots between the plug-in and plug-off times for each EV (each row of the matrix). The dark blue fields denote unavailable time slots. The distributions of energy demand (ED) after using the correction procedure and available time (AT), which were used to obtain the results in Figure 2, are illustrated in Figure 3a in the form of time slot bars. Both ED and AT are expressed in terms of the number of time slots. The charging rates in this case are shown in Figure 3b, separately for each phase.

Similar charging schedules are shown in Figure 4 for the largest analyzed farm. For this case, we set $N = 256$ and a constant charging rate of 16 A for each phase line. The distributions of the ED and AT parameters are shown in Figure 3c. Note that a small number of EVs had no assigned ED. This results from using the correction algorithm to guarantee feasibility.

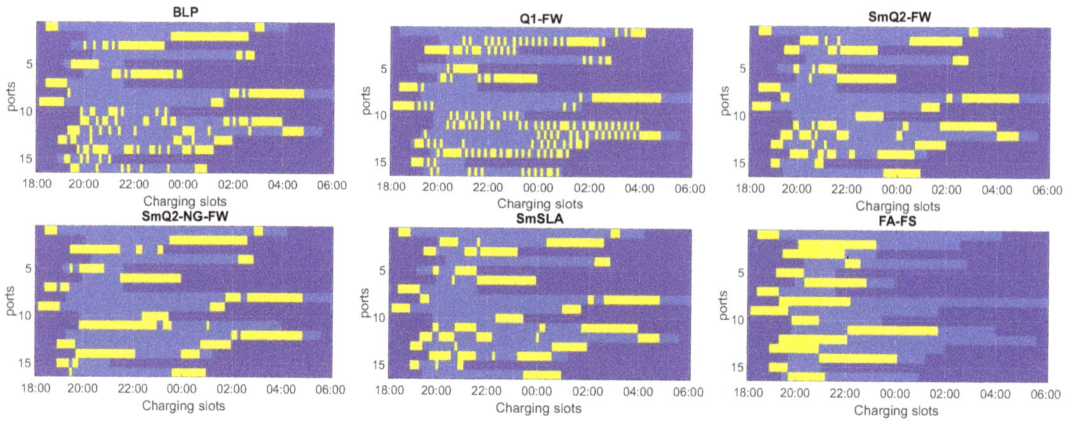

Figure 2. One MC run of charging schedules obtained with the tested algorithms for $N = 16$ and random charging rates. Charging slots are in yellow, available time slots are in light blue, and unavailable time slots are in dark blue.

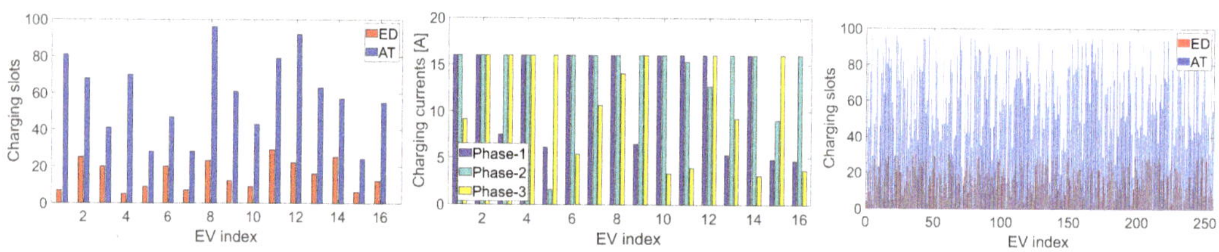

(a) (b) (c)

Figure 3. Charging settings: (**a**) ED and AT parameters for $N = 16$ and random charging rates; (**b**) charging rates for $N = 16$; (**c**) ED and AT parameters for $N = 256$ and a constant charging rate.

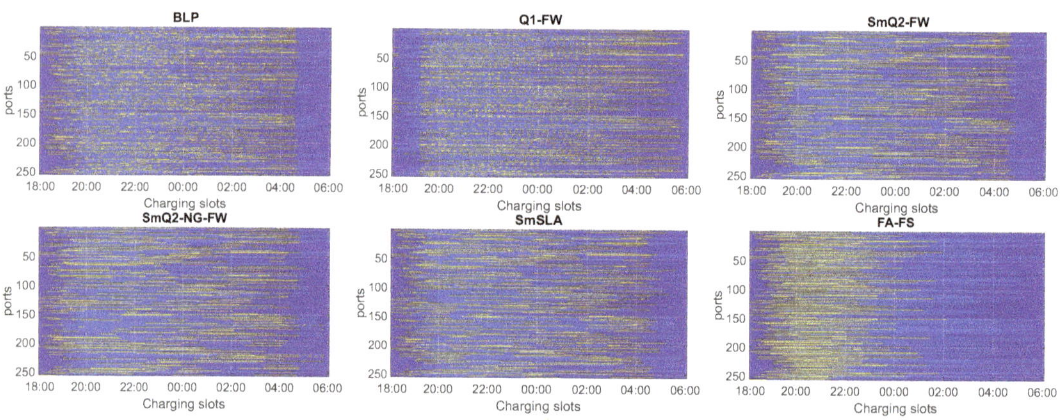

Figure 4. One MC run of charging schedules obtained with the tested algorithms for $N = 256$ and a constant charging rate.

The algorithms were also quantitatively validated using various metrics, which are listed in Tables 2–4. The results obtained for the smallest farm with $N = 16$ and random charging rates are listed in Table 2. Tables 3 and 4 contain the results obtained for the largest farm with constant and random charging rates, respectively.

One of the most important criteria for validating the correctness of the algorithm is to check if the constraints are satisfied. The equality constraints for the EDs were validated with the residual error $r_c = ||c - Ue_T||_2$. Tables 2–4 contain both the mean and median values of the residual error r_c obtained from the MC runs.

Table 2. Results obtained for the smallest farm with 16 ports, 16 EVs, and random charging rates. The maximal number of EVs that can be charged in one time slot is restricted to five owing to the power limit.

Algorithm	Mean (r_c)	Median (r_c)	F_{smooth}	S_x	T_x	L_x	L_f [%]
BLP	0.088	1.98×10^{-14}	187.7 (37.7)	107	76	4	96.7
Q1-FW	1.09	2.54×10^{-14}	191.7 (72)	105	85	4	76.7
SmQ2-FW	3.07	3.61	82.27 (38.8)	39	71	5	43.3
SmQ2-NG-FW	2.54	2.67×10^{-14}	71.33 (28.58)	41	72	5	51
SmSLA	9.68×10^{-14}	2.53×10^{-14}	100.38 (60.57)	92	81	4	100
FA-FS	0	0	31.13 (1.45)	15	53	8	100

Table 3. Results obtained for the largest farm with 256 ports, 252 EVs, and constant charging rate. The maximal number of EVs that can be charged in one time slot is restricted to 50 owing to the power limit.

Algorithm	Mean (r_c)	Median (r_c)	F_{smooth}	S_x	T_x	L_x	L_f [%]
BLP	1.17×10^{-13}	8.05×10^{-14}	3505 (238)	1909	84	50	100
Q1-FW	1.48×10^{-13}	9.12×10^{-14}	4259 (416)	2152	94	50	100
SmQ2-FW	9.33×10^{-14}	8.73×10^{-14}	924.6 (27.68)	644	87	50	100
SmQ2-NG-FW	1.1×10^{-13}	9.09×10^{-14}	772.9 (18.44)	515	93	50	100
SmSLA	2.15×10^{-13}	1.48×10^{-13}	892.6 (22.66)	621	90	50	100
FA-FS	0	0	502.3 (3.57)	242	66	123	100

Table 4. Results obtained for the largest farm with 256 ports, 196 EVs, and random charging rates. The maximal number of EVs that can be charged in one time slot is restricted to 80 owing to the power limit.

Algorithm	Mean (r_c)	Median (r_c)	F_{smooth}	S_x	T_x	L_x	L_f [%]
BLP	2.97	2.92	848.4 (240.6)	395	65	78	0
Q1-FW	0.141	5.3×10^{-14}	1322 (178)	745	94	61	90
SmQ2-FW	2.61	2.65	614 (34.17)	304	65	70	0
SmQ2-NG-FW	2.48	2.44	557 (25.4)	268	65	72	0
SmSLA	6.29×10^{-14}	5.74×10^{-14}	454.5 (38.1)	335	92	62	100
FA-FS	0	0	394.6 (12.15)	188	65	105	100

We also verified that $\max\{SU\} = 1$ for each proposed algorithm and for each analyzed scenario. This observation leads to the conclusion that each algorithm satisfies the two-link constraint in (9), which is necessary to guarantee that only one port in each charging station is active in one time slot.

Another criterion is the smoothness measure expressed by the function $F_{smooth} = u^T Q_L u$, where $u = \text{vec}(U)$ and $Q_L = L^T L \otimes I_N$ is given in (28). The mean values and standard deviations (in parentheses) of F_{smooth} are also listed in the tables. Parameter S_x denotes the total number of switching-on operations. The lower bound of this parameter was N. The tardiness, expressed by the number of time slots wherein at least one charging station is switched on, is referred to as T_x. Parameter L_x denotes the maximal number of EVs charged in any time slot. The percentage of MC runs wherein all EVs are charged up to their desired SoC level over the entire time horizon is expressed by L_f. If $L_f = 0$, it means that in each MC run, at least one EV is not charged according to the desired level. All parameters S_x, T_x, L_x, and L_f are expressed as median values.

The total power consumed by the entire farm versus time slots is illustrated in Figure 5a for the smallest farm with $N = 16$ and random charging rates, and in Figure 5b for the largest farm with $N = 256$ and a constant charging rate. The power in the t-th time slot is calculated as $P_t^{(max)} = U_s \sum_{n=1}^{N} u_{nt} \sum_{l=1}^{3} r_n^{(l)}$, where $U_s = 230\ V$ is the phase voltage. The red horizontal line determines the maximal load owing to the limit of I_{max}.

The computational complexity of the proposed algorithmic strategies was evaluated in terms of the elapsed time (ET) in seconds. The plots of the averaged ET for running the proposed algorithms on the scheduling problem in the farm with $N \in \{16, 32, 64, 128, 256\}$ are shown in Figure 6a for random charging rates and in Figure 6b for a constant charging rate. The whiskers determined the standard deviation.

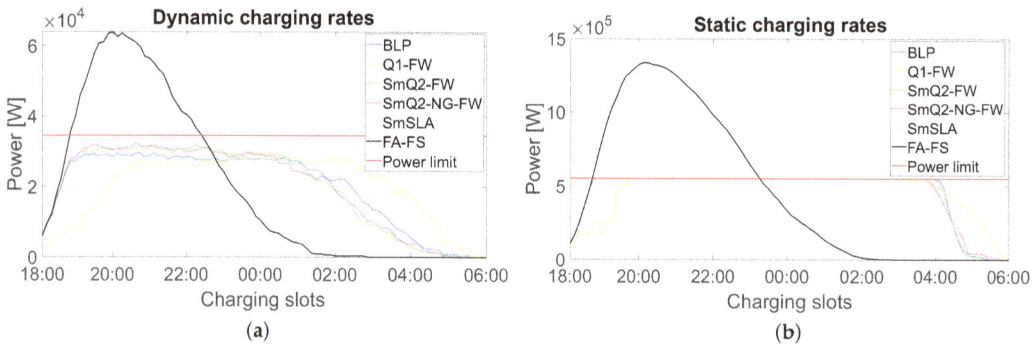

Figure 5. Averaged total power consumed by the whole farm versus the time slots: (**a**) small farm with $N = 16$ and random charging rates; (**b**) large farm with $N = 256$ and a constant charging rate.

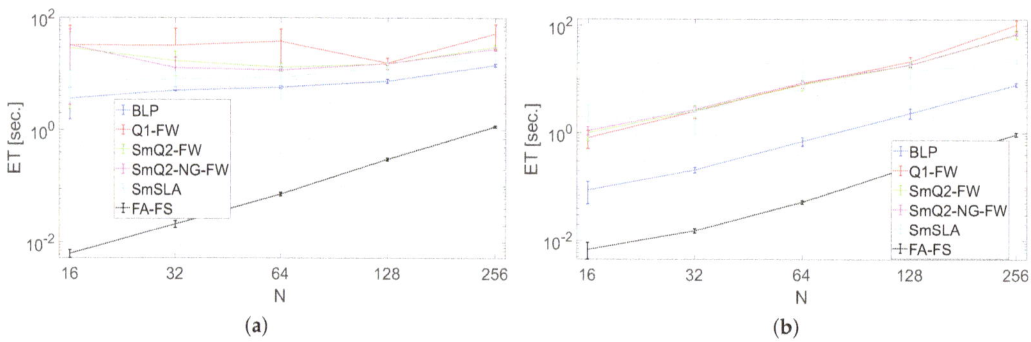

Figure 6. Averaged elapsed time (ET) for running the schedulers with (**a**) random charging rates and (**b**) a constant charging rate.

4.3. Discussion

Experiments were conducted for a variety of scheduling problems. Figure 3a,c show that ED and AT parameters are highly diversified in a wide range of their possible values in our tests. This reflects the real charging scenarios. Moreover, the charging rates considerably change with time and phase lines. In practice, the charging currents can be different for each phase line in the same time slot. This case was considered in our simulations, as shown in Figure 3b.

The experiments demonstrated that the most challenging scheduling problems, due to computational issues, were scenarios with random charging rates. Such problems were more difficult to tackle even for a small-scale farm ($N = 16$) than for a large-scale farm (for $N = 256$) but with a constant charging rate. The residual errors and the L_f measure presented in Tables 3 and 4 clearly confirm this statement. This observation is theoretically justified as the feasible region determined by the inequality constraints in (23) has the form of a complex polytop, with more vertices when $r_n^{(l)}$ varies with n and l. Our experiments showed that for a practical case with varying charging rates, we have a robust algorithm for scheduling problems, even when the number of EVs is large. It is the SmSLA strategy, which always (for each MC run) yields the solution fully satisfying all constraints, that is, the mean r_c is marginally small and $L_f = 100\%$. However, this algorithm wins only with respect to this criterion, and it is not the best one in each analyzed competition.

Regarding all the criteria used for validating the algorithm, one can conclude that there is no candidate that wins in all tested categories. The FA-FS provides the results with the shortest tardiness (lowest T_x in the tables) and the shortest ET (see Figure 6). Unfortunately, it does not have any embedded power constraints. This leads to a strong overload effect, as clearly shown in Figure 5. Hence, this algorithm is not acceptable for solving practical scheduling problems. FA-FS was used in our experiments only to demonstrate the load peak effect. All other algorithms prevent the overload effect and provide uniform energy consumption over a wide time window, as depicted in Figure 5 for each analyzed scenario. For a constant charging rate, BLP, SmQ2-FW, SmQ2-NG-FW, and SmSLA behaved similarly, and only Q1-FW provided a slightly delayed energy consumption profile. In this scenario, the farm works with the maximum acceptable power in a wide time window (from approximately 6.30 p.m. to 4 a.m.). When the charging rates are not constant, the farm works with a power slightly below the limit (the red line in Figure 5), which is still acceptable. Moreover, the energy consumption is approximately uniform from 7 p.m. to 1 a.m. for nearly all algorithms. Subsequently, a decreasing trend in energy consumption was observed until 6 a.m.

With regard to tardiness (T_x), Q1-FW and SmSLA are the least competitive. However, this criterion is correlated with the L_x measure, and it is thus obvious that fast charging is related to higher energy consumption. Q1-FW and SmSLA have a lower number of EVs in the most energy-consuming period. Hence, their schedules are safer with respect to energy consumption and maintain a higher margin to the power limit.

Another crucial criterion is the smoothness of the charging profile for each EV, which is related to the number of switching on/off operations at each charging station. The results presented in the tables demonstrate that SmQ2-NG-FW provides the smoothest charging profiles with a relatively low number of switching operations (parameter S_x). The worst results in this category were obtained by the BLP and Q1-FW. This observation can be justified by the fact that both BLP and Q1-FW do not involve any procedures for enforcing smoothness, whereas the others (excluding FA-FS) are based on the BQP strategy, where the quadratic term contains the smoothness penalty. Moreover, the quadratic II approach seems to be more favorable with respect to smoothness than the quadratic I approach. Furthermore, the results listed in the tables can be confirmed by the single-run charging matrices presented in Figures 2 and 4. The charging profiles of SmQ2-NG-FW are the smoothest, but SmQ2-FW and SmSLA are only slightly worse, which is in favor of the latter considering all the criteria.

Figures 2 and 4 also show that SmQ2-FW, SmQ2-NG-FW, and SmSLA better satisfy the rule of early charging than BLP and Q1-FW. Obviously, FA-FS is the best in this category; however, it is disqualified due to the power limit criterion (as discussed above). This rule is enforced by the matrix $\boldsymbol{D}_w$ in (12), (13), and (16), and it aims to enforce charging as early as possible. Obviously, the similar rule is incorporated in (11) via vector $\boldsymbol{w}$ but BLP seems to tackle worse in this aspect, as demonstrated in Figures 2 and 4.

The algorithms were evaluated with respect to the computational complexity expressed by the ET criterion. In this category, BLP is substantially faster than the others (excluding FA-FS) when a constant charging rate is applied (see Figure 6). For the random charging rates, the difference in ET was not very large. Interestingly, an increase in the number of EVs scales linearly for BLP, and there is no simple relation between ET and N for Q1-FW, SmQ2-FW, and SmQ2-NG-FW. This is probably owing to the high computational cost of processing a much more complicated polytop determined by (23) when random charging rates are used. For $N = 256$, Q1-FW, SmQ2-FW, SmQ2-NG-FW, and SmSLA have approximately similar ET values for random and constant charging rates. BLP is much faster for each scenario and provides satisfactory results for meeting the constraints. Hence, it can be used in the precomputing stage to check if a feasible solution exists, given the input data.

4.4. Engineering Aspects

The above experiments allowed us to select SmSLA as the most optimal algorithm for the scheduler. The charging matrix $\boldsymbol{U}$, together with the real-time intervals of the time slots, is passed to the CCU where real-time decisions on the current charging status of EVSEs are taken for each time slot. If $u_{nt} = 1$, the n-th charging point is switched on in the tth time slot. Otherwise, it is switched off. The scheduler is run each time a new EV is connected or disconnected to/from an EVSE, the scheduling period is over, or the CCU detects a considerable change in the charging currents. When a scheduler is called, the initialization procedures (which are not described herein) determine the current state of charging for each active EV. The EDs were updated using the previous charging matrices $\{\boldsymbol{U}\}$. For a new EV, the ED is evaluated using its individual charging curve, user demands, and historical charging data.

The charging problem may not have any solution, given real input data. This is a normal situation, not resulting from an algorithmic issue but rather due to a lack of a feasible solution to a given problem. To tackle this problem, BLP is used in the initialization procedure to detect this case, and if it occurs, the correction procedure is applied to decrease the EDs for the selected group of already charging EVs or for a new EV. This procedure was governed by the ED correction procedure.

5. Conclusions

In this study, we proposed a new model for on–off scheduling of EV charging, assuming that each three-phase charger is equipped in two ports that can be alternately served. The scheduler considers individual charging rates and maximal currents that supply the entire farm separately for each phase. For this model, we analyzed various binary algorithm approaches. All algorithms were validated with respect to multiple criteria, including constraint satisfying conditions, energy limit, tardiness, and charging profile smoothness. The experiments demonstrated that only SmSLA can yield the correct solution that satisfies all constraints for each MC run of each testing scenario. It also ensures smooth charging profiles. Unfortunately, this algorithm has a relatively long tardiness and is not the fastest. BLP is much faster than the others; however, it does not enforce smoothness. However, BLP could be a good choice for the preprocessing stage to guarantee feasibility.

The presented results, despite being completed, suggest that the issue of the algorithmic approach is still open, and further research in this area will be performed in the future. We still admit the possibility of designing a more robust algorithm than SmSLA, which would provide smoother charging profiles and lower tardiness. Furthermore, the topic of fast checking the feasibility (without running the scheduler) would be very interesting, with a high potential for practical applications.

In summary, we proposed a new model for the on–off scheduling of two-link EV chargers and experimentally evaluated the effectiveness of various binary algorithmic approaches with regard to multiple criteria. We statistically demonstrated that the best choice is the computational algorithm based on SLA with smoothness constraints (SmSLA). It satisfies the most important criteria and constraints in all statistical tests performed.

Author Contributions: Conceptualization, R.Z.; Methodology, R.Z.; Software, R.Z.; Investigation, R.Z., A.G., J.W.; Validation, R.Z., A.G., J.W., R.I.G.; Visualization, R.Z.; Writing—original draft, R.Z.; Writing—review and editing, R.Z.; Funding acquisition, A.G., J.W., R.I.G.; project administration, R.I.G. All authors have read and agreed to the published version of the manuscript.

Funding: This research was supported by grant RPMP.01.02.01-12-0333/19, funded by the Regional Operational Program for the Malopolska Region for 2014–2020.

Institutional Review Board Statement: Not applicable.

Informed Consent Statement: Not applicable.

Data Availability Statement: The data presented in this study and the Matlab code of the discussed algorithms are available from https://github.com/RafalZdunek/On-off-scheduler.git.

Conflicts of Interest: The authors declare no conflict of interest.

Appendix A

Let $\boldsymbol{U} = [\boldsymbol{u}_1, \ldots, \boldsymbol{u}_T] \in \mathbb{B}^{N \times T}$, $\boldsymbol{B} = \begin{bmatrix} \underline{\boldsymbol{b}}_1 \\ \vdots \\ \underline{\boldsymbol{b}}_T \end{bmatrix} \in \mathbb{B}^{T \times N}$, and $\boldsymbol{I}_T = [\boldsymbol{i}_1, \ldots, \boldsymbol{i}_T] \in \mathbb{R}^{T \times T}$ be the identity matrix. Formula (4) can be reformulated as follows:

$$\begin{aligned} \boldsymbol{c} &= \operatorname{diag}\{\boldsymbol{U}\boldsymbol{B}\} = \begin{bmatrix} \underline{\boldsymbol{b}}_1\boldsymbol{u}_1 & & 0 \\ & \ddots & \\ 0 & & \underline{\boldsymbol{b}}_T\boldsymbol{u}_T \end{bmatrix} = \begin{bmatrix} \underline{\boldsymbol{b}}_1 & & 0 \\ & \ddots & \\ 0 & & \underline{\boldsymbol{b}}_T \end{bmatrix} \begin{bmatrix} \boldsymbol{u}_1 \\ \vdots \\ \boldsymbol{u}_T \end{bmatrix} \\ &= \begin{bmatrix} \underline{\boldsymbol{b}}_1^T & & 0 \\ & \ddots & \\ 0 & & \underline{\boldsymbol{b}}_T^T \end{bmatrix}^T \begin{bmatrix} \boldsymbol{u}_1 \\ \vdots \\ \boldsymbol{u}_T \end{bmatrix} = \left[\underline{\boldsymbol{b}}_1^T \otimes \boldsymbol{i}_1, \cdots, \underline{\boldsymbol{b}}_T^T \otimes \boldsymbol{i}_T\right]^T \begin{bmatrix} \boldsymbol{u}_1 \\ \vdots \\ \boldsymbol{u}_T \end{bmatrix} \\ &= \left(\boldsymbol{B}^T \odot \boldsymbol{I}_T\right)^T \operatorname{vec}(\boldsymbol{U}). \end{aligned} \tag{A1}$$

Appendix B

Function (12) can be reformulated as follows:

$$\begin{aligned} \Psi &= \tfrac{1}{2}\textstyle\sum_{l=1}^{3} \left\| \underline{\boldsymbol{I}}^{(max,l)} - \left(\boldsymbol{r}^{(l)}\right)^T \boldsymbol{U}\boldsymbol{D}_w \right\|_2^2 = \tfrac{1}{2}\sum_{l=1}^{3} \left\| \underline{\boldsymbol{I}}^{(max,l)} - \left(\boldsymbol{D}_w \otimes (\boldsymbol{r}^{(l)})^T\right)\boldsymbol{u} \right\|_2^2 \\ &= \tfrac{1}{2}\textstyle\sum_{l=1}^{3} \left[\boldsymbol{u}^T \left(\boldsymbol{D}_w \otimes (\boldsymbol{r}^{(l)})^T\right)^T \left(\boldsymbol{D}_w \otimes (\boldsymbol{r}^{(l)})^T\right)\boldsymbol{u} - \underline{\boldsymbol{I}}^{(max,l)} \left(\boldsymbol{D}_w \otimes (\boldsymbol{r}^{(l)})^T\right)\boldsymbol{u} \right] + \text{const} \\ &= \tfrac{1}{2}\boldsymbol{u}^T \left(\boldsymbol{D}_w^2 \otimes \textstyle\sum_{l=1}^{3} \boldsymbol{r}^{(l)} (\boldsymbol{r}^{(l)})^T\right)\boldsymbol{u} - \boldsymbol{e}_3^T \begin{bmatrix} \underline{\boldsymbol{I}}^{(max,1)} \left(\boldsymbol{D}_w \otimes (\boldsymbol{r}^{(1)})^T\right) \\ \underline{\boldsymbol{I}}^{(max,2)} \left(\boldsymbol{D}_w \otimes (\boldsymbol{r}^{(2)})^T\right) \\ \underline{\boldsymbol{I}}^{(max,3)} \left(\boldsymbol{D}_w \otimes (\boldsymbol{r}^{(3)})^T\right) \end{bmatrix}^T \boldsymbol{u} + \text{const} \\ &= \tfrac{1}{2}\boldsymbol{u}^T \boldsymbol{Q}\boldsymbol{u} + \boldsymbol{d}^T\boldsymbol{u} + \text{const}. \end{aligned} \tag{A2}$$

Appendix C

Function (16) can be presented in the following form:

$$\begin{aligned} \Psi &= \tfrac{1}{2}\|\boldsymbol{c} - \boldsymbol{U}\boldsymbol{D}_w\boldsymbol{e}_T\|_2^2 + \tfrac{\alpha}{2}\|\boldsymbol{U}\boldsymbol{L}^T\|_F^2 \\ &= \tfrac{1}{2}\|\boldsymbol{c} - (\boldsymbol{e}_T^T\boldsymbol{D}_w \otimes \boldsymbol{I}_N)\boldsymbol{u}\|_2^2 + \tfrac{\alpha}{2}\|(\boldsymbol{L} \otimes \boldsymbol{I}_N)\boldsymbol{u}\|_F^2 \\ &= \tfrac{1}{2}\boldsymbol{u}^T(\boldsymbol{e}_T^T\boldsymbol{D}_w \otimes \boldsymbol{I}_N)^T(\boldsymbol{e}_T^T\boldsymbol{D}_w \otimes \boldsymbol{I}_N)\boldsymbol{u} - \boldsymbol{c}^T(\boldsymbol{e}_T^T\boldsymbol{D}_w \otimes \boldsymbol{I}_N)\boldsymbol{u} \\ &+ \tfrac{1}{2}\boldsymbol{u}^T(\boldsymbol{L} \otimes \boldsymbol{I}_N)^T(\boldsymbol{L} \otimes \boldsymbol{I}_N)\boldsymbol{u} + \text{const} \\ &= \tfrac{1}{2}\boldsymbol{u}^T(\boldsymbol{D}_w\boldsymbol{e}_T\boldsymbol{e}_T^T\boldsymbol{D}_w \otimes \boldsymbol{I}_N + \alpha(\boldsymbol{L}^T\boldsymbol{L} \otimes \boldsymbol{I}_N))\boldsymbol{u} - \boldsymbol{c}^T(\boldsymbol{e}_T^T\boldsymbol{D}_w \otimes \boldsymbol{I}_N)\boldsymbol{u} + \text{const} \\ &= \tfrac{1}{2}\boldsymbol{u}^T\boldsymbol{Q}\boldsymbol{u} + \boldsymbol{d}^T\boldsymbol{u} + \text{const}. \end{aligned} \tag{A3}$$

References

1. Karmaker, A.K.; Hossain, M.A.; Manoj Kumar, N.; Jagadeesan, V.; Jayakumar, A.; Ray, B. Analysis of Using Biogas Resources for Electric Vehicle Charging in Bangladesh: A Techno-Economic-Environmental Perspective. *Sustainability* **2020**, *12*, 2579. [CrossRef]
2. Samadi, M.; Fattahi, J.; Schriemer, H.; Erol-Kantarci, M. Demand Management for Optimized Energy Usage and Consumer Comfort Using Sequential Optimization. *Sensors* **2021**, *21*, 130. [CrossRef] [PubMed]
3. Qureshi, K.N.; Alhudhaif, A.; Jeon, G. Electric-vehicle energy management and charging scheduling system in sustainable cities and society. *Sustain. Cities Soc.* **2021**, *71*, 102990. [CrossRef]

4. Frank, M.; Wolfe, P. An algorithm for quadratic programming. *Nav. Res. Logist. Q.* **1956**, *3*, 95–110. [CrossRef]
5. Telli, M.; Bentobache, M.; Mokhtari, A. A successive linear approximation algorithm for the global minimization of a concave quadratic program. *Comp. Appl. Math.* **2020**, *39*, 272. [CrossRef]
6. Al-Ogaili, A.S.; Hashim, T.J.T.; Rahmat, N.A.; Ramasamy, A.; Marsadek, M.; Faisal, M.; Hannan, M. Review on Scheduling, Clustering, and Forecasting Strategies for Controlling Electric Vehicle Charging: Challenges and Recommendations. *IEEE Access* **2019**, *7*, 128353–128371. [CrossRef]
7. Amin, A.; Tareen, W.U.K.; Usman, M.; Ali, H.; Bari, I.; Horan, B.; Mekhilef, S.; Asif, M.; Ahmed, S.; Mahmood, A. A Review of Optimal Charging Strategy for Electric Vehicles under Dynamic Pricing Schemes in the Distribution Charging Network. *Sustainability* **2020**, *12*, 10160. [CrossRef]
8. Rahman, I.; Vasant, P.M.; Singh, B.S.; Abdullah-Al-Wadud, M.; Adnan, N. Review of recent trends in optimization techniques for plug-in hybrid, and electric vehicle charging infrastructures. *Renew. Sustain. Energy Rev.* **2016**, *58*, 1039–1047. [CrossRef]
9. Rahman, I.; Vasant, P.; Singh, B.; Abdullah-Al-Wadud, M. Novel metaheuristic optimization strategies for plug-in hybrid electric vehicles: A holistic review. *Intell. Decis. Technol.* **2016**, *10*, 149–163. [CrossRef]
10. Gan, L.; Topcu, U.; Low, S.H. Optimal decentralized protocol for electric vehicle charging. *IEEE Trans. Power Syst.* **2013**, *28*, 940–951. [CrossRef]
11. Gan, L.; Topcu, U.; Low, S.H. Stochastic distributed protocol for electric vehicle charging with discrete charging rate. In Proceedings of the 2012 IEEE Power and Energy Society General Meeting, San Diego, CA, USA, 22–26 July 2012; pp. 1–8.
12. Ma, W.J.; Gupta, V.; Topcu, U. On distributed charging control of electric vehicles with power network capacity constraints. In Proceedings of the 2014 American Control Conference, Portland, OR, USA, 4–6 June 2014; pp. 4306–4311.
13. Xu, S.; Feng, D.; Yan, Z.; Zhang, L.; Li, N.; Jing, L.; Wang, J. Ant-Based Swarm Algorithm for Charging Coordination of Electric Vehicles. *Int. J. Distrib. Sens. Netw.* **2013**, *9*, 268942. [CrossRef]
14. Karfopoulos, E.; Hatziargyriou, N. Distributed coordination of electric vehicles for conforming to an energy schedule. *Electr. Power Syst. Res.* **2017**, *151*, 86–95. [CrossRef]
15. Ardakanian, O.; Keshav, S.; Rosenberg, C. Real-Time Distributed Control for Smart Electric Vehicle Chargers: From a Static to a Dynamic Study. *IEEE Trans. Smart Grid* **2014**, *5*, 2295–2305. [CrossRef]
16. Yang, Q.; An, D.; Yu, W.; Tan, Z.; Yang, X. Towards Stochastic Optimization-Based Electric Vehicle Penetration in a Novel Archipelago Microgrid. *Sensors* **2016**, *16*, 907. [CrossRef]
17. Ma, Z.; Callaway, D.S.; Hiskens, I.A. Decentralized Charging Control of Large Populations of Plug-in Electric Vehicles. *IEEE Trans. Control Syst. Technol.* **2013**, *21*, 67–78. [CrossRef]
18. Zhang, T.; Chen, W.; Han, Z.; Cao, Z. Charging Scheduling of Electric Vehicles With Local Renewable Energy Under Uncertain Electric Vehicle Arrival and Grid Power Price. *IEEE Trans. Veh. Technol.* **2014**, *63*, 2600–2612. [CrossRef]
19. Chen, N.; Tan, C.W.; Quek, T.Q.S. Electric Vehicle Charging in Smart Grid: Optimality and Valley-Filling Algorithms. *IEEE J. Sel. Top. Signal Process.* **2014**, *8*, 1073–1083. [CrossRef]
20. Zhang, W.; Zhang, D.; Mu, B.; Wang, L.Y.; Bao, Y.; Jiang, J.; Morais, H. Decentralized Electric Vehicle Charging Strategies for Reduced Load Variation and Guaranteed Charge Completion in Regional Distribution Grids. *Energies* **2017**, *10*, 147. [CrossRef]
21. Khan, S.U.; Mehmood, K.K.; Haider, Z.M.; Rafique, M.K.; Khan, M.O.; Kim, C.H. Coordination of Multiple Electric Vehicle Aggregators for Peak Shaving and Valley Filling in Distribution Feeders. *Energies* **2021**, *14*, 352. [CrossRef]
22. Richardson, P.; Flynn, D.; Keane, A. Optimal Charging of Electric Vehicles in Low-Voltage Distribution Systems. *IEEE Trans. Power Syst.* **2012**, *27*, 268–279. [CrossRef]
23. He, Y.; Venkatesh, B.; Guan, L. Optimal Scheduling for Charging and Discharging of Electric Vehicles. *IEEE Trans. Smart Grid* **2012**, *3*, 1095–1105. [CrossRef]
24. de Hoog, J.; Alpcan, T.; Brazil, M.; Thomas, D.A.; Mareels, I. Optimal Charging of Electric Vehicles Taking Distribution Network Constraints Into Account. *IEEE Trans. Power Syst.* **2015**, *30*, 365–375. [CrossRef]
25. Jian, L.; Zheng, Y.; Shao, Z. High efficient valley-filling strategy for centralized coordinated charging of large-scale electric vehicles. *Appl. Energy* **2017**, *186*, 46–55. [CrossRef]
26. Foster, J.M.; Caramanis, M.C. Optimal Power Market Participation of Plug-In Electric Vehicles Pooled by Distribution Feeder. *IEEE Trans. Power Syst.* **2013**, *28*, 2065–2076. [CrossRef]
27. Sundstom, O.; Binding, C. Optimization Methods to Plan the Charging of Electric Vehicle Fleets. *ACEEE Int. J. Commun.* **2010**, *1*, 28–29.
28. Ma, C.; Rautiainen, J.; Dahlhaus, D.; Lakshman, A.; Toebermann, J.C.; Braun, M. Online Optimal Charging Strategy for Electric Vehicles. *Energy Procedia* **2015**, *73*, 173–181. [CrossRef]
29. Hernández-Arauzo, A.; Puente, J.; Varela, R.; Sedano, J. Electric vehicle charging under power and balance constraints as dynamic scheduling. *Comput. Ind. Eng.* **2015**, *85*, 306–315. [CrossRef]
30. Clement, K.; Haesen, E.; Driesen, J. Coordinated charging of multiple plug-in hybrid electric vehicles in residential distribution grids. In Proceedings of the 2009 IEEE/PES Power Systems Conference and Exposition, Seattle, WA, USA, 15–18 March 2009; pp. 1–7.
31. Khaki, B.; Chung, Y.W.; Chu, C.; Gadh, R. Hierarchical Distributed EV Charging Scheduling in Distribution Grids. In Proceedings of the 2019 IEEE Power Energy Society General Meeting (PESGM), Atlanta, GA, USA, 4–8 August 2019; pp. 1–5.

32. Sundstrom, O.; Binding, C. Flexible Charging Optimization for Electric Vehicles Considering Distribution Grid Constraints. *IEEE Trans. Smart Grid* **2012**, *3*, 26–37. [CrossRef]
33. Binetti, G.; Davoudi, A.; Naso, D.; Turchiano, B.; Lewis, F.L. Scalable Real-Time Electric Vehicles Charging with Discrete Charging Rates. *IEEE Trans. Smart Grid* **2015**, *6*, 2211–2220. [CrossRef]
34. Baek, S.; Kim, D.; Oh, S.; Jun, J. A queuing model with random interruptions for electric vehicle charging systems. In Proceedings of the 2011 IEEE International Conference on Consumer Electronics (ICCE), Las Vegas, NV, USA, 9–12 January 2011; pp. 679–680.
35. Harris, C.; Dusparic, I.; Galván-López, E.; Marinescu, A.; Cahill, V.; Clarke, S. Set point control for charging of electric vehicles on the distribution network. In Proceedings of the ISGT 2014, Washington, DC, USA, 19–22 February 2014; pp. 1–5.
36. Nguyen, V.L.; Tran-Quoc, T.; Bacha, S.; Nguyen, B. Charging strategies to minimize the peak load for an electric vehicle fleet. In Proceedings of the IECON 2014—40th Annual Conference of the IEEE Industrial Electronics Society, Dallas, TX, USA, 29 October–1 November 2014; pp. 3522–3528.
37. Fernandes, X.; Rebelo, J.; Gouveia, J.; Maia, R.; Silva, N.B. On-off scheduling schemes for power-constrained electric vehicle charging. *Q. J. Oper. Res.* **2017**, *15*, 163–181. [CrossRef]
38. Sun, B.; Huang, Z.; Tan, X.; Tsang, D.H.K. Optimal Scheduling for Electric Vehicle Charging With Discrete Charging Levels in Distribution Grid. *IEEE Trans. Smart Grid* **2018**, *9*, 624–634. [CrossRef]
39. Jawad, M.; Qureshi, M.B.; Ali, S.M.; Shabbir, N.; Khan, M.U.S.; Aloraini, A.; Nawaz, R. A Cost-Effective Electric Vehicle Intelligent Charge Scheduling Method for Commercial Smart Parking Lots Using a Simplified Convex Relaxation Technique. *Sensors* **2020**, *20*, 4842. [CrossRef]
40. Bashash, S.; Moura, S.J.; Forman, J.C.; Fathy, H.K. Plug-in hybrid electric vehicle charge pattern optimization for energy cost and battery longevity. *J. Power Sources* **2011**, *196*, 541–549. [CrossRef]
41. Peterson, S.B.; Apt, J.; Whitacre, J. Lithium-ion battery cell degradation resulting from realistic vehicle and vehicle-to-grid utilization. *J. Power Sources* **2010**, *195*, 2385–2392. [CrossRef]
42. Trippe, A.E.; Arunachala, R.; Massier, T.; Jossen, A.; Hamacher, T. Charging optimization of battery electric vehicles including cycle battery aging. In Proceedings of the IEEE PES Innovative Smart Grid Technologies, Istanbul, Turkey, 12–15 October 2014; pp. 1–6.
43. De Vroey, L.; Jahn, R.; Omar, N.; Van Mierlo, J. Impact of smart charging on the EV battery ageing—Discussion from a 3 years real life experience. *World Electr. Veh. J.* **2015**, *7*, 613–620. [CrossRef]
44. van Kooten Niekerk, M.E.; Van den Akker, J.M.; Hoogeveen, J.A. Scheduling electric vehicles. *Public Transp.* **2017**, *9*, 155–176. [CrossRef]
45. Umetani, S.; Fukushima, Y.; Morita, H. A linear programming based heuristic algorithm for charge and discharge scheduling of electric vehicles in a building energy management system. *Omega* **2017**, *67*, 115–122. [CrossRef]
46. Shrestha, G.B.; Ang, S.G. A study of electric vehicle battery charging demand in the context of Singapore. In Proceedings of the 2007 International Power Engineering Conference (IPEC 2007), Singapore, 3–6 December 2007; pp. 64–69.
47. Sassi, O.; Oulamara, A. Electric vehicle scheduling and optimal charging problem: Complexity, exact and heuristic approaches. *Int. J. Prod. Res.* **2017**, *55*, 519–535. [CrossRef]
48. Franco, J.F.; Rider, M.J.; Romero, R. An MILP model for the plug-in electric vehicle charging coordination problem in electrical distribution systems. In Proceedings of the 2014 IEEE PES General Meeting | Conference Exposition, National Harbor, MD, USA, 27–31 July 2014; pp. 1–5.
49. Nizami, M.S.H.; Hossain, M.J.; Mahmud, K. A Coordinated Electric Vehicle Management System for Grid-Support Services in Residential Networks. *IEEE Syst. J.* **2021**, *15*, 2066–2077. [CrossRef]
50. Ma, T.Y. Two-stage battery recharge scheduling and vehicle-charger assignment policy for dynamic electric dial-a-ride services. *PLoS ONE* **2021**, *16*, e0251582. [CrossRef]
51. García-Álvarez, J.; González, M.A.; Vela, C.R. A Genetic Algorithm for Scheduling Electric Vehicle Charging. In Proceedings of the GECCO '15: Proceedings of the 2015 Annual Conference on Genetic and Evolutionary Computation, Madrid, Spain, 11–15 July 2015; pp. 393–400.
52. Wu, H.; Pang, G.K.H.; Choy, K.L.; Lam, H.Y. Dynamic resource allocation for parking lot electric vehicle recharging using heuristic fuzzy particle swarm optimization algorithm. *Appl. Soft Comput.* **2018**, *71*, 538–552. [CrossRef]
53. Yin, W.; Ming, Z.; Wen, T. Scheduling strategy of electric vehicle charging considering different requirements of grid and users. *Energy* **2021**, *232*, 121118. [CrossRef]
54. Arias, N.B.; Franco, J.F.; Lavorato, M.; Romero, R. Metaheuristic optimization algorithms for the optimal coordination of plug-in electric vehicle charging in distribution systems with distributed generation. *Electr. Power Syst. Res.* **2017**, *142*, 351–361. [CrossRef]
55. García-Álvarez, J.; González, M.A.; Vela, C.R. Metaheuristics for solving a real-world electric vehicle charging scheduling problem. *Appl. Soft Comput.* **2018**, *65*, 292–306. [CrossRef]
56. García Álvarez, J.; González, M.Á; Rodríguez Vela, C.; Varela, R. Electric Vehicle Charging Scheduling by an Enhanced Artificial Bee Colony Algorithm. *Energies* **2018**, *11*, 2752. [CrossRef]
57. Usman, M.; Tareen, W.U.K.; Amin, A.; Ali, H.; Bari, I.; Sajid, M.; Seyedmahmoudian, M.; Stojcevski, A.; Mahmood, A.; Mekhilef, S. A Coordinated Charging Scheduling of Electric Vehicles Considering Optimal Charging Time for Network Power Loss Minimization. *Energies* **2021**, *14*, 5336. [CrossRef]

58. Alonso, M.; Amaris, H.; Germain, J.G.; Galan, J.M. Optimal Charging Scheduling of Electric Vehicles in Smart Grids by Heuristic Algorithms. *Energies* **2014**, *7*, 2449–2475. [CrossRef]
59. Azar, A.G.; Jacobsen, R.H. Agent-based charging scheduling of electric vehicles. In Proceedings of the 2016 IEEE Online Conference on Green Communications (OnlineGreenComm), Piscataway, NJ, USA, 14–17 November 2016; pp. 64–69.
60. Xydas, E.; Marmaras, C.; Cipcigan, L.M. A multi-agent based scheduling algorithm for adaptive electric vehicles charging. *Appl. Energy* **2016**, *177*, 354–365. [CrossRef]
61. Zhang, C.; Liu, Y.; Wu, F.; Tang, B.; Fan, W. Effective Charging Planning Based on Deep Reinforcement Learning for Electric Vehicles. *IEEE Trans. Intell. Transp. Syst.* **2021**, *22*, 542–554. [CrossRef]
62. Zhong, J.; Xiong, X. An Orderly EV Charging Scheduling Method Based on Deep Learning in Cloud-Edge Collaborative Environment. *Adv. Civ. Eng.* **2021**, *2021*, 6690610.
63. Li, B.; Coutiño, M.; Giannakis, G.B.; Leus, G. A Momentum-Guided Frank-Wolfe Algorithm. *IEEE Trans. Signal Process.* **2021**, *69*, 3597–3611. [CrossRef]
64. Ferreira, O.P.; Sosa, W.S. On the Frank–Wolfe algorithm for non-compact constrained optimization problems. *Optimization* **2021**, 1–15. in press. [CrossRef]
65. Marra, F.; Yang, G.; Træholt, C.; Larsen, E.; Rasmussen, C.; You, S. Demand Profile Study of Battery Electric Vehicle under Different Charging Options. In Proceedings of the 2012 IEEE Power and Energy Society General Meeting, San Diego, CA, USA, 22–26 July 2012.
66. Miao, Y.; Hynan, P.; von Jouanne, A.; Yokochi, A. Current Li-Ion Battery Technologies in Electric Vehicles and Opportunities for Advancements. *Energies* **2019**, *12*, 1074. [CrossRef]

sensors

MDPI

Article

Advanced Driver Assistance Systems (ADAS) Based on Machine Learning Techniques for the Detection and Transcription of Variable Message Signs on Roads

Gonzalo De-Las-Heras [1], Javier Sánchez-Soriano [2] and Enrique Puertas [2,*]

1 SICE Canada Inc., Toronto, ON M4P 1G8, Canada; gdelasheras@sice.com
2 Department of Science, Computing and Technology Department, Universidad Europea de Madrid, 28670 Madrid, Spain; javier.sanchez@universidadeuropea.es
* Correspondence: enrique.puertas@universidadeuropea.es

Abstract: Among the reasons for traffic accidents, distractions are the most common. Although there are many traffic signs on the road that contribute to safety, variable message signs (VMSs) require special attention, which is transformed into distraction. ADAS (advanced driver assistance system) devices are advanced systems that perceive the environment and provide assistance to the driver for his comfort or safety. This project aims to develop a prototype of a VMS (variable message sign) reading system using machine learning techniques, which are still not used, especially in this aspect. The assistant consists of two parts: a first one that recognizes the signal on the street and another one that extracts its text and transforms it into speech. For the first one, a set of images were labeled in PASCAL VOC format by manual annotations, scraping and data augmentation. With this dataset, the VMS recognition model was trained, a RetinaNet based off of ResNet50 pretrained on the dataset COCO. Firstly, in the reading process, the images were preprocessed and binarized to achieve the best possible quality. Finally, the extraction was done by the Tesseract OCR model in its 4.0 version, and the speech was done by the cloud service of IBM Watson Text to Speech.

Keywords: VMS; machine learning; ADAS; image processing; environment perception

Citation: De-Las-Heras, G.; Sánchez-Soriano, J.; Puertas, E. Advanced Driver Assistance Systems (ADAS) Based on Machine Learning Techniques for the Detection and Transcription of Variable Message Signs on Roads. *Sensors* **2021**, *21*, 5866. https://doi.org/10.3390/s21175866

Academic Editor: Juan A. Cabrera

Received: 28 July 2021
Accepted: 28 August 2021
Published: 31 August 2021

1. Introduction

1.1. Motivation

Since the democratization of the private car, the world's fleet has continued to grow [1,2] (in Spain, each household has almost two vehicles [3]). This increase has brought with it the problem of traffic accidents. Data from the World Health Organization (WHO) estimate that during the period 2011–2020, 1.1 million people died due to traffic accidents and between 20 and 50 million were injured [4].

In Spain, the Dirección General de Tráfico (DGT) has produced a series of statistical yearbooks, which illustrate the evolution from 1960 to 2018 [5,6]. Generally speaking, the number of casualties has increased in recent years. The number of fatalities and hospitalized victims has decreased while the number of non-hospitalized injured victims has increased. Accidents are still occurring, but the probability of death is decreasing.

The causes of traffic accidents can be classified according to the risk factor that causes them. They are distinguished by human, mechanical and environmental factors (the state of the asphalt or traffic signs and weather conditions). According to the DGT, in 2018, 88% of accidents were the result of inappropriate driver behaviors [7] (similar conclusion to study [8], which states that 90% are due to human causes). In first place were distractions (33%), followed by speeding (29%) and alcohol consumption (26%) [7]. The same organization has prepared a document that lists the main distractions and explains how they affect accidents [9]. It shows that actions such as using a cell phone, eating or smoking are activities that require time and attention, reducing concentration while driving. The driver's physical condition also affects his reaction time and ability to be distracted. This has a direct impact on braking distance, which is a serious risk. Many of these behaviors are known to drivers and many are declared offenders [10].

The WHO, in its report on 2011–20, proposes five action points to improve safety [4]. Examples of the third (safer vehicles) are initiatives such as Prometheus [11,12], created by an association of vehicle manufacturers and researchers, or DRIVE (Dedicated Road Infrastructure for Vehicle Safety in Europe), funded by the EU (European Union) [12], which has promulgated a large number of papers on fundamental and practical problems, such as GIDS (Generic Intelligent Driver Support) [13]. Its aim was to "to determine the requirements and design standards for a class of intelligent co-driver (GIDS) systems that are maximally consistent with the information requirements and performance capabilities of the human driver" [13]. It was the beginning of what we know today as ADASs (advanced driver assistance systems), successors to basic safety systems and enablers of autonomous driving in the future [14].

Variable message signs (VMSs) are roadside ATIS (advanced traveler information system) devices consisting of LEDs (light-emitting diodes) that stand out against a black background (Figure 1). They are the mechanism used by traffic agencies to communicate useful information to drivers in order to improve their safety. These messages convey information by means of personalized text and/or traffic sign pictograms [15].

Figure 1. VMS example [16].

Several studies indicate that VMSs have a positive impact on driving by reducing speed [17] and relieving congestion caused by accidents or other events [18]. The very act of reading the VMS itself causes a reduction in speed while approaching it [19]. However, the act of investing attention and time into reading the message and understanding it is in itself a distraction and therefore a risk. Additionally, if we add to a main task, such as driving, the task of reading and understanding the information, we obtain a decrease in the effectiveness of both tasks [20]. There are approaches to reduce the attention required, simplifying the information by means of pictograms or messages consisting of a single word. The latter are more effective in understanding the message than even pictograms, because comprehension does not depend on prior knowledge of the pictogram [21]. There are conventions, such as the Vienna Convention [22], but each country is free to alter their signs, which makes it difficult to recognize them quickly.

There are solutions such as READit VMS [23], which through a client–server architecture and the geolocation of the user performs a locution of the content of the sign or displays a pictogram on an internal screen of the vehicle. These applications require constant connectivity to geolocation and the Internet to check the nearest VMS and may suffer from latency issues. They are also limited to the VMSs registered in the system. Due to these dependencies, they are not autonomous systems that allow the vehicle to be independent wherever it travels. The most similar ADAS are traffic signal recognition systems that, using sophisticated computer vision and machine learning techniques, display the signal to the driver on a screen located on the dashboard.

The motivation of this project is to provide solutions to the challenge of road fatalities by developing an ADAS that intervenes in the major cause of accidents, the distractions [7,10]. On the road we find panels with information that many studies have reported to cause a reduction in vehicle speed. However, the cause of this is the attention that is required to read and understand the message [24]. This results in less efficient driving [20]. This issue has been addressed by client–server software [23], but not by machine learning and computer vision techniques. This ADAS will allow the vehicle to be independent from network latency, geopositioning and the sign database. The solution will consist of a VMS recognizer that reproduces the signal content using a synthetic voice. To do so, it recognizes and trims the VMS from the road images, delivers it to the OCR (optical character recognition) subsystem that transcribes the panel content and announces it via the IBM Watson Text to Speech cloud service [25].

1.2. Vehicle Safety Systems

The report [14] carried out by The Boston Consulting Group (BCG) for The Motor & Equipment Manufacturers Association (MEMA) describes the evolution of safety systems in three periods: assistance and comfort systems, ADASs and semi/autonomous vehicles.

First assistants. In the first period, the first projects were developed to improve vehicle safety. Although they may seem simple, they are very useful, since they not only help the driver, but also provide greater comfort (an aspect closely related to safety [26]).

Some of these systems are cruise control, ABS (antilock braking system), ESP (electronic stability program), etc.

ADAS. As technology developed, more advanced systems emerged that operated in increasingly complex situations. The report [27] proposes a taxonomy based on the type of sensor used:

- Vision systems. These have cameras (monocular, stereo and infrared) placed at strategic points of the vehicle that provide images of the environment from which knowledge of the scene is extracted. These kinds of systems have problems with depth and lens obstructions; however, they are affordable [27];
- LiDAR (light detection and ranging). This is a technology that generates a 3D environment by projecting rays and measuring the distance to different objects. This allows the vehicle to know the elements around it in high resolution. It is a cutting-edge technology, but at the same time expensive. There is currently a debate between LiDAR and conventional cameras. Companies on a par with Tesla bet on the extraction of knowledge through multiple cameras plus other devices, such as radars. Others, for instance Waymo, believe that LiDAR is the solution of the future [28];
- Radars. These systems measure the speed and distance of objects in the environment (thanks to the Doppler effect). They emit a series of microwaves and measure the change in wave frequency. One case of use is adaptive cruise control [27];
- Ultrasound. Using a series of sound waves, these systems measure the distance to nearby objects. An example is the parking collision warning device [27];
- All these ADASs are complemented with other functionalities to improve their accuracy. For example, IMUs (inertial measurement units) or GPSs (global positioning systems) are auxiliary systems for distance measurement [27].

Semi/autonomous vehicles. In the latest era, which comes up to the present day, the challenge is to create cars that can drive themselves. With the help of new ADAS, such as the autopilot for traffic jams or the automatic lane change, this is possible. By 2025, it is expected that there will be 8 million autonomous and semi-autonomous vehicles worldwide [29–31].

The J3016 standard "Levels of Driving Automation" of the Society of Automotive Engineers (SAE) established six levels with which to define the autonomy of a vehicle. They range from 0 (fully manual) to 5 (fully autonomous) [32].

1.3. Recognition Systems

1.3.1. Object Recognition

The history of object recognizers is divided into two periods: traditional models and, since 2014, those based on deep learning [33].

First-generation detectors had to deal with a lack of computational and feature representation resources. For this reason, these algorithms contained hand-crafted features and methods that took full advantage of machine power [33].

- **Viola Jones** [34,35]. This is an extremely fast face recognizer, which slides a window over the entire image until a face is identified in one of the subsections.
- **HOG (histogram of oriented gradients)** [36]. This detector is designed to work on a uniform grid. Although it can be used to detect a variety of objects, it was primarily motivated for pedestrian detection [33].
- **DPM (deformable part-based model)** [37]. This method is an extension of the HOG detector, which applies the divide and conquer strategy. For example, the problem of recognizing a car can be decomposed into locating parts such as wheels or windows. It consists of a main filter and several secondary filters configured by supervised learning as if they were latent variables [33].

With the evolution of machine learning techniques, artificial neural networks (ANNs) emerged and within them, deep convolutional neural networks (CNNs) have improved image classification [38,39] and object detection [39–41] accuracy. Within CNNs, those dedicated to object detection are divided into two groups: one-stage and two-stage. The first ones treat the task as a regression problem by learning the probabilities of a class and the coordinates of the bounding box. The second ones group a series of regions of interest (first step) that are sent to the object classifier and the coordinate delimiter (second step). Each strategy has advantages and disadvantages. For example, one-step ones are faster, but have less accuracy [42].

Two-stage models:

- **R-CNN** [40]. This system takes the image and divides it into about 2000 regions on which the features are computed by a CNN. Finally, each region is classified by linear one-vs-rest SVMs (support vector machines) [40];
- **Fast R-CNN** [39]. Based on the previous model, fast R-CNN directly extracts features from the entire image, which are sent to the CNN for classification and localization at the same time. Thanks to this improvement, training time decreases while accuracy increases [39];
- **Faster R-CNN** [43]. This model eliminates the bottleneck that fast R-CNN had when selecting the region of interest (RoI) [33] by using a CNN called a region proposal network (RPN) to predict it. Faster R-CNN merges the RPN and fast R-CNN into a single network, so that the first one tells the second one where to focus. This is achieved by sharing their convolutional characteristics. This way, the RoI selection is practically zero cost, and the system is very close to real time [43].

Single-stage models:

- **YOLO** (You Only Look Once v1 [44], v2/9000 [45], v3 [46], v4 [47]). This is a real-time object recognition system thanks to the fact that the entire detection process is done by a single network. The process consists of a phase in which the system resizes the image to 488 x 488 and then executes a single CNN that returns the confidence of the detected object [44]. There are several enhancements to this model that are focused on increasing the accuracy but keeping the fast execution. The most recent version is v4 [45–47];
- **SSD (single shot detector)** [48]. This model's main contribution is the introduction of multi-reference and multi-resolution detection techniques, which significantly improve detection accuracy, especially for some small objects [33];
- **RetinaNet** [49]. Thanks to the authors of [49], it was found that the extreme imbalance of the foreground class is the main cause of their lower accuracy. To solve it, they

introduced a new loss function called "focal loss" to make the classifier focus on the most difficult examples of the misclassified ones. This brings this model up to the accuracy of the two-stage models.

There are several surveys in the literature that compare these object recognition models by measuring accuracy and speed, both for training and for inference. One of the best works comparing each of these models is [50], in which a systematic review of each of the models presented above is made and they are compared in terms of different metrics such as accuracy or inference speed. It is difficult to choose a clear winner since it depends on the specific task we are performing and whether we are more interested in a fast model for inference or if we need to obtain a higher accuracy in object recognition. In our work we have chosen RetinaNet as it is a model with one of the best accuracy–FPS balances.

1.3.2. Text Recognition

As with object detection, there are two eras. A first one in which the techniques were based on "hand-made" features to discriminate the characters, and another one in which machine learning models predominate [51,52].

Pre-deep learning period:

- **Connected-component analysis (CCA).** These classifiers extract candidate components at first and then filter out non-textual components using manual rules or trained classifiers [53]. There are two methods, these being stroke width transform (SWT) and maximally stable extremal regions (MSER) [51];
- **Sliding window (SW).** This model works by sliding a small multi-scale window through all possible locations on the image, classifying whether text is present or not [51].

In the era of deep learning, [52] proposes a hierarchical taxonomy divided into text detectors, transcribers, end-to-end systems and auxiliary methods that improve the model quality:

- **Detection.** Text detection can be defined as a subset of the problem of object detection, in which there are three tendencies [52]:
- **Reduction of pipelines** to simplify the training process and reduce error. **Decomposition into subtexts** and then joining them into a complete instance. **Specific recognition** in cases such as curved text, irregularly shaped text or text with complex backgrounds;
- **Transcribers.** In traditional methods, the process consisted of preprocessing, segmentation and character recognition. However, segmentation is costly and has a longer execution time. To avoid this step, connectionist temporal classification (CCT) methods [54] and attention mechanisms [52] are used;
- **End-to-end systems.** Instead of dividing the main problem into detection and recognition subproblems, these systems integrate the entire process for reading directly from the image [52];
- **Auxiliary techniques.** An important aspect is techniques that improve training quality, such as creating synthetic examples, reducing noise in the image or incorporating information from the environment [52].

Some examples for object detection in vehicle security systems are:

Traffic light recognition [55–57]. These are assistants that detect this type of signaling, so that they can inform the driver of their current status. If they were connected directly to the vehicle control system, the vehicle could even brake automatically. The main challenges of this ADAS are related to the different types of traffic lights, since there are several models depending on the country, and the existence of intersections or multiple lanes;

Signal recognition [58,59]. Traffic sign identification is one of the tasks required for environment perception. They are the main source through which drivers receive information (maximum speed, prohibitions, intersections, etc.). Although there are currently commercialized ADAS (such as the Toyota Road Sign Assist, or RSA [60]), it is still a challenge. The main problem is the diversity in size and shapes;

Panel recognition [61,62]. Information boards are a type of signage located above the lanes, which primarily communicate information by text. Therefore, the challenge for the assistants lies in the recognition of the characters, not only in the identification of the object on the road.

2. Methodology

The processing steps are summarized in Figure 2. The images captured by the vehicle camera are initially processed by the VMS object recognition module. The next step is to normalize the section that corresponds to the VMS by cropping the image, changing the perspective and angle in addition to adjusting the color to facilitate the following task of extracting the text from the image. Finally, the text is converted to audio using a "text to speech" service in the cloud.

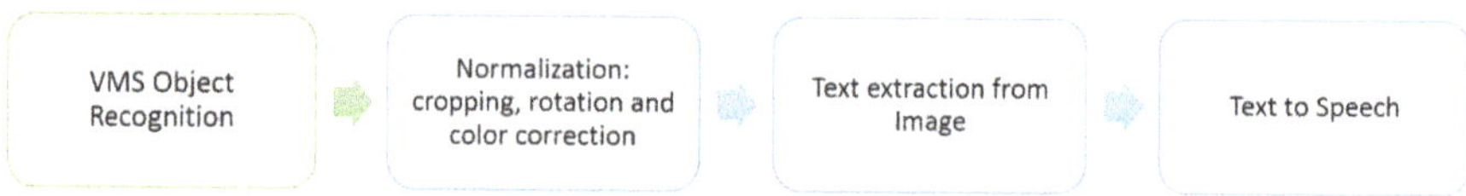

Figure 2. Processing steps.

These processing steps for the VMS speech system are divided into two subsystems combining local processing and cloud services: a VMS recognizer and a content extractor and speaker (Figure 3).

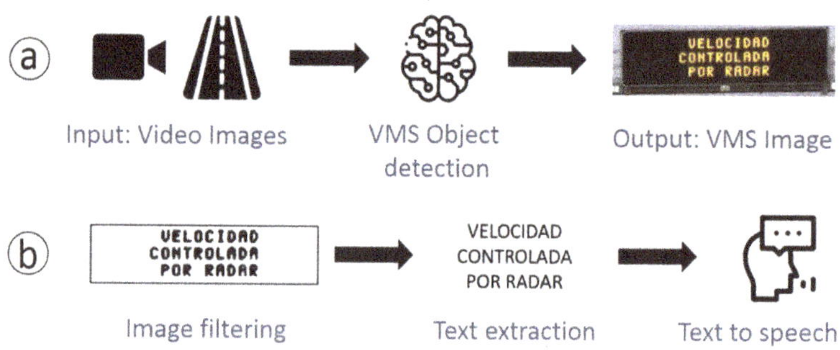

Figure 3. The VMS reading process consists of 2 steps. (**a**) VMS extraction and (**b**) processes the image to extract the content and speak it.

2.1. VMS Recognizer

From a picture of the environment taken by a camera located on the front of the vehicle, it recognizes the VMS and produces another image as an output, consisting only of the sign itself. This task is carried out by a deep CNN, a machine learning model that gives great results in image classification and object detection [38–41]. In order to do so, it is necessary to build a set of labeled images to train and evaluate the model.

2.2. Content Extractor and Speaker

Taking as an input the image produced by the VMS recognizer, it processes it to obtain the text of the panel and reproduces it using a synthetic voice. The process is as follows.

First, it is necessary to preprocess the image to make it easier to extract the text. The steps to follow are: (1) Angle correction. Straightens the orientation of the VMS. (2) Cropping of the VMS. Generates an image with only the content of the panel by eliminating margins that do not correspond to the VMS. (3) Color adjustment. Transforms the previous image into another one with black text over a white background; this will make the extraction task easier.

Then, using an OCR model, it transcribes the text contained in the panel. Finally, the system makes a call to the IBM Watson Text to Speech cloud service, which returns a sound file with the spoken text.

3. Variable Message Sign Recognition

3.1. Dataset

Labeled Image Collection

The strategy is to join different sources to maximize the number of examples with the least manual work. This is a key point, since each image must be annotated individually, which is very time-consuming. Therefore, a process has been designed to obtain a minimal dataset and to create a basic model with which to label the images iteratively. Thus, although the first search will be completely manual, subsequent searches will consist of small adjustments on images extracted from videos (Table 1), which would otherwise involve a lot of work. The initial acquisition can be divided into three steps:

1. **Collection.** By searching Google Images, YouTube, several websites and manual clippings combined with scraping scripts.
2. **Labeling.** Each image is manually annotated using the software in [63], which generates an XML (Extensible Markup Language) file in PASCAL VOC (Visual Object Classes) format.
3. **Data augmentation.** Data augmentation is a widespread method that consists of applying modifications to the image (rotations, cropping, translations, etc.) in order to create apparently new instances. For this project, since the VMS will always be in the top position of the image, we have chosen to flip the image on the y-axis. That way, the signs on one side will be placed on the opposite side, generating a new instance.

Once the first version of the dataset (134 VMS examples) was obtained, a RetinaNet [49] was trained with it on a ResNet50 model [64] pretrained on COCO [65]. This model has been selected due to the fact that even though it is a single-stage model, it achieves results very close to those of two stages, maintaining the advantages of the single-stage models [49]. Results are shown on Table 1.

Table 1. Basic model training parameters and results.

Epochs	25
No. of training images	134
Time	≈01:30:00
Learning rate	10^{-5}
Loss	0.174

Thanks to this model, an iterative process begins in which new labeled images are obtained more quickly. There are two methods with which to do so:

- **Manual.** As in the first acquisition, the VMS images are manually selected. The difference is that the labeling is performed by the basic model;
- **Semiautomatic.** In this case, we select videos to be analyzed by the basic model in order to extract a set of labeled candidate images from hours of footage, which would otherwise be much more tedious.

Since this first model is not perfect (nor is it intended to be), it is necessary to check the automatic selection and detection. Finally, once the images have been validated with their annotations, data augmentation (flipping on the y-axis) is applied.

3.2. Final Dataset

Every machine learning algorithm is sensitive to overfitting its parameters to the data with which it has been trained. In this situation, the model memorizes this information, which prevents it from generalizing and, therefore, from performing well in real situations. To avoid this situation, the dataset has been divided into two portions, one exclusively for training and another for validation. This method is a popular practice for correctly measuring the quality of a model.

At a certain epoch, generalization is transformed into memorization of the training set. This manifests itself as an increase in the validation error after a downward trend, while the training error decreases until it almost disappears. The best model is found just before this occurs.

The training set contains 706 (324 with VMSs) images extracted partially from 19 YouTube videos with a total duration of 05:19:27. The test set contains 153 (56 with VMSs) images that were manually reviewed to ensure the best comparison.

3.3. VMS Recognizer

Next, the training process performed to obtain the final model is detailed. A public distribution called Keras RetinaNet [66] has been used, which works on TensorFlow 2.0 [67]. Table 2 shows hardware specifications of on-board PC used for training and deployment.

Table 2. Hardware used for training.

Processor	Intel i7 9800K 3.6 GHz
RAM	32 GBs
Graphics card	Nvidia RTX 2080 Ti
Hard disk	1 Tb SSD M2

It has been established as an indicator to maximize the AP (average precision), which is the area under the coverage–precision curve given an IoU (Intersection over Union). The IoU indicates the amount of overlap between the recognized area and the real area. It is used as a threshold to find the true positives (TP), false positives (FP) and false negatives (FN) that define the accuracy and coverage value.

The training parameters and results (Table 3) are as follows.

Table 3. Training parameters and results.

Epochs	16	Best epoch	7
Loss	0.008	Loss (epoch 7)	0.024
lr	10^{-5}	AP (epoch 7)	0.703
IoU	0.5	Time	01:20:00

Once the first training is finished, it can be resumed by reducing the learning rate (lr) to slightly improve the model. This is because the lr guides the gradient descent through the error space until the local minimum (or in the optimal case, the absolute minimum) is reached. A high value of the lr causes the network to diverge, while a low value, even though it requires more time, will converge to the local minimum (or in the optimal case, the absolute minimum).

The parameters and results of the training continuation are shown on Table 4.

Table 4. Retraining parameters and results.

Epochs	14	Best epoch	7
Loss	0.009	Loss (epoch 7)	0.024
lr	10^{-7}	AP (epoch 7)	0.703
IoU	0.5	Time	01:15:00

Observing the retraining results, it is concluded that the model with the best AP is still the one achieved at epoch 7. The lr reduction did not produce the desired effect.

4. Text Extraction

4.1. Preprocessing

4.1.1. Image Straightening

The VMS image may have a small rotation that affects the OCR. In order to correct it, a procedure based on the Canny algorithm [68–71] and the Hough transform [69,72–75] has been designed.

- **Edge detection.** This task is carried out by the Canny algorithm on a grayscale image, on which a 5×5 Gaussian filter has been previously applied to reduce noise (although the Canny algorithm already applies one by default). The parameterization used is inspired by [76]. Thresholds are automatically calculated as follows:
    1. Obtain the average pixel intensity, v;
    2. Apply the following formulas with $\sigma = 0.33$ to find the lower and upper thresholds:
        - $T_l = \max(0,\ (1-\sigma)\cdot v)$
        - $T_H = \min(255,\ (1+\sigma)\cdot v)$
- **Straight line recognition within the image.** The Hough transform is applied on the output image of the Canny algorithm, obtaining a list of $(\rho,\ \theta)$ pairs. The parameters established are:
    - Accumulator distance on the axis $\rho = 1$;
    - Accumulator distance on the axis $\theta = \frac{\pi}{180}\ radians = 1^{\circ}$;
    - Threshold $T = 100$.
- **Calculation of the rotation angle, $\boldsymbol{\theta}$.** For each pair $(\rho,\ \theta)$, Equation (1) is applied to find the equation of the line in the xy plane. From it, the slope, a, required to transform it into degrees using Equation (2) is obtained and entered into a list. The rotation angle, θ, is estimated by the arithmetic mean of all the slopes of the detected lines.

$$y = \left(-\frac{\cos\theta}{\sin\theta}\right)x + \left(\frac{r}{\sin\theta}\right) \tag{1}$$

$$degrees = \frac{a\ 180}{\pi} \tag{2}$$

- **Calculation of the rotation matrix, $\boldsymbol{R}$.** Finally, by applying a rotation matrix, R (3), to the original image, the straightened image is obtained. For this, it is necessary to calculate α and β by means of Equations (4) and (5), knowing that $center = \left(\frac{width}{2}, \frac{altura}{2}\right)$, $scale = 1$ and θ is the value obtained in step three.

$$R = \begin{bmatrix} \alpha & \beta & (1-\alpha)\cdot center\cdot x - \beta\cdot center\cdot y \\ -\beta & \alpha & \beta\cdot center\cdot x + (1-\alpha)\cdot center\cdot y \end{bmatrix} \tag{3}$$

$$\alpha = scale\cdot \cos\theta \tag{4}$$

$$\beta = scale\cdot \sin\theta \tag{5}$$

4.1.2. Image Cropping

Once the slope has been adjusted, the next step is to crop the image so that only the inside of the VMS is shown. The objective is to identify the lines that delimit the panel and mark the cut points. The following algorithm details the procedure.

1. Find the equations of the lines on the image.

Through steps one, two and three of the above procedure, $(\rho,\ \theta)$ of the horizontal (between 0° and 1° slope), r_{hi}, and vertical (between 88° and 92°), r_{vj}, lines in the image are obtained. Then, the equations in the xy plane are calculated.

2. Calculate the intersection point with the image limits.

Side limits. For each straight line, r_{hi}, the intersection with the vertical limits $x = 0$ (22) and $x = w$ (23), where w is the width of the image, is calculated to store the y coordinate of each slice in the list, l_h. This way, each element of l_h is a candidate to be the limit of the horizontal slice.

$$Left\ side\ cut = \begin{cases} x = 0 \\ y = a * 0 + b = b \end{cases} \tag{6}$$

$$Right\ side\ cut = \begin{cases} x = w \\ y = aw + b \end{cases} \tag{7}$$

Upper and lower limits. For each straight line, r_{vj}, the intersection with the horizontal limits $y = 0$ (8) and $y = h$ (9), where h is the height of the image, is calculated to store the x coordinate of each slice in the list, l_v. This way, each element of l_v is a candidate to be the limit of the vertical slice.

$$Upper\ cut = \begin{cases} x = \left(\frac{0-b}{a}\right) \\ y = 0 \end{cases} \tag{8}$$

$$Lower\ cut = \begin{cases} x = \left(\frac{h-b}{a}\right) \\ y = h \end{cases} \tag{9}$$

3. Identify the cutting points and extract the subsection.
   - **Horizontal cut.** Identify the upper, I_{Hh}, and lower, I_{Lh}, cut-off points of l_h that satisfy:

$$I_{Hh} = \max(p) \text{ being } p \in l_h \text{ and } 0 \leq p \leq \frac{h}{6} \tag{10}$$

$$I_{Lh} = \min(p) \text{ being } p \in l_h \text{ and } \left(\frac{5}{6}h\right) \leq p \leq h. \tag{11}$$

   - **Vertical cut.** Identify the left, I_{Lv}, and right, I_{Rv}, cut-off points of l_v that satisfy:

$$I_{Lv} = \max(p) \text{ being } p \in l_v \text{ and } 0 \leq p \leq \frac{h}{10}. \tag{12}$$

$$I_{Rv} = \min(p) \text{ being } p \in l_v \text{ and } \left(\frac{9}{10}h\right) \leq p \leq h. \tag{13}$$

The range of p values for I_{Hh} and I_{Lh} in addition to I_{Lv} and I_{Rv}, as well as the following increments have been experimentally established:

$$I_{Hh} = I_{Hh} + 0.03 I_{Lh} \text{ and } I_{Lh} = I_{Lh} + 0.05 I_{Lh}. \tag{14}$$

$$I_{Lv} = I_{Lv} + 0.03 I_{Rv} \text{ and } I_{Rv} = I_{Rv} + 0.05 I_{Rv}. \tag{15}$$

4.1.3. Color Adjustment for OCR

Once the VMS content has been isolated, the image is ready for OCR. The objective is to create a new binarized picture, i.e., black text on a white background.

1. **Binarize the image.**
   - **Convert to grayscale.** By applying the formula presented in [77], the gray value is obtained (R, G and B being the values of the red, green and blue channels, respectively).
   - **Apply Otsu's method.** Otsu binarization [69,78–80] is an unsupervised parameterless method that consists of automatically finding a threshold, T, that minimizes the intraclass variance in black and white pixels. This way, a binary image is left.

- **Reverse the image color.** The output of Otsu's method is an image with white text on a black background. Therefore, it is necessary to apply the NOT logic gate on each value.

2. **Join discontinuous strokes.**

The binarized image may have small discontinuities in the letter strokes. To correct these imperfections that affect recognition, the closing morphological transformation [69,81,82] has been used to solve this problem.

Morphological transformations are operations that usually work on a binarized image by moving a kernel over it (similar to 2D convolution). The closing one (16) consists of a dilation that fills the small holes in the stroke, followed by an erosion that corrects the unwanted pixels that the first operation has enlarged.

$$A \cdot B = (A \oplus B) \ominus B \tag{16}$$

Dilation transforms the value of a pixel to 1 if all pixels below the kernel are 1, and erosion when at least one has the value 1.

3. **Histogram equalization.** Finally, it is necessary to increase the contrast so that the subsequent OCR model will be able to recognize the text. For this purpose, the histogram [69,83] of the image, $H(i)$, has been equalized by mapping it to the normalized cumulative distribution, $H'(i)$, q, which is more uniform.

4.2. Recognition and Speech

Once the VMS image has been preprocessed, it is ready to be transcribed using the Tesseract OCR model, and then spoken by the IBM Watson Text to Speech cloud service [25]. Tesseract [84–86] is an optical character recognition engine. The version used in this project is Tesseract 4.0, which implements LSTM (long short-term memory) recurrent neural networks, resulting in better and much faster results.

The last step in the pipeline is the voice-over of the content. This task is very easy thanks to the IBM Watson Text to Speech cloud service [25]. It provides the user with a REST API that receives the text and returns an audio file.

5. Results and Discussion

The presentation of the results has been divided into two parts, according to the subsystems of the project. All the results have been obtained with the same hardware with which the VMS recognizer model has been trained.

5.1. VMS Detector

An average precision of 0.7 has been achieved on 153 test images. These are some examples of the VMS detector. As can be seen, the detector confuses some static signals as if they were VMSs (Figure 4). This is a reasonable error due to the small number of images used to train the model and the similarity between both types of signals. However, this problem could be solved by adding another machine learning model that classifies between VMSs and non-VMSs. Additionally, different types of VMSs affect processing differently. Basic panels, with road signs and logos on the sides, can be found, as can LED matrices with higher or lower resolutions.

5.2. Image Preprocessing and Text Extraction

Qualitative results of the preprocessing and text extraction are presented below. As can be seen, the quality of the image and the resolution of the VMS affect both preprocessing (Figure 5) and transcription (Figure 6). Images with very low resolution are especially complicated. In addition, signs with pictograms affect the image processing and text extraction in the same negative way.

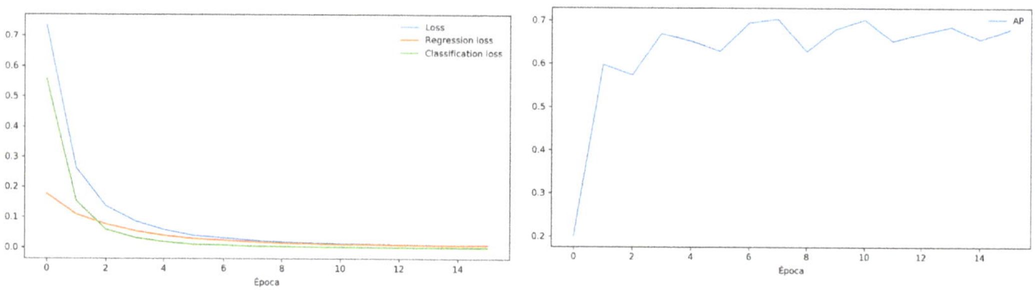

(a) (b)

Figure 4. (**a**) Loss, regression loss and classification loss for each training epoch; (**b**) AP by epoch.

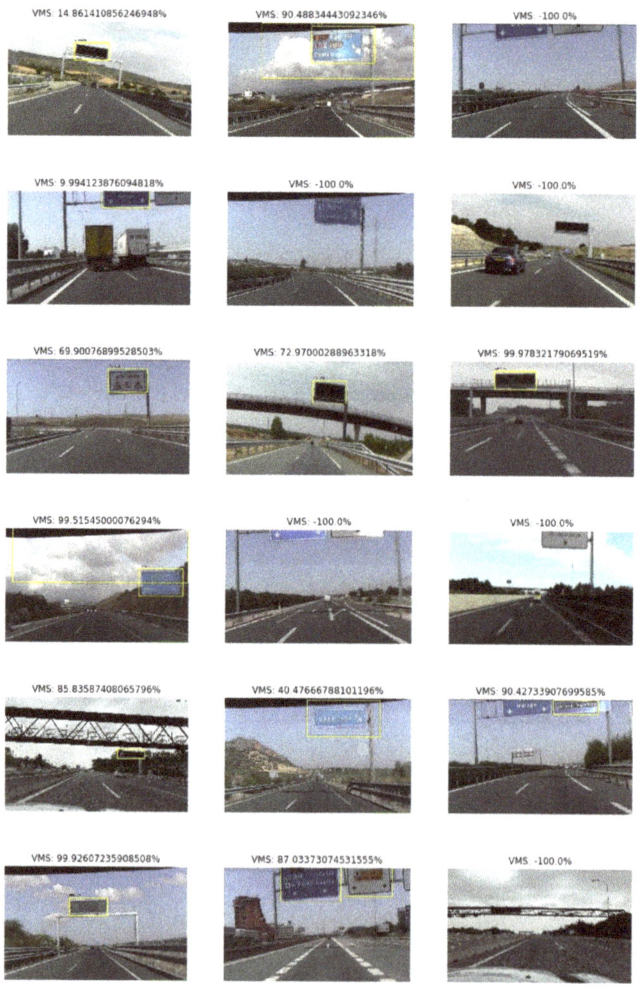

Figure 5. Some examples of VMS detection.

Figure 6. Image preprocessing examples.

It has been detected that, in some images such as the following two (Figure 7), the OCR model performs much better without the last steps of the preprocessing algorithm (Figure 8). In particular, without the last color adjustment, lower resolution instances have a better transcription.

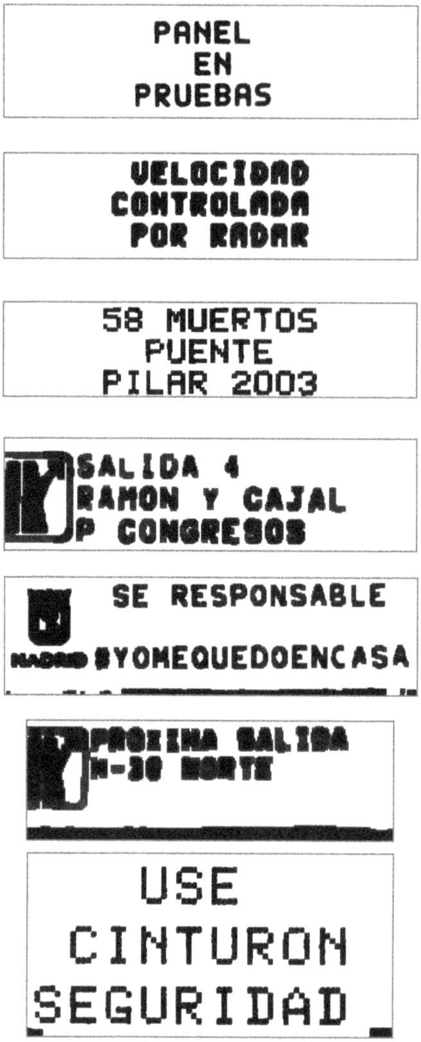

PANEL EN PRUEBAS

VELOCIDAD CONTROLADA
POR RADAR

58 MUERTOS PUENTE PILAR 2003

SALIDA Il jearos Y casal
PP" CONGRESOS

SE RESPONSABLE
maprio Y Y OMEQUEDOENCASA

PANA

USE CINTURON DE SEGURIDAD

Figure 7. Some examples of OCR (part 1).

SALIDA 4 I RAMON Y CAJAL
P CONGRESOS

PROXIMA SALIDA M-30 NORTE

Figure 8. Some examples of OCR (part 2).

6. Conclusions

As a result of the research, a prototype ADAS for reading variable message signs has been obtained. It works with a RetinaNet, a type of neural network based on ResNet50 with an average accuracy of 0.703, which recognizes the VMS in an image and indicates the location of it with a confidence percentage. Next, the section of the image with the VMS is processed to extract the count with an OCR model called Tesseract.

Author Contributions: G.D.-L.-H. was responsible for creating the dataset and programming the information extraction algorithms. J.S.-S. was in charge of the general supervision of the work and the evaluation of the machine learning results. E.P. was responsible for the design of the system architecture and the validation of the final prototype. All authors contributed to writing and reviewing the final manuscript. All authors have read and agreed to the published version of the manuscript.

Funding: This work was supported in part by the National Plan for Research PN I+D+i under PID2019-104793RB-C32, and the Comunidad de Madrid through SEGVAUTO-4.0-CM (P2018/EMT-4362) grants.

Conflicts of Interest: The authors declare no conflict of interest.

Appendix A. Dataset Sources

Table 1. Sources of the videos used for the preparation of the dataset.

URL	Duration	URL	Duration
youtu.be/MFzuxq4V0XI	00:06:59	youtu.be/dJcH8YFuvY4	00:25:12
youtu.be/GREMRp7rvoY	00:08:16	youtu.be/M2rvG-e04HE	00:16:22
youtu.be/lNhy2mT94Ao	00:05:22	youtu.be/8ifk3BHz1_c	00:13:35
youtu.be/37YgdfidwkA	00:19:58	youtu.be/6bkOZcBECsk	00:12:37
youtu.be/JctnDDdoy0A	00:22:56	youtu.be/afCyj52txC0	00:06:40
youtu.be/H1gxWeWsa_E	01:20:25	youtu.be/vU82-jnUi_E	00:07:57
youtu.be/UZFLDp_LLj4	00:27:23	youtu.be/D5RHKJNhw7I	00:08:01
youtu.be/tz8bEIirIx4	00:10:50	youtu.be/S1DE3pvnG8s	00:12:33
youtu.be/QJ_XSlOeCBw	00:10:48	youtu.be/XLQbclKjNrw	00:03:36
youtu.be/4s-WfvYUbPM	00:19:57		

References

1. Dargay, J.; Gately, D.; Sommer, M. Vehicle ownership and income growth, Worldwide: 1960–2030. *Energy J.* **2007**, *28*, 143–170. [CrossRef]
2. Dirección General de Tráfico—Ministerio del Interior. Anuario Estadístico General. 2018. Available online: http://www.dgt.es/es/seguridad-vial/estadisticas-e-indicadores/publicaciones/anuario-estadistico-general/ (accessed on 7 May 2021).
3. Instituto Nacional de Estadística. Encuesta Continua de Hogares. 2018. Available online: https://www.ine.es/prensa/ech_2018.pdf (accessed on 7 May 2021).
4. Organización Mundial de la Salud. Decade of Action for Road Safety 2011–2020. 2011. Available online: https://www.who.int/publications/i/item/decade-of-action-for-road-safety-2011-2020 (accessed on 7 May 2021).
5. Dirección General de Tráfico—Ministerio del Interior. Las Principales Cifras de la Siniestralidad Vial. 2018. Available online: http://www.dgt.es/es/seguridad-vial/estadisticas-e-indicadores/publicaciones/principales-cifras-siniestralidad/ (accessed on 7 May 2021).
6. Dirección General de Tráfico—Ministerio del Interior. Anuario Estadístico de Accidentes. 2018. Available online: http://www.dgt.es/es/seguridad-vial/estadisticas-e-indicadores/publicaciones/anuario-estadistico-accidentes/ (accessed on 7 May 2021).
7. Dirección General de Tráfico—Ministerio del Interior. Las distracciones Causan Uno de Cada Tres Accidentes Mortales. 2018. Available online: http://www.dgt.es/es/prensa/notas-de-prensa/2018/20180917_campana_distracciones.shtml (accessed on 7 May 2021).
8. Smiley, A.; Brookhuis, K.A. Alcohol, drugs and traffic safety. In *Road Users and Traffic Safety*; Transportation Research Board: Washington, DC, USA, 1987; pp. 83–104.
9. Dirección General de Tráfico—Ministerio del Interior. Distracciones al Volante. Available online: http://www.dgt.es/PEVI/documentos/catalogo_recursos/didacticos/did_adultas/Distracciones_al_volante.pdf (accessed on 7 May 2021).

10. Dirección General de Tráfico—Ministerio del Interior. Las Distracciones Son la Causa de Uno de Cada Cuatro Accidentes. 2019. Available online: http://www.dgt.es/es/prensa/notas-de-prensa/2019/Las-distracciones-son-la-causa-de-uno-de-cada-cuatro-accidentes.shtml (accessed on 7 May 2021).
11. Billington, J. The Prometheus Project: The Story behind One of AV's Greatest Developments. 2018.Available online: https://www.autonomousvehicleinternational.com/features/the-prometheus-project.html (accessed on 8 May 2021).
12. Brookhuis, K.A.; de Waard, D.; Janssen, W.H. Behavioural impacts of advanced driver assistance systems—An overview. *Eur. J. Transp. Infrastruct. Res.* **2001**, *1*, 246–253.
13. Michon, J.A. *Generic Intelligent Driver Support: A Comprehensive Report on GIDS*; CRC Press: Boca Raton, FL, USA, 1993; pp. 3–18.
14. BCG. A Roadmap to Safer Driving through Advanced Driver Assistance Systems. 2015. Available online: https://image-src.bcg.com/Images/MEMA-BCG-A-Roadmap-to-Safer-Driving-Sep-2015_tcm9-63787.pdf (accessed on 8 May 2021).
15. Nygårdhs, S.; Helmers, G. *VMS—Variable Message Signs: A Literature Review*; Transportation Research Board: Washington, DC, USA, 2007.
16. Autopistas.com. Paneles de Mensajería Variable. Available online: https://www.autopistas.com/blog/paneles-de-mensajeria-variable/ (accessed on 8 May 2021).
17. Kolisetty, V.G.B.; Iryo, T.; Asakura, Y.; Kuroda, K. Effect of variable message signs on driver speed behavior on a section of expressway under adverse fog conditions—A driving simulator approach. *J. Adv. Transp.* **2006**, *40*, 47–74. [CrossRef]
18. Peeta, S.; Ramos, J.L. Driver response to variable message signs-based traffic information. *IEEE Proc. Intell. Transp. Syst.* **2006**, *153*, 2–10. [CrossRef]
19. Guattari, C.; De Blasiis, M.R.; Calvi, A. The effectiveness of variable message signs information: A driving simulation study. *Procedia Soc. Behav. Sci.* **2012**, *53*, 692–702. [CrossRef]
20. Gopher, G. *Attentional Allocation in Dual Task Environments, Attention and Performance III*; Elsevier: Amsterdam, The Netherlands, 1990.
21. Roca, J.; Insa, B.; Tejero, P. Legibility of text and pictograms in variable message signs: Can single word messages outperform pictograms? *J. Hum. Factors Ergon. Soc.* **2018**, *60*, 384–396. [CrossRef]
22. Simlinger, P.; Egger, S.; Galinski, C. Proposal on Unified Pictograms, Keywords, Bilingual Verbal Messages and Typefaces for VMS in the TERN, International Institute for Information Design. 2008. Available online: https://ec.europa.eu/transport/road_safety/sites/roadsafety/files/pdf/projects_sources/in-safety_d2_3.pdf (accessed on 8 May 2021).
23. Universitat de València. READit VMS. 2018. Available online: https://www.uv.es/uvweb/estructura-investigacion-interdisciplinar-lectura/es/productos-tecnologicos/productos-tecnologicos/readit-vms-1286067296453.html (accessed on 8 May 2021).
24. Erke, A.; Sagberg, F.; Hagman, R. Effects of route guidance variable message signs (VMS) on driver behaviour. *Transp. Res. Part F Traffic Psychol. Behav.* **2007**, *10*, 447–457. [CrossRef]
25. IBM. Watson Speech to Text. Available online: https://www.ibm.com/es-es/cloud/watson-speech-to-text (accessed on 11 July 2021).
26. Dirección General de Tráfico—Ministerio del Interior. Cuestiones De Seguridad Vial. 2018. Available online: http://www.dgt.es/Galerias/seguridad-vial/formacion-vial/cursos-para-profesores-y-directores-de-autoescuelas/XXI-Cuso-Profesores/Manual-II-Cuestiones-de-Seguridad-Vial-2018.pdf (accessed on 7 May 2021).
27. Kukkala, V.K.; Tunnell, J.; Pasricha, S.; Bradley, T. Advanced driver-assistance systems: A path toward autonomous vehicles. *IEEE Consum. Electron. Mag.* **2018**, *7*, 18–25. [CrossRef]
28. O'Kane, S. How Tesla and Waymo Are Tackling a Major Problem for Self-Driving Cars: Data, The Verge. 2018. Available online: https://www.theverge.com/transportation/2018/4/19/17204044/tesla-waymo-self-driving-car-data-simulation (accessed on 8 May 2021).
29. Bay, O. ABI Research Forecasts 8 Million Vehicles to Ship with SAE Level 3, 4 and 5 Autonomous Technology in 2025. 2018. Available online: https://www.abiresearch.com/press/abi-research-forecasts-8-million-vehicles-ship-sae-level-3-4-and-5-autonomous-technology-2025/ (accessed on 8 May 2021).
30. Stoma, M.; Dudziak, A.; Caban, J.; Droździel, P. The future of autonomous vehicles in the opinion of automotive market users. *Energies* **2021**, *14*, 4777. [CrossRef]
31. Silva, D.; Csiszar, C.; Földes, D. Autonomous vehicles and urban space management. *Sci. J. Sil. Univ. Technol. Ser. Transp.* **2020**, *110*, 13.
32. Synopsys. The 6 Levels of Vehicle Autonomy Explained. Available online: https://www.synopsys.com/automotive/autonomous-driving-levels.html (accessed on 8 May 2021).
33. Zou, Z.; Shi, Z.; Guo, Y.; Ye, J. Object detection in 20 years: A survey. *arXiv* **2019**, arXiv:1905.05055v2.
34. Viola, P.; Jones, M.J. Robust real-time face detection. *Int. J. Comput. Vis.* **2004**, *57*, 137–154. [CrossRef]
35. Viola, P.; Jones, M.J. Rapid object detection using a boosted cascade of simple features. In Proceedings of the 2001 IEEE Computer Society Conference, Kauai, HI, USA, 8–14 December 2001.
36. Dalal, N.; Triggs, B. Histograms of Oriented Gradients for Human Detection. In Proceedings of the 2005 IEEE Computer Society Conference on Computer Vision and Pattern Recognition, San Diego, CA, USA, 20–25 June 2005.
37. Felzenszwalb, P.; McAllester, D.; Ramanan, D. A discriminatively trained, multiscale, deformable part mode, computer vision and pattern recognition. In Proceedings of the 2008 IEEE Conference on Computer Vision and Pattern Recognition, Anchorage, AK, USA, 23–28 June 2008.

38. Krizhevsky, A.; Sutskever, I.; Hinton, G.E. ImageNet classification with deep convolutional neural networks. *Commun. ACM* **2017**, *60*, 84–90. [CrossRef]
39. Girshick, R. Fast R-CNN. In Proceedings of the 2015 IEEE International Conference on Computer Vision (ICCV), Santiago, Chile, 7–13 December 2015; pp. 1440–1448.
40. Girshick, R.; Donahue, J.; Darrell, T.; Malik, J. Rich feature hierarchies for accurate object detection and semantic segmentation. In Proceedings of the 2014 IEEE Conference on Computer Vision and Pattern Recognition, Columbus, OH, USA, 23–28 June 2014.
41. Sermanet, P.; Eigen, D.; Zhang, X.; Mathieu, M.; Fergus, R.; Lecun, Y. Over Feat: Integrated Recognition, Localization and Detection using Convolutional Networks. *arXiv* **2013**, arXiv:1312.6229.
42. Soviany, P.; Ionescu, R.T. Optimizing the trade-off between single-stage and two-stage deep object detectors using image difficulty prediction. In Proceedings of the 2018 20th International Symposium on Symbolic and Numeric Algorithms for Scientific Computing, Timisoara, Romania, 20–23 September 2018.
43. Ren, S.; He, K.; Girshick, R.; Sun, J. Faster R-CNN: Towards real-time object detection with region proposal networks. *IEEE Trans. Pattern Anal. Mach. Intell.* **2017**, *39*, 1137–1149. [CrossRef]
44. Redmon, J.; Divvala, S.; Girshick, R.; Farhadi, A. You only look once: Unified, real-time object detection. In Proceedings of the 2016 IEEE Conference on Computer Vision and Pattern Recognition, Las Vegas, NV, USA, 27–30 June 2016; pp. 779–788.
45. Redmon, J.; Farhadi, A. YOLO9000: Better, faster, stronger. *arXiv* **2017**, arXiv:1612.08242v1.
46. Redmon, J.; Farhadi, A. YOLOv3: An incremental improvement. *arXiv* **2018**, arXiv:1804.02767.
47. Bochkovskiy, A.; Wang, C.-Y.; Liao, H.-Y.M. YOLOv4: Optimal Speed and Accuracy of Object Detection. *arXiv* **2020**, arXiv:2004.10934.
48. Liu, W.; Anguelov, D.; Erhan, D.; Szegedy, C.; Reed, S.; Fu, C.-Y.; Berg, A.C. Ssd: Single shot multibox detector. In Proceedings of the Computer Vision—ECCV 2016, Amsterdam, The Netherlands, 11–14 October 2016; pp. 21–37.
49. Lin, T.-Y.; Goyal, P.; Girshick, R.; He, K.; Dollár, P. Focal loss for dense object detection. *IEEE Trans. Pattern Anal. Mach. Intell.* **2018**, *42*, 318–327. [CrossRef]
50. Hui, J. Object Detection: Speed and Accuracy Comparison (Faster R-CNN, R-FCN, SSD, FPN, RetinaNet and YOLOv3). Available online: https://jonathan-hui.medium.com/object-detection-speed-and-accuracy-comparison-faster-r-cnn-r-fcn-ssd-and-yolo-5425656ae359 (accessed on 25 May 2021).
51. Lin, H.; Yang, P.; Zhang, F. Review of scene text detection and recognition. *Arch. Comput. Methods Eng.* **2020**, *27*, 433–454. [CrossRef]
52. Long, S.; He, X.; Yao, C. Scene text detection and recognition: The deep learning era. *Int. J. Comput. Vis.* **2020**, *129*, 161–184. [CrossRef]
53. Zhu, Y.; Yao, C.; Bai, X. Scene text detection and recognition: Recent advances and future trends. *Front. Comput. Sci.* **2015**, *10*, 19–36. [CrossRef]
54. Graves, A.; Fernández, S.; Gomez, F.; Schmidhuber, J. Connectionist temporal classification: Labelling unsegmented sequence data with recurrent neural networks. In Proceedings of the 23rd International Conference on Machine Learning, New York, NY, USA, 25–29 June 2006; Volume 2006, pp. 369–376.
55. Charette, R.; Nashashibi, F. Real time visual trafric lights recognition based on spot light detection and adaptive traffic lights templates. In Proceedings of the 2009 IEEE Intelligent Vehicles Symposium, Xi'an, China, 3–5 June 2009; pp. 358–363.
56. Fairfield, N.; Urmson, C. Traffic light mapping and detection. In Proceedings of the 2011 IEEE International Conference on Robotics and Automation, Shanghai, China, 9–13 May 2011; pp. 5421–5426.
57. Chung, Y.-C.; Wang, J.-M.; Chen, S.-W. A vision-based traffic light detection system at intersections. *J. Taiwan Norm. Univ. Math. Sci. Technol.* **2002**, *47*, 67–86.
58. Lu, Y.; Lu, J.; Zhang, S.; Hall, P. Traffic signal detection and classification in street views using an attention model. *Comput. Vis. Media* **2018**, *4*, 253–266. [CrossRef]
59. Zhu, Z.; Liang, D.; Zhang, S.; Huang, X.; Li, B.; Hu, S. Traffic-sign detection and classification in the wild. In Proceedings of the 2016 IEEE Conference on Computer Vision and Pattern Recognition, Las Vegas, NV, USA, 27–30 June 2016; pp. 2110–2118.
60. Toyota Motor Sales. Toyota Safety Sense™ Comes Standard on Many New Toyotas. Available online: https://www.toyota.com/safety-sense/animation/drcc (accessed on 8 May 2021).
61. González, A.; Bergasa, L.M.; Yebes, J.J. Text detection and recognition on traffic panels from street-level imagery using visual appearance. *IEEE Trans. Intell. Transp. Syst.* **2014**, *15*, 228–238. [CrossRef]
62. Vazquez-Reina, A.; Sastre, R.; Arroyo, S.; Gil-Jiménez, P. Adaptive traffic road sign panels text extraction. In Proceedings of the 5th WSEAS International Conference on Signal Processing, Robotics and Automation, Madrid, Spain, 15–17 February 2006; pp. 295–300.
63. Tzutalin, LabelImg, Github. Available online: https://github.com/tzutalin/labelImg (accessed on 14 May 2021).
64. He, K.; Zhang, X.; Ren, S.; Sun, J. Deep residual learning for image recognition. In Proceedings of the 2016 IEEE Conference on Computer Vision and Pattern Recognition (CVPR), Las Vegas, NV, USA, 27–30 June 2016; pp. 770–778.
65. Lin, T.-Y.; Maire, M.; Belongie, S.; Hays, J.; Perona, P.; Ramanan, D.; Dollár, P.; Zitnick, C. Microsoft coco: Common objects in context. *arXiv* **2014**, arXiv:1405.0312v3.
66. Fizyr, Keras-Retinanet, Github. Available online: https://github.com/fizyr (accessed on 14 May 2021).
67. Google. TensorFlow. Available online: https://www.tensorflow.org/tutorials/quickstart/beginner (accessed on 14 May 2021).

68. Canny, J. A computational approach to edge detection. *IEEE Trans. Pattern Anal. Mach. Intell.* **1986**, *8*, 679–698. [CrossRef]
69. Gonzalez, R.C.; Woods, R.E. *Digital Image Processing*; Prentice Hall: Hoboken, NJ, USA, 2008.
70. Rebaza, J.V. Detección de Bordes Mediante el Algoritmo de Canny. Master's Thesis, Universidad Nacional de Trujillo, Trujillo, Peru, 2007.
71. OpenCV. Open Source Computer Vision, Canny Edge Detector. Available online: https://docs.opencv.org/master/da/d5c/tutorial_canny_detector.html (accessed on 30 May 2021).
72. OpenCV. Open Source Computer Vision, Hough Line Transform. Available online: https://docs.opencv.org/master/d9/db0/tutorial_hough_lines.html (accessed on 30 May 2021).
73. Shehata, A.; Mohammad, S.; Abdallah, M.; Ragab, M. A survey on Hough transform, theory, techniques and applications. *arXiv* **2015**, arXiv:1502.02160.
74. Hough, P.V.C. Methods and Means for Recognizing Complex Patterns. U.S. Patent US3069654A, 18 December 1962.
75. Ballard, D.H.; Brown, C.M. *Computer Vision*; Prentice Hal: Englewood Cliffs, NJ, USA, 1982.
76. Rosebrock, A. Zero-Parameter, Automatic Canny Edge Detection with Python and OpenCV. 2015. Available online: https://www.pyimagesearch.com/2015/04/06/zero-parameter-automatic-canny-edge-detection-with-python-and-opencv/ (accessed on 30 May 2021).
77. OpenCV. Open Source Computer Vision, Color Conversions. Available online: https://docs.opencv.org/3.4/de/d25/imgproc_color_conversions.html (accessed on 30 May 2021).
78. Otsu, N. Fronts propagating with curvature-dependent speed: Algorithms based on Hamilton-Jacobi formulations. *J. Comput. Phys.* **1988**, *79*, 12–49.
79. OpenCV. Open Source Computer Vision, Image Thresholding. Available online: https://docs.opencv.org/master/d7/d4d/tutorial_py_thresholding.html (accessed on 30 May 2021).
80. Morse, S. Lecture 4: Thresholding (Brigham Young University). Available online: http://homepages.inf.ed.ac.uk/rbf/CVonline/LOCAL_COPIES/MORSE/threshold.pdf (accessed on 30 May 2021).
81. OpenCV. Open Source Computer Vision, Morphological Transformations. Available online: https://docs.opencv.org/trunk/d9/d61/tutorial_py_morphological_ops.html (accessed on 30 May 2021).
82. Sreedhar, K.; Panlal, B. Enhancement of images using morphological transformations. *Int. J. Comput. Sci. Inf. Technol.* **2012**, *4*, 33–50. [CrossRef]
83. OpenCV. Open Source Computer Vision, Histogram Equalization. Available online: https://docs.opencv.org/3.4/d4/d1b/tutorial_histogram_equalization.html (accessed on 30 May 2021).
84. Smith, R. An overview of the tesseract OCR engine. In Proceedings of the Ninth International Conference on Document Analysis and Recognition (ICDAR 2007), Curitiba, Brazil, 23–26 September 2007.
85. Tesseract-Ocr, Tesseract, GitHub. Available online: http://code.google.com/p/tesseract-ocr (accessed on 3 June 2021).
86. Kumar, A. Performing OCR by Running Parallel Instances of Tesseract 4.0: Python. 2018. Available online: https://appliedmachinelearning.blog/2018/06/30/performing-ocr-by-running-parallel-instances-of-tesseract-4-0-python/ (accessed on 3 June 2021).

sensors

Article

A Bidirectional Versatile Buck–Boost Converter Driver for Electric Vehicle Applications

Catalina González-Castaño [1,†], Carlos Restrepo [2,†], Samir Kouro [3,†], Enric Vidal-Idiarte [4,*,†] and Javier Calvente [4,†]

1 Department of Engineering Sciences, Universidad Andres Bello, Santiago 7500971, Chile; inv.cet@unab.cl
2 Department of Electromechanics and Energy Conversion, Universidad de Talca, Curicó 3340000, Chile; crestrepo@utalca.cl
3 Electronics Engineering Department, Universidad Técnica Federico Santa María, Valparaíso 2390123, Chile; samir.kouro@usm.cl
4 Departament d'Enginyeria Electrònica, Elèctrica i Automàtica, Escola Tècnica Superior d'Enginyeria, Universitat Rovira i Virgili, 43007 Tarragona, Spain; javier.calvente@urv.cat
* Correspondence: enric.vidal@urv.cat; Tel.: +34-977-559622
† These authors contributed equally to this work.

Abstract: This work presents a novel dc-dc bidirectional buck–boost converter between a battery pack and the inverter to regulate the dc-bus in an electric vehicle (EV) powertrain. The converter is based on the versatile buck–boost converter, which has shown an excellent performance in different fuel cell systems operating in low-voltage and hard-switching applications. Therefore, extending this converter to higher voltage applications such as the EV is a challenging task reported in this work. A high-efficiency step-up/step-down versatile converter can improve the EV powertrain efficiency for an extended range of electric motor (EM) speeds, comprising urban and highway driving cycles while allowing the operation under motoring and regeneration (regenerative brake) conditions. DC-bus voltage regulation is implemented using a digital two-loop control strategy. The inner feedback loop is based on the discrete-time sliding-mode current control (DSMCC) strategy, and for the outer feedback loop, a proportional-integral (PI) control is employed. Both digital control loops and the necessary transition mode strategy are implemented using a digital signal controller TMS320F28377S. The theoretical analysis has been validated on a 400 V 1.6 kW prototype and tested through simulation and an EV powertrain system testing.

Keywords: noninverting buck–boost converter; high efficiency; wide bandwidth control; discrete-time sliding-mode current control (DSMCC); electric vehicle (EV); driver vehicle system; energy management

Citation: González-Castaño, C.; Restrepo, C.; Kouro, S.; Vidal-Idiarte, E.; Calvente, J. A Bidirectional Versatile Buck–Boost Converter Driver for Electric Vehicle Applications. *Sensors* **2021**, *21*, 5712. https://doi.org/10.3390/s21175712

Academic Editors: Arturo de la Escalera Hueso and Omprakash Kaiwartya

Received: 21 June 2021
Accepted: 19 August 2021
Published: 25 August 2021

1. Introduction

Electric vehicles (EVs) are an essential part of meeting global goals on reducing the carbon footprint of vehicle emissions that contribute to climate change [1,2]. All the EVs powertrain configurations shown in Figure 1 have a common system that is formed by the battery, the power converter, and the electric motor. Each of these components has been the subject of extensive research in recent years and a high level of development to improve the performance of the automotive traction systems. However, these three components represent a tremendous research challenge given the complexity of integrating these elements in EVs application.

In EVs, the battery is generally sized by the energy requirements to allow a specific range to be reached. Still, there is not a linear relationship between car range and battery capacity because adding the weight of the battery reduces the efficiency on the road [2,3]. The battery cells for EVs are usually connected in series to meet the voltage requirements of the power converter (inverter). The connection of cells in series exponentially increases the probability of failure of the battery pack. The performance of the whole pack is limited by the weakest cell and the oversizing of the power inverter and the electric motor to ensure peak power delivery at a low state of charge (SoC) of a battery pack with a wide voltage variation at different SoC [4,5]. Thus, there is a limitation of the maximum number of battery cells that could be connected in series, and a step-up dc-dc power converter is required to reach the requirements of the inverter converter. Therefore, the power converter shown in Figure 1 is implemented using a step-up DC-DC in cascade with a DC-AC traction inverter (see Figure 2 DC-DC + DC-AC block) [6].

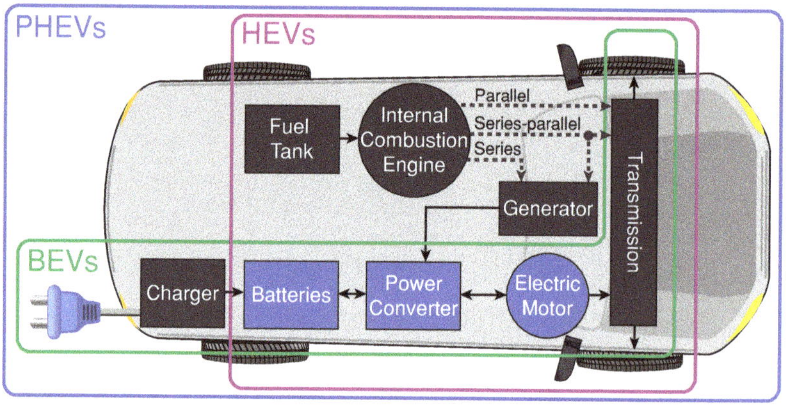

Figure 1. EVs powertrain configurations: hybrid electric vehicles (HEVs), plug-in hybrid electric vehicles (PHEVs) and battery electric vehicles (BEVs).

The induction machine (IM) and the permanent magnet synchronous machine (PMSM) are the most used [7–9] electric motors in EV traction applications. In the constant torque operation region (Figure 2), the maximum torque capability is determined by the current rating of the inverter [7,10]. The maximum torque at base speed (point A in Figure 2) defines the vehicle performance at starting or climbing hills [8]. The available torque at maximum speed (point B in Figure 2) in the constant power region limits the vehicle speed highways. In the last region, the torque and power reduction are due to the back-electromotive force's increasing influence [7,10].

Connecting a DC-DC converter between the battery and the inverter allows optimizing the inverter's DC input voltage, improving power capability, and maximizing the electric motor efficiency [11]. A bidirectional DC-DC converter can be used to control the voltage at the input of the inverter according to the motor speed. In this way, the converter can optimize the efficiency of the inverter (modulation index $MI = 1$ achieves it) in a wider range of operating speeds, as can be seen in Figure 2 (see DC-DC + DC-AC block) [12,13].

In [12], the authors used an interleaving half-bridge bidirectional converter to regulate a variable DC-bus voltage, showing the efficiency improvement both in the step-up converter and inverter. A detailed inverter loss model is developed in [13], where a variable DC-bus voltage closely related to the rotational motor speed significantly improves the inverter efficiency for voltages above the battery voltage. Despite proposals of using composite topologies for high step-up gain [14] or flying capacitors topologies to reduce inductor size [15,16], the most commonly used converter for this application has been the bidirectional half-bridge [12,13,17–24] and boost [25] converters. In [18], the authors proposed a three-level version of this converter to use lower breakdown voltage MOSFETs. A coupled inductor in each phase is proposed to increase the power [19].

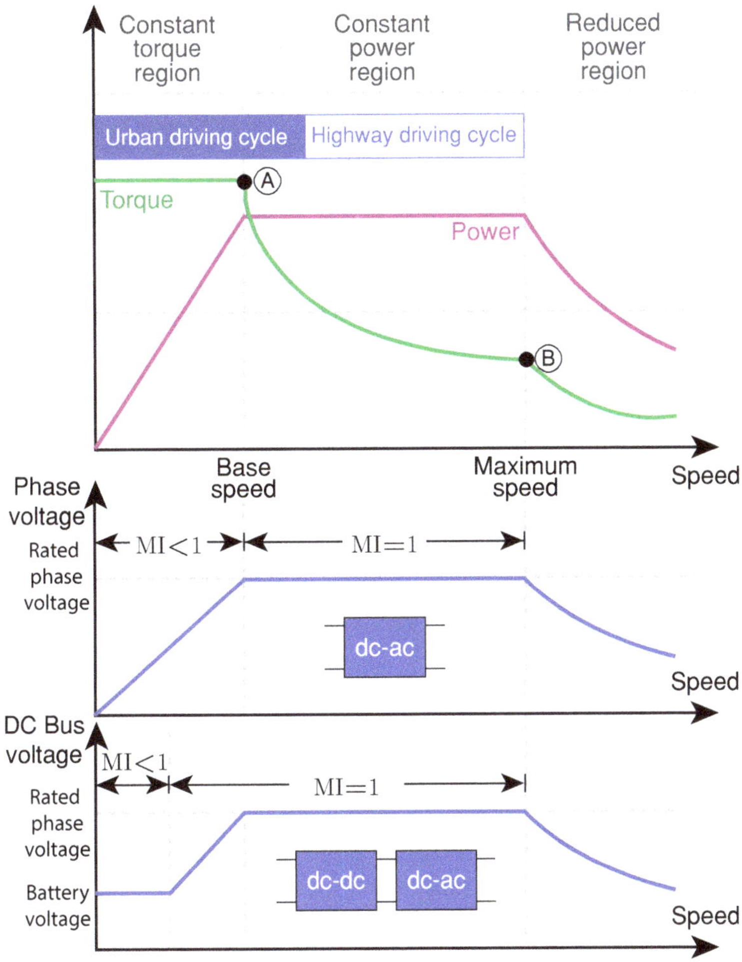

Figure 2. Torque and power requirements for the EV drive systems.

An interleaved zero voltage switching (ZVS) version, included in multifunctional power electronic interface and operating at 60 kHz, is presented in [20], achieving high-efficiency measurements. Integration of this bidirectional converter in a new topology is proposed in [24] for a hybrid electric vehicle system. This converter interfaces between two different voltage values corresponding to the battery system and a DC-bus. It is worth noting that this bidirectional converter operates as a boost- or buck-converter depending on whether the motor is in driving or regenerative mode [26]. Therefore, this electric drive topology is more suitable for highway driving cycles (see Figure 2), reducing the system's efficiency under an urban driving cycle. The latter is because the inverter efficiency cannot be guaranteed under low speeds since the boost converter cannot reduce the DC-bus voltage below the battery voltage [6,27]. A converter with step-up and step-down characteristics, not only will extend the efficient range to urban driving cycle, but will also add more flexibility in designing both battery and inverter.

The noninverting buck–boost converter with coupled inductors known as the versatile buck–boost converter and shown in Figure 3a, could be an excellent candidate to optimize the global efficiency of the system. It has many advantages, such as noninverting voltage step-up and step-down characteristic in both motor operating modes, high efficiency, wide bandwidth [28], and input or output currents regulation because of their low ripple values [29,30]. It provides smooth transitions between buck and boost operating modes due an hysteresis PWM control strategy used to activate the controlled switches [31]. In addition, the introduction of an RC damping network in parallel with the intermediate capacitor, combined with the coupled-inductors, eliminates the right half-plane zero that limits the closed-loop bandwidth of the step-up converters [28]. All the advantages mentioned above have allowed its use in different fuel cell hybrid power systems [32–34] and deepen on various digital current control techniques [35–37].

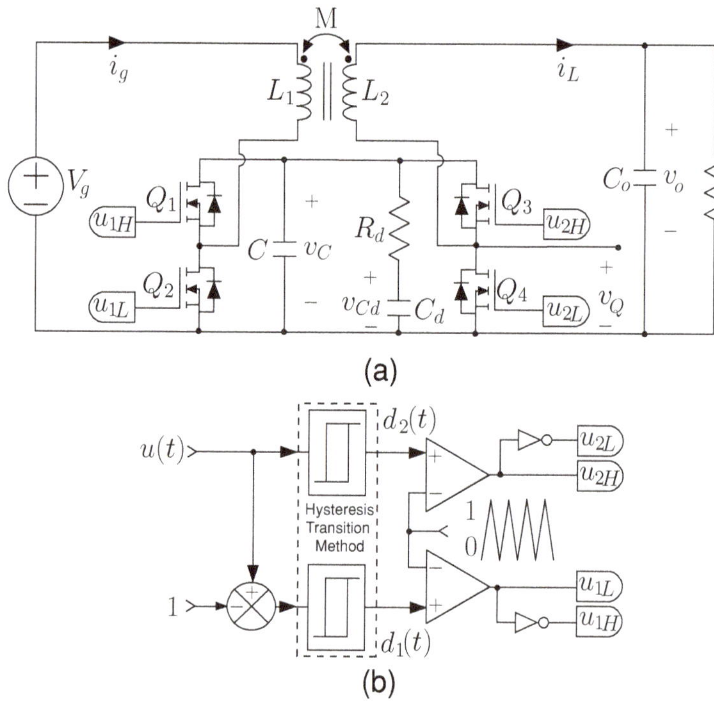

Figure 3. Schemes of (**a**) the buck–boost converter, (**b**) switch signals generation.

A novel bidirectional version of the versatile buck–boost converter is presented to extend its use in electric vehicle applications. This new converter shares some similarities with their previous ones (see Figure 4) to preserve all the advantages of the versatile converter. However, two significant changes to match the hard-switching high-voltage bidirectional EV application are included in this new version. The first one corresponds the use of the Silicon Carbide (SiC) devices that extend the operation at high-voltage with low switching losses [23,38–40]. The second one corresponds with a redesign of the coupled inductors to reduce the parasitic winding-to-winding capacitance [41]. Other important aspects that differentiate the converter presented in this work from the existing ones are summarized in Figure 4.

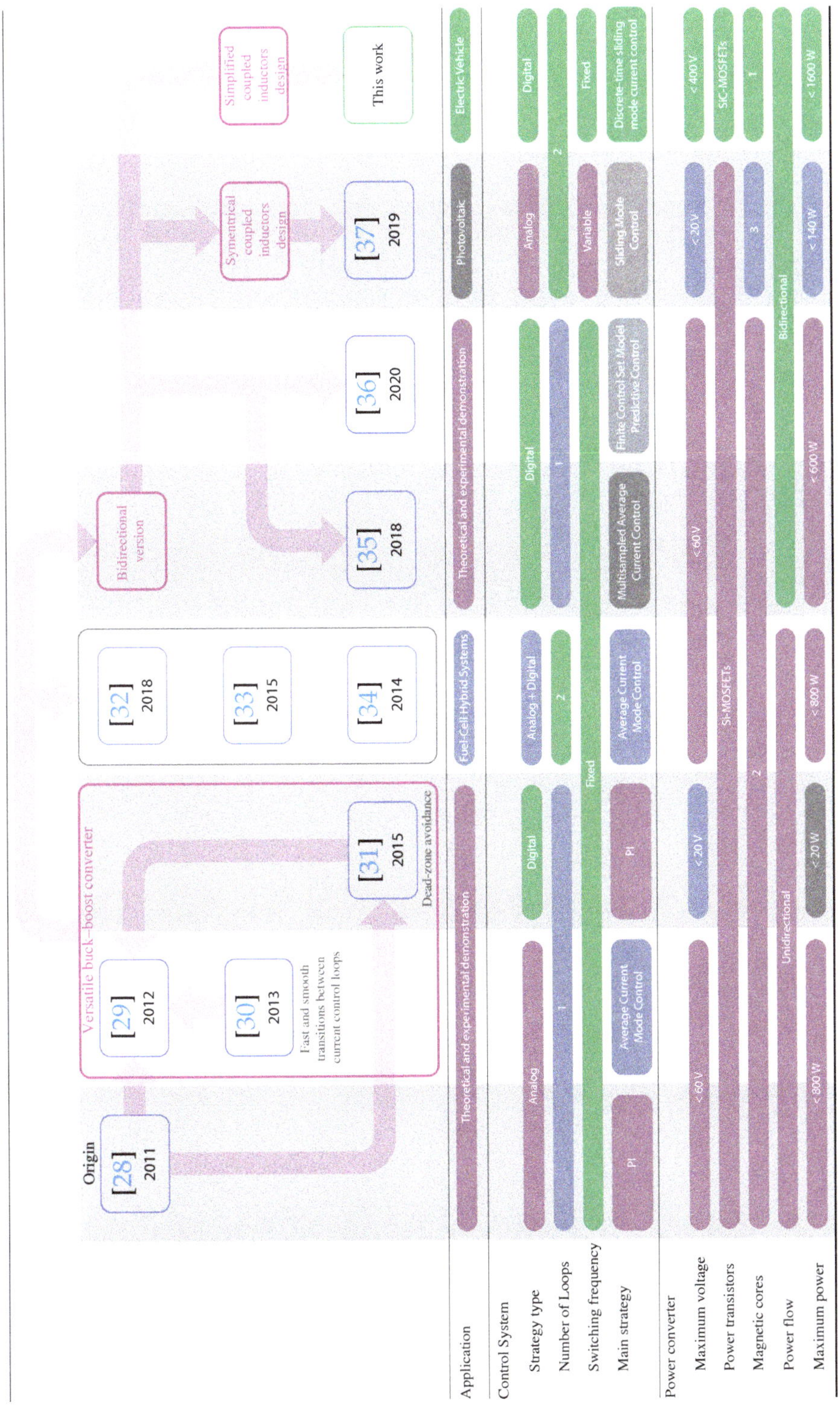

Figure 4. Versatile buck–boost converter evolution throughout all the investigations carried out to date.

This work presents a novel high-voltage bidirectional buck–boost converter with digital control that allows the regulation of the high-voltage DC-bus for EV applications. This voltage regulation and the energy flow between the battery and the motor drive are managed by means of a two-loop digital current control strategy, which facilitates the hysteresis transition between voltage step-up and step-down modes. The resulting control provides output voltage regulation in the presence of variations in output voltage and load power. The whole system is tested experimentally in a 1.6 kW prototype applied to a resistive load and an EV hardware emulation platform. Based on Figure 4 and the state of the art, the main contributions of this paper can be summarized as follows:

- A novel high-voltage high-switching bidirectional converter is presented. This new converter has step-up and step-down characteristics in both current directions to extend the EV traction inverter efficiency under a wide range of speeds. This converter guarantees a high power conversion efficiency for EV powertrain applications due to silicon carbide (SiC) devices and the design with a low winding-to-winding parasitic capacitance of the coupled inductor. It can operate in boost or buck mode.
- A two-loop digital control design with a current (inner loop) controller and a voltage (outer loop) controller regulate the DC-bus voltage during traction and regenerative modes. The proposed controller ensures zero steady-state voltage error and fast transient responses to the voltage reference and power variations.
- A DSMCC control is proposed for the inner loop of the voltage feedback outer loop. The proposed controller ensures fast-tracking of the control set-points and low steady-state error under demanding tests that include system start-up and dc bus voltage reference with small and large variations. It is the first time that the DSMCC control strategy is used for the versatile buck–boost converter.

This paper is organized as follows: Section 2 presents an analysis of the coupled inductors buck–boost converter with the goal of obtaining the inductor current slope equations. Section 3 describes the DSMCC technique implemented for the inner control loop. This section also includes the outer voltage feedback loop analysis, which is based on a PI controller. Simulations and experimental results of the current control technique under startup, small and large variations, and using an EV emulator are presented and discussed in Section 4. Finally, the main conclusions and the remaining challenges for the future are summarized in Section 5.

2. Bidirectional Noninverting Coupled-Inductor Buck–Boost Converter

The converter scheme depicted in Figure 3a is composed of two half-bridge MOSFETs, an R_dC_d damping network connected in parallel with the intermediate capacitor C, a constant input voltage V_g, and a resistance load R_o. In addition, the coupled inductor has a unitary ideal turns ratio N_2/N_1, a coupled coefficient $k = 0.5$, a mutual inductance M and equal values for the primary (L_1) and secondary (L_2) self-inductances ($L = L_1 = L_2$). In the analysis, a continuous conduction mode (CCM) operation is considered, with no parasitic effects and a switching frequency much higher than the converter's natural frequencies. The use of the state-space averaging (SSA) method to model the converter leads to the following set of differential equations [37]:

$$\frac{di_g(t)}{dt} = \frac{L(V_g - v_c(1 - u_{1L})) - M(v_o - v_c u_{2H})}{L^2 - M^2} \tag{1}$$

$$\frac{di_L(t)}{dt} = \frac{M(V_g - v_c(1 - u_{1L})) - L(v_o - v_c u_{2H})}{L^2 - M^2} \tag{2}$$

$$\frac{dv_c(t)}{dt} = \frac{1}{C}\left(-i_L u_{2H} + i_g(-u_{1L} + 1) - \frac{1}{R_d}(v_c - v_{cd})\right) \tag{3}$$

$$\frac{dv_{cd}(t)}{dt} = \frac{v_c - v_{cd}}{C_d R_d} \tag{4}$$

$$\frac{dv_o(t)}{dt} = \frac{i_L}{C_o} - \frac{v_o}{R_o C_o} \tag{5}$$

In the scheme of Figure 3, the duty cycle $d_1(t)$ is used to activate the switch Q_1 and Q_2 for boost mode. Q_3 and Q_4 are switched with the duty cycle $d_2(t)$ for buck mode. The activation signals u_{1H} and u_{1L} are for the half-bridge composed of Q_1 and Q_2, and the activation signals u_{2H} and u_{2L} are for the half-bridge composed by Q_3 and Q_4. u_{1H} and u_{1L} operate in a complementary manner while u_{2H} is set at 1 and u_{2L} is set at 0, in boost mode. Otherwise, u_{2H} and u_{2L} operate in a complementary manner while u_{1H} is set at 1 and u_{1L} is set at 0, in buck mode. The duty cycles are computed considering a variable control $u(t)$, where $u(t) = 1 + d_1(t)$ in boost mode and $u(t) = d_2(t)$ for buck mode [28]. Figure 3b shows the hysteresis transition method avoids oscillations in the transitions between buck, boost, and buck–boost working modes [31]. The aim of this analysis is to find the converter's current output slope $\frac{di_L}{dt}$ in each operation mode (buck or boost) to design the digital inner current programmed controller. The output current has a periodic triangular waveform where the current rises with a slope of m_1 and falls with a slope $-m_2$. Table 1 presents the converter output current waveform slopes based on Equation (2) for the boost and buck modes.

Table 1. Slope of the output current waveform.

Mode	m_1	$-m_2$
Buck	$\frac{M(V_g - v_c) - L(v_o - v_c)}{L^2 - M^2}$	$\frac{M(V_g - v_c) - Lv_o}{L^2 - M^2}$
Boost	$\frac{MV_g - L(v_o - v_c)}{L^2 - M^2}$	$\frac{M(V_g - v_c) - L(v_o - v_c)}{L^2 - M^2}$

3. Digital Control for Output Voltage Regulation

The control method implemented to regulate the converter's output voltage is a two-loop digital control. This strategy allows smooth transitions between motoring and regenerative braking operations and during the DC-bus voltage reference changes. The digital control has the advantage of simplifying the implementation of complex control strategies, the soft start of the converter, higher robustness to noise, and flexibility in design without the need to make any component or hardware changes [42]. In addition, it allows the integration of the hysteresis mode transition strategy in the digital controller, making the implementation and tuning of this transition strategy easier. The digital control proposed has an inner current programmed controller with an outer voltage feedback loop (PI compensator).

The current control loop must present a fast dynamic response to reduce the transient response between buck and boost modes. This can be achieved using discrete-time sliding-mode current control (DSMCC) for the output current i_L, taking into account the converter dynamics.

3.1. Discrete-Time Sliding-Mode Current Control

This work presents a fixed switching DSMCC control for the bidirectional noninverting buck–boost converter. This discrete sliding control has been presented for a boost converter in [43] and a buck converter [44]. In this control strategy, the DSMCC aims to compute the variable control $u[n]$ in the n-th time sample period that ensures the control surface (Equation 6)) is reached in the next sampling period ($f_{samp} = f_s$).

$$s[n] = i_{Lref}[n-1] - i_L[n]. \tag{6}$$

The Euler approximation leads to the following discrete-time output current expression, assuming the averaged model that the converter's current output slope $\frac{di_L}{dt} \approx \frac{i_L[n+1]-i_L[n]}{T}$

$$i_L[n+1] = i_L[n] + T(m_1+m_2)d_x[n] - m_2T. \tag{7}$$

Hence, the resulting expression of the duty cycle is

$$d_x[n] = \frac{1}{(m_1+m_2)T}[i_{Lref}[n] - i_L[n]] + \frac{m_2}{m_1+m_2} \tag{8}$$

where x in Equation (8) corresponds to the operating mode of the bidirectional buck–boost converter (x = 1 for boost mode, x = 2 for buck mode), and $i_{Lref}[n] = i_L[n+1]$, using the expressions for m_1 and $-m_2$ for the output current slopes from Table 1 in Equation (8). The expression $m_1 + m_2$ is obtained from Table 1 for each converter operation mode, yielding

$$m_1 + m_2 = \begin{cases} \frac{MV_c}{L^2-M^2} & \text{for boost mode} \\ \frac{LV_c}{L^2-M^2} & \text{for buck mode.} \end{cases} \tag{9}$$

For $m_2/(m_1+m_2)$, it is given by

$$\frac{m_2}{m_1+m_2} = \begin{cases} \frac{-M(V_g-V_c)+L(V_o-V_c)}{MV_c} & \text{for boost mode} \\ \frac{-M(V_g-V_c)+LV_o}{LV_c} & \text{for buck mode.} \end{cases} \tag{10}$$

In this control method, the output current $i_L(t)$ and voltages are sampled at the beginning of each switching period, then, at the end of the switching cycle, $i_L[n] = i_{Lref}[n-1]$. The steady-state duty cycle from the equivalent control law (Equation 8)) can be obtained by substituting the voltage of the intermediate capacitor v_c by V_g for buck mode and by $v_c = V_o$ for boost mode in Equation (10). In steady-state, the duty cycle is $U = m_2/(m_1+m_2)$, thus, the variable control $u[n]$ can be written as

$$u[n] = \frac{1}{(m_1+m_2)T}\left(i_{Lref}[n] - i_L[n]\right) + U_n \tag{11}$$

where $U_n = U$ for buck mode and $U_n = 1 + U$ for boost mode. The schematic diagram of the DSMCC is depicted in Figure 5.

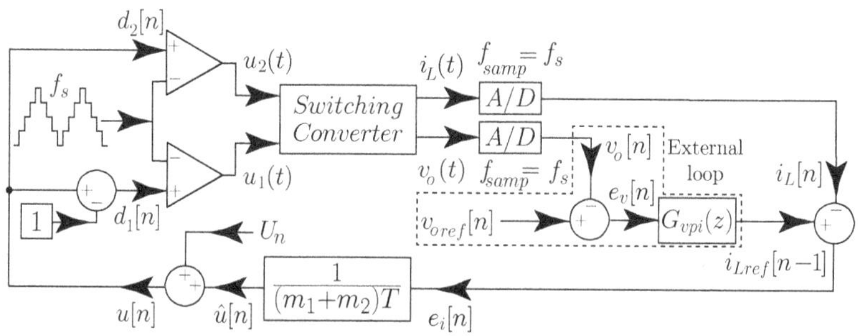

Figure 5. Schematic diagram of the two-loop control using DSMCC method.

3.2. Digital Proportional-Integral Voltage Control

In order to guarantee dc bus voltage regulation, it is necessary to add a slower outer voltage control loop. With this new loop, the switching converter can be operated as a controlled current source due to the control DSMCC ensures the load current will follow the current reference. Therefore, the current-controlled buck–boost converter is operated as a current source that allows driving the energy consumption of the load. From the point of view of the dc voltage control loop, voltage variations with power constant should be compensated charging or discharging the dc bus capacitor [45]. Hence, this PI control is designed to consider the filter output capacitor value C_o or the capacitor in the dc-link for an EV powertrain C_{bus}. The transfer function of the PI voltage controller in the Laplace domain can be written as follows

$$G_{vpi}(s) = K_{pv} + \frac{K_{iv}}{s}. \tag{12}$$

The output current reference to output voltage transfer function is obtained from Equation (5)

$$H_{voiL}(s) = \frac{v_o(s)}{i_L(s)} = \frac{R_o}{R_o C_o s + 1} \tag{13}$$

The loop-gain of the external closed loop voltage can be written as:

$$G(s) = H_{voiL}(s) G_{vpi}(s) H_v(s) e^{-sT_m}, \tag{14}$$

where $H_v(s)$ represents the sensor gain. The term e^{-sT_m} represents half switching cycle delay, $T_m = T/2$. Then, the controller transfer function (Equation (12)) can be expressed in the z domain using the forward Euler method, as follows

$$G_{vpi}(z) = K_{pv} + \frac{K_{iv} T_{samp}}{z - 1}. \tag{15}$$

where $T_{samp} = 1/f_{samp}$. The forward Euler method is used to find the recurrence equation of the discrete-time integral PI control

$$\begin{aligned} i_{Lp}[n] &= K_{pv} e_v[n] \\ i_{Li}[n] &= K_{iv} T_{samp} e_v[n] + i_{Li}[n-1] \\ i_{Lref}[n] &= i_{Lp}[n] + i_{Li}[n]. \end{aligned} \tag{16}$$

where

$$K_{pv} = C_o 2\pi f_c \tag{17}$$

$$K_{iv} = \frac{K_{pv}}{T_i} \tag{18}$$

$$T_i \geq \frac{10}{2\pi f_c} \tag{19}$$

can be obtained from Equations (12) and (14), taking into account that the zero of Equation (12) is placed lower than one decade below f_c, which represents the crossover frequency (CF). The value of the crossover frequency for the voltage loop (f_c) should be lower than that of the current loop. Hence, a $f_c = 2500$ Hz was selected for the voltage feedback loop.

Figure 6 depicts the Bode plots of simulated (PSIM) and experimental voltage loop gain under different operation modes (boost and buck) for the versatile buck–boost converter with a gain of the measurement system $H_v(s) = 0.044$. These Bode plots show a similar behavior at a low frequency of the magnitude plot for the experiment and the simulation. For a quantitative evaluation, the CF and the phase margin (PM) are listed in Table 2. From this table, it can be concluded that the closed-loop system is stable.

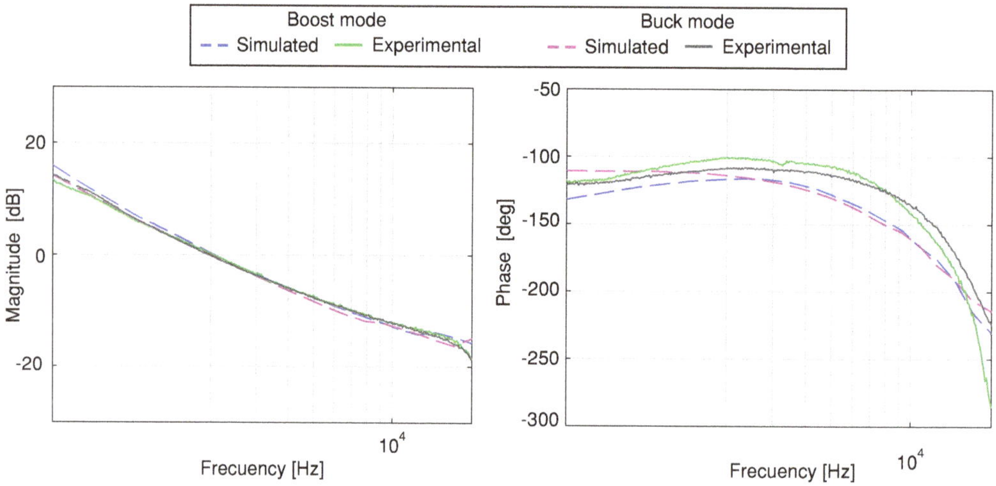

Figure 6. Simulated and experimental voltage loop gain Bode plots of the buck–boost converter.

Table 2. CF and PM of voltage loop gain.

Mode	Simulated		Experimental	
	CF [kHz]	PM [deg]	CF [kHz]	PM [deg]
Boost	1.99	63.79	2.03	79.4
Buck	1.99	66.52	1.94	71.2

4. Simulation and Experimental Results

Validation of the proposed current control strategy is performed on a 1.6 kW versatile buck–boost converter. A Texas Instruments TMS320F28377S Digital Signal Processor (DSP) is used to implement the proposed control algorithm to calculate the variable control u and the hysteresis buck–boost transition method introduced in [31], which was employed to compute the duty cycle values. These duty cycles allow the PWM generation using a symmetric triangular signal to get the activation signals of the MOSFETs switches. In addition, a direct voltage source AMREL SPS800X13-K02D is used as a power supply for the input voltage of the buck–boost converter shown in Figure 3 and whose parameters are listed in Table 3. The design guides of the versatile converter are described in detail in [41].

4.1. System Startup

The simulated and experimental results for the system startup in closed-loop are given in Figure 7. The voltage reference V_{oref} is increased from 0 V in each switching period during 12 ms until 293 V with an input voltage (V_g) of 200 V and 350 V for boost and buck mode, respectively. It must also be noted that during the startup in boost mode, the system begins in buck mode and ends in boost mode in steady-state. Therefore, this experiment exhibits a smooth transition between the buck and boost operating modes. The experimental results demonstrate that the voltage output is well regulated in all the operation modes. In addition, a good agreement can be observed between the experimental measurement and the simulated with a fast and soft startup.

Table 3. Selected components and parameters for the buck–boost converter.

Parameter	Value or Type
Input voltage V_g	200–400 V
Output voltage V_o	100–400 V
Rated Power	1.6 kW
Switching frequency f_s	100 kHz
Output capacitor C_o	6× R75PW44704030J, 28 μF, 630 V
Damping capacitor C_d	MKP1848S62070JP2F, 20 μF, 700 V
Intermediate capacitor C	4× R76PN33304030J, 1.32 μF, 630 V
Coupled inductor	M =135 μH and L =270 μH, Core: 77,908 Magnetics, Number turns: 80, Wire size: 18 AWG.
Damping resistance R_d	2× BPR10100J in parallel, 5 Ω, 10 W, 500 V
MOSFET Driver	UCC27714D
Power semiconductors $Q_1 - Q_4$	SCT2450KEC

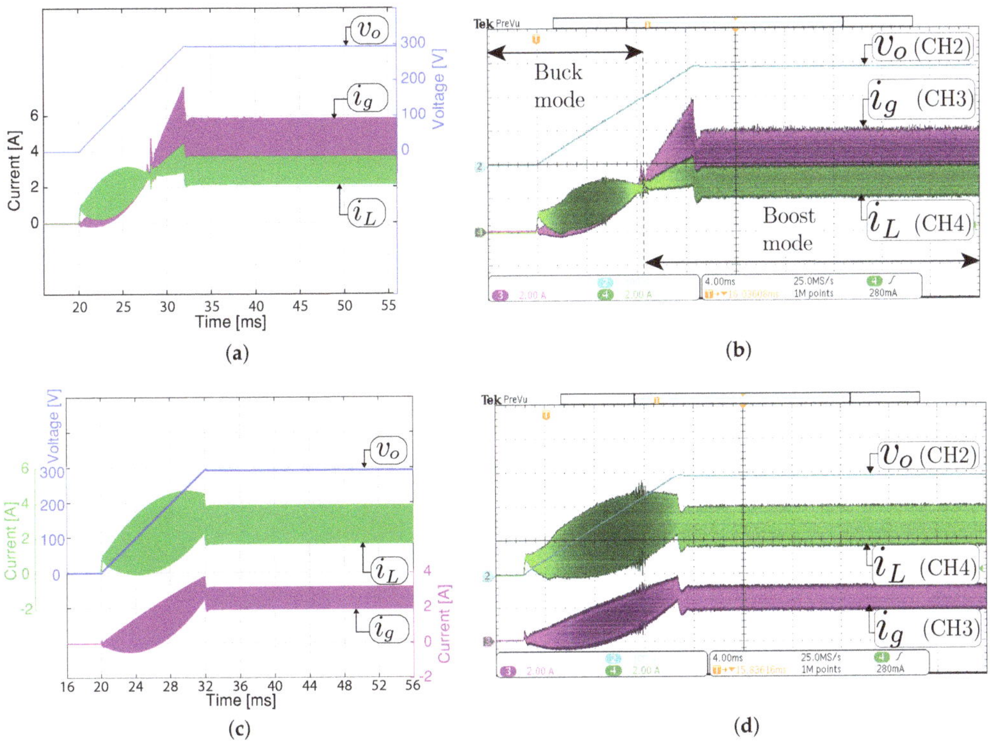

Figure 7. System startup with a constant resistive load. Simulated (**a**,**c**) and experimental (**b**,**d**). Two operation modes in steady-state are shown: (**a**,**b**) boost mode ($V_g = 200$ V, $v_o = 293$ V and and $R_o = 200$ Ω) and (**c**,**d**) buck mode ($V_g = 350$ V, $v_o = 293$ V and $R_o = 32.3$ Ω). CH1 or CH2: v_o (100 V/div), CH3: i_g (2 A/div), CH4: i_L (2A/div), and time base of 4 ms.

4.2. Small-Signal Response to Output Voltage Reference Variation

Figures 8 and 9 show the small-signal control loop response to small output voltage reference changes during the boost and buck operation, respectively. The input voltage is set at 200 V with a constant resistive load $R_o = 200\ \Omega$ for all the study cases. In boost mode (Figure 8), the output voltage reference changed between 294 V and 296 V. While in buck mode, the output voltage reference changes between the values of 98 V and 100 V, as shown in Figure 9. The dc component in Figures 8 and 9 have been removed to comprise the ± 2 V step change in the output voltage reference. These results show that the output voltage is well regulated to its desired reference, and the output and input current are increased or decreased when the voltage reference changes to recover the converter's operating point. The figures also demonstrate a good agreement between the experimental and simulation results with a short outer voltage transient of around 400 µs, which validates the proposed control method's satisfactory operation.

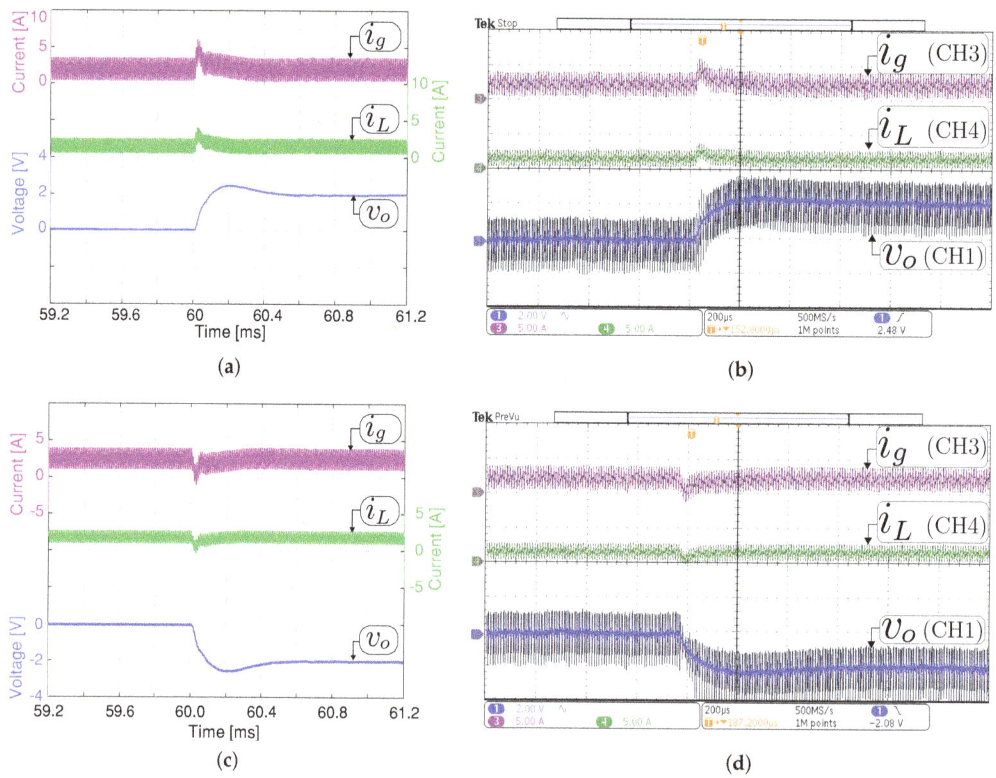

Figure 8. Small signal transient response with a constant resistive load $R_o = 200\ \Omega$ in boost mode ($V_g = 200$ V). Simulated (**a**,**c**) and experimental (**b**,**d**). Transient response when the output voltage reference changes from 294 to 296 V (**a**,**b**), and from 296 V to 294 V (**c**,**d**). CH1: v_o (2 Vac/div), CH3: i_g (5 A/div), CH4: i_L (5 A/div), and time base of 200 µs.

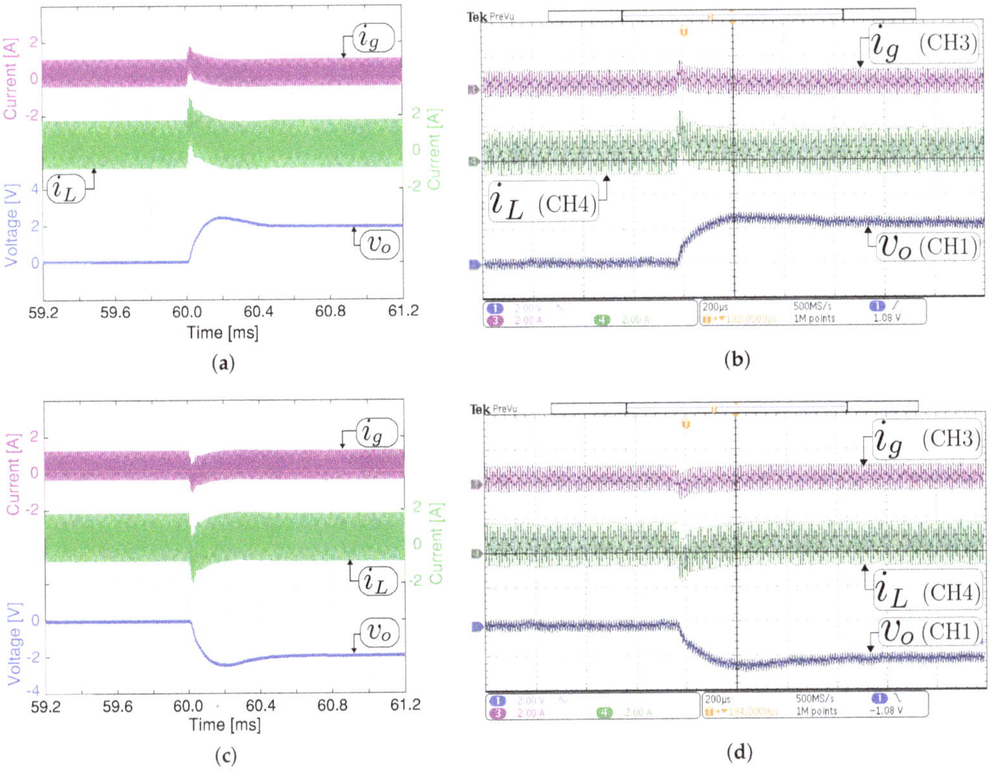

Figure 9. Small signal transient response with a constant resistive load $R_o = 200\ \Omega$ in buck mode ($V_g = 200$ V). Simulated (**a**,**c**) and experimental (**b**,**d**). Transient response when the output voltage reference changes from 98 to 100 V (**a**,**b**), and from 100 V to 98 V (**c**,**d**). CH1: v_o (2 Vac/div), CH3: i_g (2 A/div), CH4: i_L (2 A/div), and time base of 200 µs.

4.3. Large-Signal Response to Output Voltage Reference Variation

Figures 10 and 11 compare the large-signal response when the output voltage has a ±20 V step change. The figures show simulation and experimental waveforms of the input and output current and output voltage. Figure 10 depicts the response when the converter operates in boost mode. The output voltage reference has been changed between 293 V and 313 V. The dc component in the experimental and simulation results has been removed to appreciate the output voltage variation in boost mode. The results for buck mode are shown in Figure 11, where the output voltage reference has been changed between 100 V and 120 V. From these figures, for both control methods, the transient average current output value was successfully limited to ±4 A, which is the rated output current of the converter. These results confirm the direct relationship between the output voltage response time with the output filter capacitor value. It should be remembered that the slew-rate (SR) is defined in this case as $SR = i/C_o$ [V/µs], where i is the instantaneous current through the capacitor C_o. Therefore, the response time to step output variation depends on the SR parameter. Note that the measured current i_L follows the current reference accurately. Some differences are presented between the simulated and experimental results concerning the input current. These differences are because the converter does not control this current and its dynamic depends on the dc power supply internal control. Again, a good agreement between the experimental and simulation results is observed.

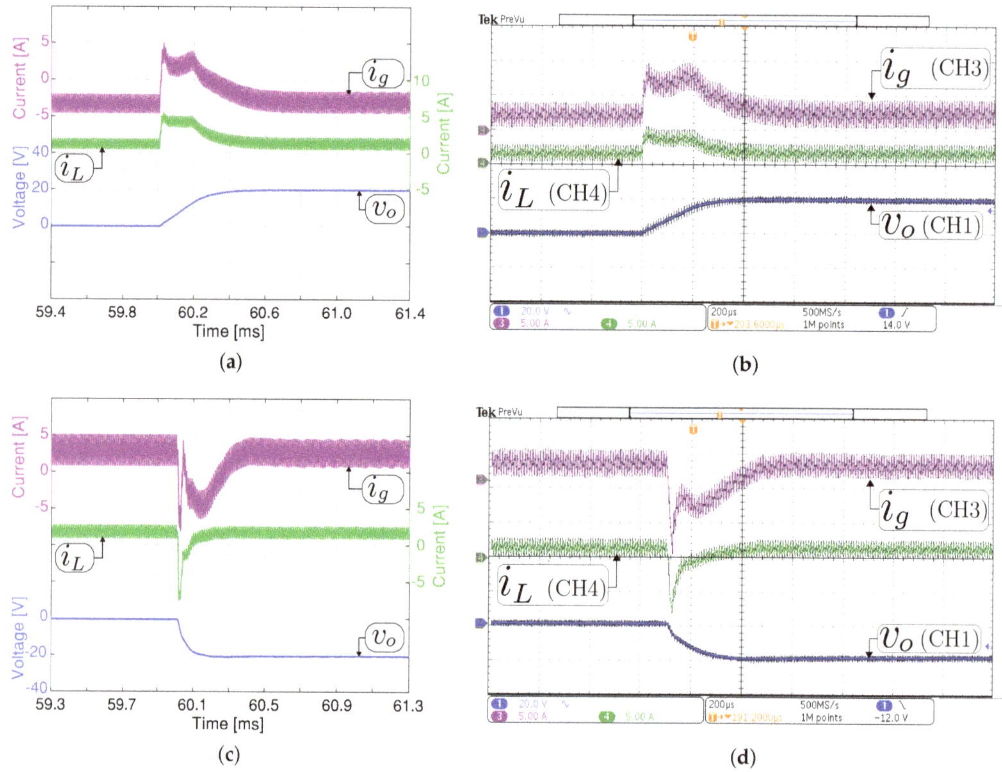

Figure 10. Large signal transient response with a constant resistive load $R_o = 200\ \Omega$ in boost mode ($V_g = 200$ V). Simulated (**a**,**c**) and experimental (**b**,**d**). Transient response when the output voltage reference changes from 294 to 314 V (**a**,**b**), and from 314 to 294 V (**c**,**d**). CH1: v_o (20 Vac/div), CH3: i_g (5 A/div), CH4: i_L (5 A/div), and time base of 200 µs.

4.4. Experiments with an EV Powertrain System Emulation

Energy management system in auxiliary supply in EV topologies and the use of dc-dc converter as an interface between the primary energy source and high-voltage powertrain are some of the applications for EVs. This application can be studied using a powertrain emulation system or simulation [46]. Some experiments are carried out considering the experimental PMSM platform described in Figure 12 that emulates an EV powertrain. This system is composed of two permanent-magnet synchronous motors (PMSMs). One of them (LSRPM 100 L) works as a traction motor with a maximum power of 4.5 kW, and the other one (LSRPM 90 SL) works as a controlled torque load with a maximum power of 3 kW. In order to verify the correct operation of the whole system shown in Figure 12, a test with a third of the total power is tested in this work. To increase the system's operating power, it will be necessary to connect two more converters in parallel which is possible since they are current controlled modules. The traction part is controlled by a universal variable speed ac drive (SP2202), and it is fed using the buck–boost converter described in this work. The battery is emulated using a DC power supply (AMREL SPS800X13-K02D) connected in parallel with an electronic load to absorb the current in the case of regenerative mode. This converter is connected between the battery emulator and the output filter capacitor C_o (R75PW44704030J). Subsequently, the traction motor is mechanically coupled to the motor that emulates the load (EV behavior). This traction motor is controlled according to a speed profile provided by a specific driving cycle. On the other hand, the load motor is controlled by a universal variable speed ac drive (SP1405) to follow a torque reference based on the vehicle dynamics. An EV powertrain system model has been implemented in the PSIM software with the parameters of the PMSMs listed in Table 4. To startup, the system with an initial voltage to fed the unidrive SP2202, a soft-starting of the dc-dc converter is implemented by the algorithm as it was previously described, and the simulated and experimental results are shown in Figure 13. During the startup, the reference voltage v_{oref} changed from 0 V to the final desired output voltage value with a short transient around 0.9 s. The switching frequency for the inverter (SP2202) is 16 kHz. Figure 13a,b shows the startup response in boost mode with V_g = 200 V and steady-state the output voltage value of 350 V. In this experiment, the currents have an average value of 0 A in steady-state because the motors are not operating during the startup; however, the inverter remains switched because it has a 200 V power supply terminals. Figure 13c,d shows the steady-state converter response in buck mode with an input voltage (V_g) of 400 V and output voltage (v_o) of 300 V. The figures also show good agreement between the experimental data and the simulation results.

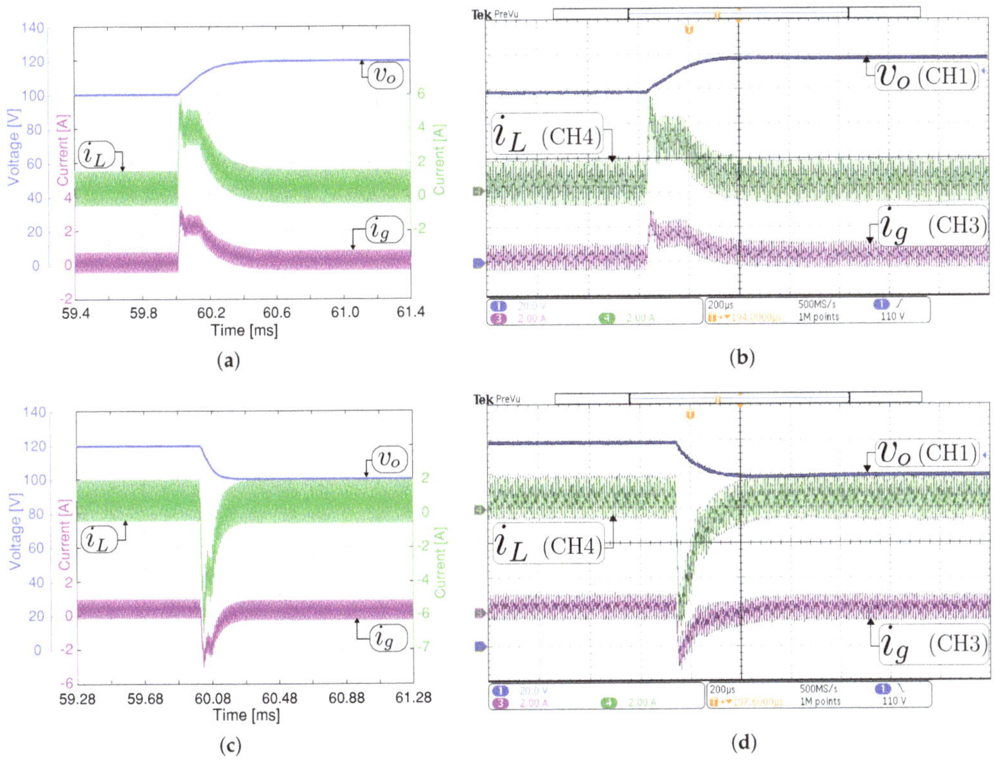

Figure 11. Large signal transient response with a constant resistive load $R_o = 200\ \Omega$ in buck mode ($V_g = 200$ V). Simulated (**a**,**c**) and experimental (**b**,**d**). Transient response when the output voltage reference changes from 100 to 120 V (**a**,**b**), and from 120 to 100 V (**c**,**d**). CH1: v_o (20 V/div), CH3: i_g (2 A/div), CH4: i_L (2 A/div), and time base of 200 µs.

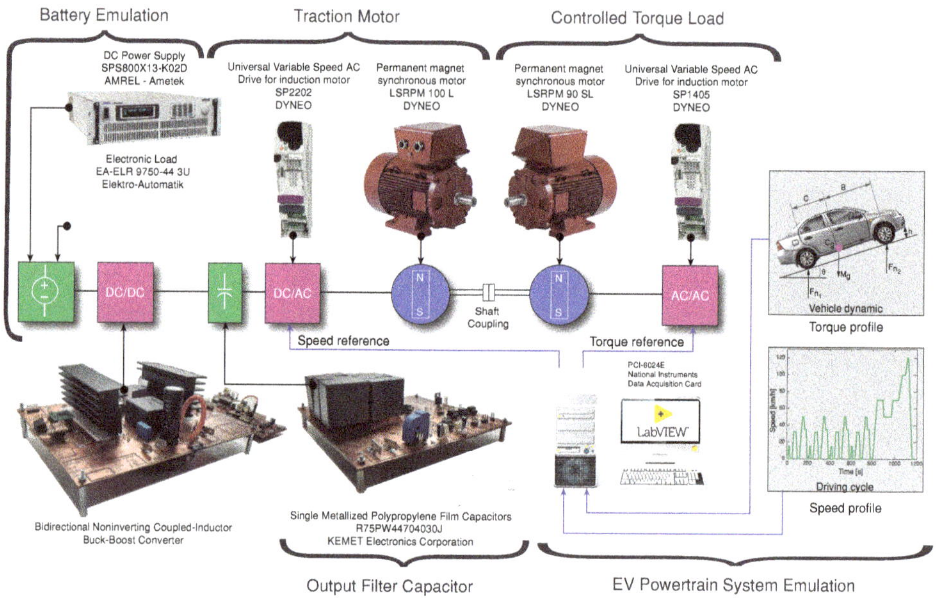

Figure 12. Diagram of the experimental setup: Converter dc-dc and EV powertrain.

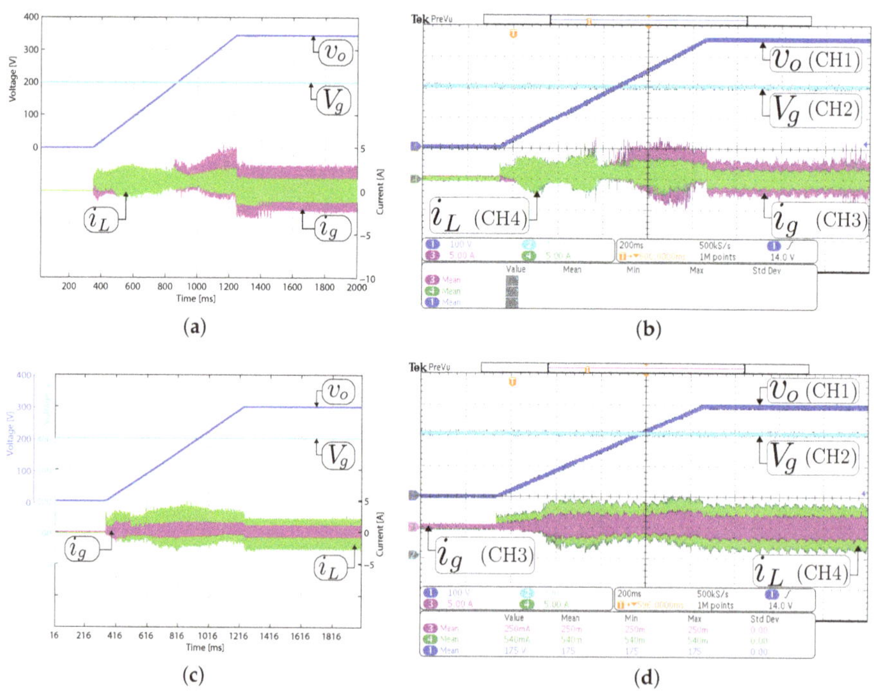

Figure 13. System startup with an EV powertrain system emulation. Simulated (**a**,**c**) and experimental (**b**,**d**). Two operation modes in steady-state are shown: (**a**,**b**) boost mode ($V_g = 200$ V, $v_o = 350$ V) and (**c**,**d**) buck mode ($V_g = 400$ V, $v_o = 300$ V). CH1: v_o (100 V/div), CH2: V_g (100 V/div), CH3: i_g (5 A/div), CH4: i_L (5 A/div), and time base of 200 ms.

Table 4. Parameters of the PMSMs.

Parameter	90 SL	100 L
Motor rated speed	1500 [rpm]	1500 [rpm]
Number of pole pairs	8	8
Stator resistance R_s	2.34 Ω	1.277 Ω
d-axis inductance L_d	50.124 mH	29.128 mH
q-axis inductance L_d	29.128 mH	19.295 mH
Moment of inertia J	0.0032 kg·m^2	0.0066 kg·m^2
Electrical constant k_e	212 V_{kp}/k_{rpm}	223 V_{kp}/k_{rpm}

Figure 14 shows the transient response with a 450 W step change in the load power, setting the speed reference in 500 rpm and the torque value in 3.77 Nm to obtain a dc bus power demand of 300 W. Later, the speed reference changes to 1250 rpm to obtain a dc bus power demand of 750 W. As a result, the output current changes from 1 A to 2.5 A gradually while the output voltage is regulated at 300 V. In the experimental and simulated results of Figure 15, the converter can be seen working in boost mode with V_g = 200 V and bidirectional power flow. The speed of the traction motor is set to 500 rpm, and the torque of the load motor to 7.53 Nm to get a dc bus demand of 600 W. The dc component (300 V) was removed to appreciate the ±20 V step change in the output voltage reference. Figure 15a,b shows the results when the output voltage is changed from 300 V to 320 V. Consequently, the current output i_L quickly goes to 4 A. Figure 15c,d shows the results when the output voltage is changed from 320 V to 300 V with a step change, the current output i_L decreases to −3 A. This current is limited above the rated current (−4 A) due to the limitation of the source is 13 A. The battery is simulated with a dc source (AMREL SPS800X13-K02D) in parallel with an electronic load (EA-ELR 9750-44 3U) (see Figure 12) in resistance mode to absorb 6 A, and it can absorb the current when the output voltage is decreased. The output voltage has a 20 V step change over 28 ms and has a −20 V step change at the output voltage over 12 ms. This time is different for each case because the unidrive SP2202 has an input filter capacitor of 2870 μF. Accordingly, this time depends on the filter capacitor value and the instantaneous current through the capacitor during the charging or discharging due to the output voltage step changes. Finally, there is a qualitatively good agreement between the simulated and the experimental results in the EV powertrain system emulation for the proposed controller and all the converter operation modes.

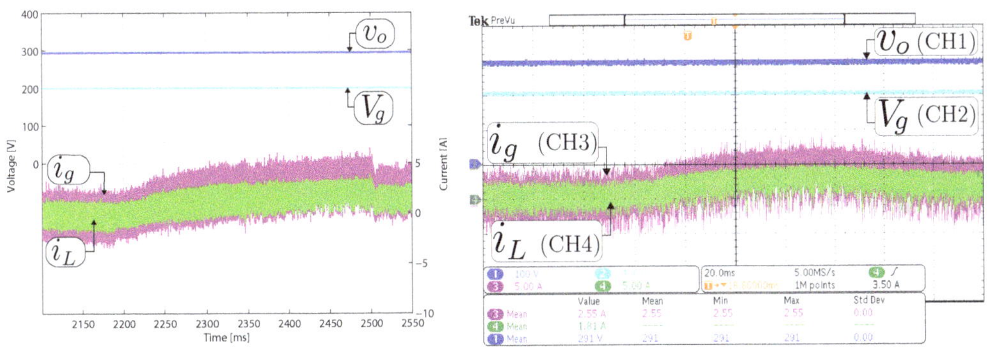

Figure 14. Boost operation mode in steady-state for a step power transition (P_o = 300 to 750 W, V_g = 200 V, V_o = 300 V) with an EV powertrain system emulation. Simulated (**a**), and experimental (**b**). CH1: V_o (100 V/div), CH2: V_g (100 V/div), CH3: i_g (5 A/div), CH4: i_L (5 A/div), and time base of 20 ms.

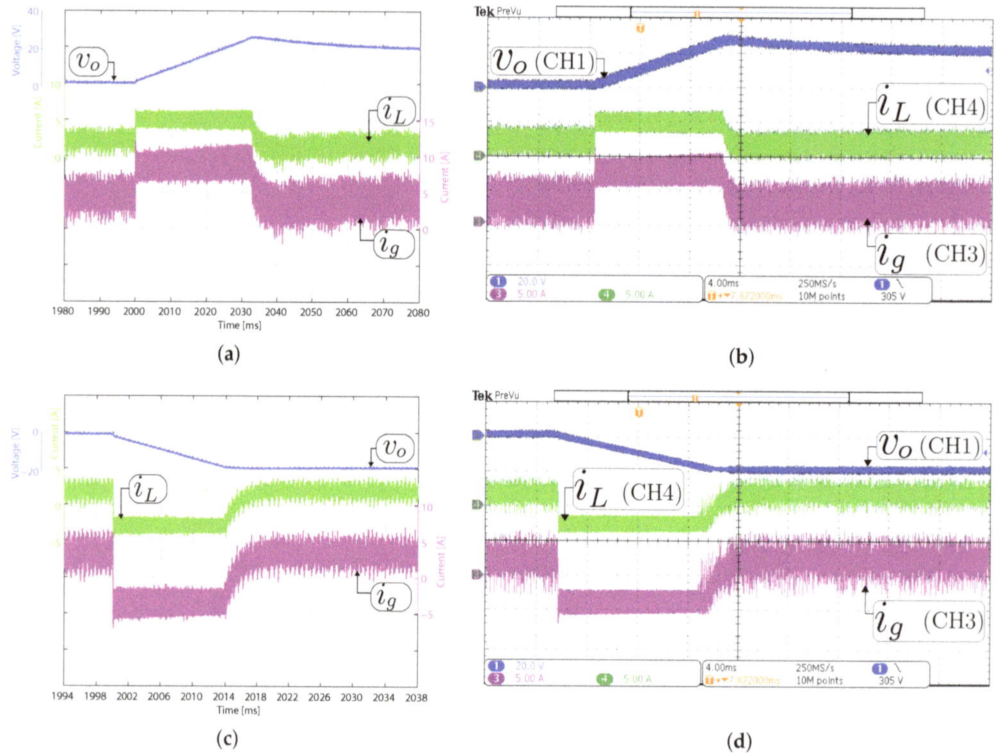

Figure 15. Boost operation mode in steady-state with an EV powertrain system emulation $V_g = 200$ V. Transient response when the output voltage reference changes from 300 to 320 V. Simulated (**a**,**c**) and experimental (**b**,**d**), time base of 10 ms (**a**,**b**), and from 320 to 300 V, time base of 4 ms (**c**,**d**). CH1: V_o (20 V/div), CH3: i_g (5 A/div), CH4: i_L (5 A/div).

5. Conclusions

This paper proposed the bidirectional versatile buck–boost converter modified to operate at high voltage. This converter is an alternative to conventional topologies based on the boost converter in electric vehicle applications. The versatile converter has been located between the battery and traction inverter to regulate the dc bus in electric vehicle powertrains. The use of a high-efficiency step-up/step-down converter can improve the performance efficiency of the EV powertrain. This improvement includes an extensive range of electric motor speeds, which comprises urban and highway driving cycles. The proposed dc-dc bidirectional buck–boost converter is responsible for the dc bus voltage regulation through an outer voltage feedback loop and an inner current programmed controller. A Texas Instruments TMS320F28377S DSC is used to implement the digital control loops. The digital implementation of this current controller has allowed to include a dead-zone avoidance technique that effectively has suppressed very effectively undesirable nonlinear phenomena in the buck–boost mode transitions such as sub-harmonics or other undesirable nonlinear phenomena. The theoretical analyses have been validated using simulations and experimental tests performed on a 400-V 1.6-kW prototype. The current controller allows regulating the traction dc bus during motoring and regenerative brake conditions. The system presents zero steady error and fast transient response in the start-up for dc bus voltage reference changes and under realistic conditions using an EV powertrain system emulation. The experimental results are in good agreement with the simulation and the theoretical predictions. Future works will address the parallelization of power converters to increase the operating power of the system.

Author Contributions: Conceptualization, J.C., C.R., S.K., E.V.-I. and C.G.-C.; methodology, J.C., C.R., S.K., E.V.-I. and C.G.-C.; software, C.R. and C.G.-C.; validation, C.R. and C.G.-C.; formal analysis, J.C., E.V.-I. and S.K.; investigation, J.C., C.R., S.K., E.V.-I. and C.G.-C.; resources, J.C., C.R., S.K., E.V.-I. and C.G.-C.; data curation, S.K., C.R. and C.G.-C.; writing—original draft preparation, S.K., C.R. and C.G.-C.; writing—review and editing, J.C., C.R., S.K., E.V.-I. and C.G.-C.; visualization, S.K., C.R. and C.G.-C.; supervision, E.V.-I., C.R. and J.C.; project administration, J.C., S.K. and C.R.; funding acquisition, S.K., C.R. and E.V.-I. All authors have read and agreed to the published version of the manuscript.

Funding: This work was supported by the Spanish Agencia Estatal de Investigación (AEI) and the Fondo Europeo de Desarrollo Regional (FEDER) under research projects DPI2016-80491-R (AEI/FEDER, UE). The work was also supported by the the Chilean Government under Project CONICYT/FONDECYT 1191680 and by SERC Chile (CONICYT/FONDAP/15110019).

Institutional Review Board Statement: Not applicable.

Informed Consent Statement: Not applicable.

Data Availability Statement: Not applicable.

Conflicts of Interest: The authors declare no conflict of interest. The funders had no role in the design of the study; in the collection, analyses, or interpretation of data; in the writing of the manuscript, or in the decision to publish the results.

References

1. Yong, J.Y.; Ramachandaramurthy, V.K.; Tan, K.M.; Mithulananthan, N. A review on the state-of-the-art technologies of electric vehicle, its impacts and prospects. *Renew. Sustain. Energy Rev.* **2015**, *49*, 365–385. [CrossRef]
2. Shi, X.; Pan, J.; Wang, H.; Cai, H. Battery electric vehicles: What is the minimum range required? *Energy* **2019**, *166*, 352–358. [CrossRef]
3. Andwari, A.M.; Pesiridis, A.; Rajoo, S.; Martinez-Botas, R.; Esfahanian, V. A review of Battery Electric Vehicle technology and readiness levels. *Renew. Sustain. Energy Rev.* **2017**, *78*, 414–430. [CrossRef]
4. Sewergin, A.; Wienhausen, A.H.; Oberdieck, K.; Doncker, R.W.D. Modular bidirectional full-SiC DC-DC converter for automotive applications. In Proceedings of the 2017 IEEE 12th International Conference on Power Electronics and Drive Systems (PEDS), Honolulu, HI, USA, 12–15 December 2017; pp. 277–281.
5. Rothgang, S.; Baumhöfer, T.; van Hoek, H.; Lange, T.; Doncker, R.W.D.; Sauer, D.U. Modular battery design for reliable, flexible and multi-technology energy storage systems. *Appl. Energy* **2015**, *137*, 931–937. [CrossRef]
6. Quraan, M.; Tricoli, P.; D'Arco, S.; Piegari, L. Efficiency Assessment of Modular Multilevel Converters for Battery Electric Vehicles. *IEEE Trans. Power Electron.* **2017**, *32*, 2041–2051. [CrossRef]
7. Zhu, Z.Q.; Howe, D. Electrical Machines and Drives for Electric, Hybrid, and Fuel Cell Vehicles. *Proc. IEEE* **2007**, *95*, 746–765. [CrossRef]
8. Yang, Z.; Shang, F.; Brown, I.P.; Krishnamurthy, M. Comparative Study of Interior Permanent Magnet, Induction, and Switched Reluctance Motor Drives for EV and HEV Applications. *IEEE Trans. Transp. Electrif.* **2015**, *1*, 245–254. [CrossRef]
9. Song, Z.; Liu, C.; Zhao, H. Investigation on Magnetic Force of a Flux-Modulated Double-Rotor Permanent Magnet Synchronous Machine for Hybrid Electric Vehicle. *IEEE Trans. Transp. Electrif.* **2019**, *5*, 1383–1394. [CrossRef]
10. Cheng, M.; Sun, L.; Buja, G.; Song, L. Advanced Electrical Machines and Machine-Based Systems for Electric and Hybrid Vehicles. *Energies* **2015**, *8*, 9541–9564. [CrossRef]
11. Zhao, N.; Schofield, N.; Yang, R.; Gu, R. Investigation of DC-Link voltage and temperature variations on EV traction system design. *IEEE Trans. Ind. Appl.* **2017**, *53*, 3707–3718. [CrossRef]
12. Lin, Y.S.; Hu, K.W.; Yeh, T.H.; Liaw, C.M. An electric-vehicle IPMSM drive with interleaved front-end DC/DC converter. *IEEE Trans. Veh. Technol.* **2015**, *65*, 4493–4504. [CrossRef]
13. Deng, W.; Zhao, Y.; Wu, J. Energy efficiency improvement via bus voltage control of inverter for electric vehicles. *IEEE Trans. Veh. Technol.* **2016**, *66*, 1063–1073. [CrossRef]
14. Chen, H.; Kim, H.; Erickson, R.; Maksimović, D. Electrified Automotive Powertrain Architecture Using Composite DC–DC Converters. *IEEE Trans. Power Electron.* **2017**, *32*, 98–116. [CrossRef]
15. Qian, W.; Cha, H.; Peng, F.Z.; Tolbert, L.M. 55-kW variable 3X DC-DC converter for plug-in hybrid electric vehicles. *IEEE Trans. Power Electron.* **2012**, *27*, 1668–1678. [CrossRef]
16. Elsayad, N.; Moradisizkoohi, H.; Mohammed, O.A. A New Hybrid Structure of a Bidirectional DC-DC Converter With High Conversion Ratios for Electric Vehicles. *IEEE Trans. Veh. Technol.* **2019**, *69*, 194–206. [CrossRef]
17. Reimers, J.; Dorn-Gomba, L.; Mak, C.; Emadi, A. Automotive traction inverters: Current status and future trends. *IEEE Trans. Veh. Technol.* **2019**, *68*, 3337–3350. [CrossRef]

18. Dusmez, S.; Hasanzadeh, A.; Khaligh, A. Comparative analysis of bidirectional three-level DC–DC converter for automotive applications. *IEEE Trans. Ind. Electron.* **2015**, *62*, 3305–3315. [CrossRef]
19. Pavlovský, M.; Guidi, G.; Kawamura, A. Assessment of coupled and independent phase designs of interleaved multiphase buck/boost DC–DC converter for EV power train. *IEEE Trans. Power Electron.* **2013**, *29*, 2693–2704. [CrossRef]
20. Lu, X.; Wang, H. A highly efficient multifunctional power electronic interface for PEV hybrid energy management systems. *IEEE Access* **2018**, *7*, 8964–8974. [CrossRef]
21. Hegazy, O.; Mierlo, J.V.; Lataire, P. Analysis, Modeling, and Implementation of a Multidevice Interleaved DC/DC Converter for Fuel Cell Hybrid Electric Vehicles. *IEEE Trans. Power Electron.* **2012**, *27*, 4445–4458. [CrossRef]
22. Onar, O.C.; Kobayashi, J.; Erb, D.C.; Khaligh, A. A Bidirectional High-Power-Quality Grid Interface With a Novel Bidirectional Noninverted Buck–Boost Converter for PHEVs. *IEEE Trans. Veh. Technol.* **2012**, *61*, 2018–2032. [CrossRef]
23. Han, D.; Noppakunkajorn, J.; Sarlioglu, B. Comprehensive efficiency, weight, and volume comparison of SiC-and Si-based bidirectional DC–DC converters for hybrid electric vehicles. *IEEE Trans. Veh. Technol.* **2014**, *63*, 3001–3010. [CrossRef]
24. Lai, C.M.; Cheng, Y.H.; Hsieh, M.H.; Lin, Y.C. Development of a Bidirectional DC/DC Converter With Dual-Battery Energy Storage for Hybrid Electric Vehicle System. *IEEE Trans. Veh. Technol.* **2018**, *67*, 1036–1052. [CrossRef]
25. Guo, J.; Rodriguez, R.; Gareau, J.; Schumacher, D.; Alizadeh, M.; Azer, P.; Bauman, J.; Bilgin, B.; Emadi, A. A Comprehensive Analysis for High-Power Density, High-Efficiency 60 kW Interleaved Boost Converter Design for Electrified Powertrains. *IEEE Trans. Veh. Technol.* **2020**, *69*, 7131–7145. [CrossRef]
26. Sun, Q.; Wu, J.; Gan, C.; Si, J.; Guo, J.; Hu, Y. Cascaded Multiport Converter for SRM-Based Hybrid Electrical Vehicle Applications. *IEEE Trans. Power Electron.* **2019**, *34*, 11940–11951. [CrossRef]
27. Estima, J.O.; Cardoso, A.J.M. Efficiency Analysis of Drive Train Topologies Applied to Electric/Hybrid Vehicles. *IEEE Trans. Veh. Technol.* **2012**, *61*, 1021–1031. [CrossRef]
28. Restrepo, C.; Calvente, J.; Cid-Pastor, A.; El Aroudi, A.; Giral, R. A noninverting buck–boost DC–DC switching converter with high efficiency and wide bandwidth. *IEEE Trans. Power Electron.* **2011**, *26*, 2490–2503. [CrossRef]
29. Restrepo, C.; Calvente, J.; Romero, A.; Vidal-Idiarte, E.; Giral, R. Current-Mode Control of a Coupled-Inductor Buck–Boost DC–DC Switching Converter. *IEEE Trans. Power Electron.* **2012**, *27*, 2536–2549. [CrossRef]
30. Restrepo, C.; Konjedic, T.; Calvente, J.; Milanovic, M.; Giral, R. Fast Transitions Between Current Control Loops of the Coupled-Inductor Buck–Boost DC–DC Switching Converter. *IEEE Trans. Power Electron.* **2013**, *28*, 3648–3652. [CrossRef]
31. Restrepo, C.; Konjedic, T.; Calvente, J.; Giral, R. Hysteretic transition method for avoiding the dead-zone effect and subharmonics in a noninverting buck–boost converter. *IEEE Trans. Power Electron.* **2015**, *30*, 3418–3430. [CrossRef]
32. Ramírez-Murillo, H.; Restrepo, C.; Konjedic, T.; Calvente, J.; Romero, A.; Baier, C.R.; Giral, R. An Efficiency Comparison of Fuel-Cell Hybrid Systems Based on the Versatile Buck–Boost Converter. *IEEE Trans. Power Electron.* **2018**, *33*, 1237–1246. [CrossRef]
33. Ramírez-Murillo, H.; Restrepo, C.; Calvente, J.; Romero, A.; Giral, R. Energy Management of a Fuel-Cell Serial–Parallel Hybrid System. *IEEE Trans. Ind. Electron.* **2015**, *62*, 5227–5235. [CrossRef]
34. Ramírez-Murillo, H.; Restrepo, C.; Calvente, J.; Romero, A.; Giral, R. Energy Management DC System Based on Current-Controlled Buck-Boost Modules. *IEEE Trans. Smart Grid* **2014**, *5*, 2644–2653. [CrossRef]
35. Restrepo, C.; Konjedic, T.; Flores-Bahamonde, F.; Vidal-Idiarte, E.; Calvente, J.; Giral, R. Multisampled Digital Average Current Controls of the Versatile Buck-Boost Converter. *IEEE J. Emerg. Sel. Top. Power Electron.* **2018**, *7*, 879–890. [CrossRef]
36. Restrepo, C.; Garcia, G.; Flores-Bahamonde, F.; Murillo-Yarce, D.; Guzman, J.I.; Rivera, M. Current Control of the Coupled-Inductor Buck–Boost DC–DC Switching Converter using a Model Predictive Control Approach. *IEEE J. Emerg. Sel. Top. Power Electron.* **2020**, *8*, 3348–3360. [CrossRef]
37. Méndez-Díaz, F.; Pico, B.; Vidal-Idiarte, E.; Calvente, J.; Giral, R. HM/PWM Seamless Control of a Bidirectional Buck–Boost Converter for a Photovoltaic Application. *IEEE Trans. Power Electron.* **2019**, *34*, 2887–2899. [CrossRef]
38. Zhang, L.; Yuan, X.; Wu, X.; Shi, C.; Zhang, J.; Zhang, Y. Performance evaluation of high-power SiC MOSFET modules in comparison to Si IGBT modules. *IEEE Trans. Power Electron.* **2018**, *34*, 1181–1196. [CrossRef]
39. Bosshard, R.; Kolar, J.W. All-SiC 9.5 kW/dm^3 on-board power electronics for 50 kW/85 kHz automotive IPT system. *IEEE J. Emerg. Sel. Top. Power Electron.* **2016**, *5*, 419–431. [CrossRef]
40. Ghazanfari, A.; Perreault, C.; Zaghib, K. EV/HEV Industry Trends of Wide-bandgap Power Semiconductor Devices for Power Electronics Converters. In Proceedings of the 2019 IEEE 28th International Symposium on Industrial Electronics (ISIE), Vancouver, BC, Canada, 12–14 June 2019; pp. 1917–1923.
41. González-Castaño, C.; Restrepo, C.; Giral, R.; García-Amoros, J.; Vidal-Idiarte, E.; Calvente, J. Coupled inductors design of the bidirectional non-inverting buck–boost converter for high-voltage applications. *IET Power Electron.* **2020**, *13*, 3188–3198. [CrossRef]
42. Kim, B.S.; Kim, H.J.; Jin, C.; Huh, D.Y. A digital controlled DC-DC converter for electric vehicle applications. In Proceedings of the 2011 International Conference on Electrical Machines and Systems (ICEMS), Beijing, China, 20–23 August 2011; pp. 1–5.
43. Vidal-Idiarte, E.; Marcos-Pastor, A.; Garcia, G.; Cid-Pastor, A.; Martinez-Salamero, L. Discrete-time sliding-mode-based digital pulse width modulation control of a boost converter. *IET Power Electron.* **2015**, *8*, 708–714. [CrossRef]
44. Vidal-Idiarte, E.; Marcos-Pastor, A.; Giral, R.; Calvente, J.; Martinez-Salamero, L. Direct digital design of a sliding mode-based control of a PWM synchronous buck converter. *IET Power Electron.* **2017**, *10*, 1714–1720. [CrossRef]

45. Teodorescu, R.; Liserre, M.; Rodriguez, P. *Grid Converters for Photovoltaic and Wind Power Systems*; John Wiley & Sons: New York, NY, USA, 2011; Volume 29.
46. Geng, Z.; Hong, T.; Qi, K.; Ambrosio, J.; Gu, D. Modular Regenerative Emulation System for DC–DC Converters in Hybrid Fuel Cell Vehicle Applications. *IEEE Trans. Veh. Technol.* **2018**, *67*, 9233–9240. [CrossRef]

MDPI
St. Alban-Anlage 66
4052 Basel
Switzerland
Tel. +41 61 683 77 34
Fax +41 61 302 89 18
www.mdpi.com

Sensors Editorial Office
E-mail: sensors@mdpi.com
www.mdpi.com/journal/sensors

www.ingramcontent.com/pod-product-compliance
Lightning Source LLC
Chambersburg PA
CBHW040745250926
51850CB00007BB/1410
9783036560090